DATE DUE

Demco, Inc. 38-293

The PSI Handbook of Virtual Environments for Training and Education

The PSI Handbook of Virtual Environments for Training and Education

DEVELOPMENTS FOR THE MILITARY AND BEYOND

Volume 3
Integrated Systems,
Training Evaluations, and Future Directions

Edited by Joseph Cohn, Denise Nicholson,
and Dylan Schmorrow

Technology, Psychology, and Health

PRAEGER SECURITY INTERNATIONAL
Westport, Connecticut · London

Library of Congress Cataloging-in-Publication Data

The PSI handbook of virtual environments for training and education : developments for the
  military and beyond.
      p. cm. – (Technology, psychology, and health, ISSN 1942–7573 ; v. 1-3)
  Includes bibliographical references and index.
  ISBN 978–0–313–35165–5 (set : alk. paper) – ISBN 978–0–313–35167–9 (v. 1 : alk. paper) –
  ISBN 978–0–313–35169–3 (v. 2 : alk. paper) – ISBN 978–0–313–35171–6 (v. 3 : alk. paper)
1. Military education–United States. 2. Human-computer interaction. 3. Computer-assisted
instruction. 4. Virtual reality. I. Schmorrow, Dylan, 1967- II. Cohn, Joseph, 1969- III. Nicholson,
Denise, 1967- IV. Praeger Security International. V. Title: Handbook of virtual environments for
training and education. VI. Title: Praeger Security International handbook of virtual environments
for training and education.
U408.3.P75 2009
355.0078'5–dc22   2008027367

British Library Cataloguing in Publication Data is available.

Library of Congress Catalog Card Number: 2008027367
ISBN-13: 978–0–313–35165–5 (set)
            978–0–313–35167–9 (vol. 1)
            978–0–313–35169–3 (vol. 2)
            978–0–313–35171–6 (vol. 3)
ISSN: 1942–7573

First published in 2009

Praeger Security International, 88 Post Road West, Westport, CT 06881
An imprint of Greenwood Publishing Group, Inc.
www.praeger.com

Printed in the United States of America

The paper used in this book complies with the
Permanent Paper Standard issued by the National
Information Standards Organization (Z39.48–1984).

10 9 8 7 6 5 4 3 2 1

To our families, and to the men and women who have dedicated their lives to educate, train, and defend to keep them safe

CONTENTS

SERIES FOREWORD

LAUNCHING THE TECHNOLOGY, PSYCHOLOGY, AND HEALTH DEVELOPMENT SERIES

The escalating complexity and operational tempo of the twenty-first century requires that people in all walks of life acquire ever-increasing knowledge, skills, and abilities. Training and education strategies are dynamically changing toward delivery of more effective instruction and practice, wherever and whenever needed. In the last decade, the Department of Defense has made significant investments to advance the science and technology of virtual environments to meet this need. Throughout this time we have been privileged to collaborate with some of the brightest minds in science and technology. The intention of this three-volume handbook is to provide comprehensive coverage of the emerging theories, technologies, and integrated demonstrations of the state-of-the-art in virtual environments for training and education.

As Dr. G. Vincent Amico states in the Preface, an important lesson to draw from the history of modeling and simulation is the importance of *process*. The human systems engineering process requires highly multidisciplinary teams to integrate diverse disciplines from psychology, education, engineering, and computer science (see Nicholson and Lackey, Volume 3, Section 1, Chapter 1). This process drives the organization of the handbook. While other texts on virtual environments (VEs) focus heavily on technology, we have dedicated the first volume to a thorough investigation of learning theories, requirements definition, and performance measurement. The second volume provides the latest information on a range of virtual environment component technologies and a distinctive section on training support technologies. In the third volume, an extensive collection of integrated systems is discussed as virtual environment use-cases along with a section of training effectiveness evaluation methods and results. Volume 3, Section 3 highlights future applications of this evolving technology that span cognitive rehabilitation to the next generation of museum exhibitions. Finally, a glimpse into the potential future of VEs is provided as an original short story entitled "Into the Uncanny Valley" from Judith Singer and Hollywood director Alex Singer.

Through our research we have experienced rapid technological and scientific advancements, coinciding with a dramatic convergence of research achievements representing contributions from numerous fields, including neuroscience, cognitive psychology and engineering, biomedical engineering, computer science, and systems engineering. Historically, psychology and technology development were independent research areas practiced by scientists and engineers primarily trained in one of these disciplines. In recent years, however, individuals in these disciplines, such as the close to 200 authors of this handbook, have found themselves increasingly working within a unified framework that completely blurs the lines of these discrete research areas, creating an almost "metadisciplinary" (as opposed to multidisciplinary) form of science and technology. The strength of the confluence of these two disciplines lies in the complementary research and development approaches being employed and the interdependence that is required to achieve useful technological applications. Consequently, with this handbook we begin a new Praeger Security International Book Series entitled *Technology, Psychology, and Health* intended to capture the remarkable advances that will be achieved through the continued seamless integration of these disciplines, where unified and simultaneously executed approaches of psychology, engineering, and practice will result in more effective science and technology applications. Therefore, the esteemed contributors to the *Technology, Psychology, and Health Development Series* strive to capture such advancements and effectively convey both the practical and theoretical elements of the technological innovations they describe.

The *Technology, Psychology, and Health Development Series* will continue to address the general themes of requisite foundational knowledge, emergent scientific discoveries, and practical lessons learned, as well as cross-discipline standards, methodologies, metrics, techniques, practices, and visionary perspectives and developments. The series plans to showcase substantial advances in research and development methods and their resulting technologies and applications. Cross-disciplinary teams will provide detailed reports of their experiences applying technologies in diverse areas—from basic academic research to industrial and military fielded operational and training systems to everyday computing and entertainment devices.

A thorough and comprehensive consolidation and dissemination of psychology and technology development efforts is no longer a noble academic goal—it is a twenty-first century necessity dictated by the desire to ensure that our global economy and society realize their full scientific and technological potentials. Accordingly, this ongoing book series is intended to be an essential resource for a large international audience of professionals in industry, government, and academia.

We encourage future authors to contact us for more information or to submit a prospectus idea.

Dylan Schmorrow and Denise Nicholson
Technology, Psychology, and Health Development Series Editors
TPHSeries@ist.ucf.edu

PREFACE

G. Vincent Amico

It is indeed an honor and pleasure to write the preface to this valuable collection of articles on simulation for education and training. The fields of modeling and simulation are playing an increasingly important role in society.

You will note that the collection is titled virtual environments for *training and education*. I believe it is important to recognize the distinction between those two terms. Education is oriented to providing fundamental scientific and technical skills; these skills lay the groundwork for training. Simulations for training are designed to help operators of systems effectively learn how to operate those systems under a variety of conditions, both normal and emergency situations. Cognitive, psychomotor, and affective behaviors must all be addressed. Hence, psychologists play a dominant role within multidisciplinary teams of engineers and computer scientists for determining the effective use of simulation for training. Of course, the U.S. Department of Defense's Human Systems Research Agencies, that is, Office of the Secretary of Defense, Office of Naval Research, Air Force Research Lab, Army Research Laboratory, and Army Research Institute, also play a primary role—their budgets support many of the research activities in this important field.

Volume 1, Section 1 in this set addresses many of the foundational learning issues associated with the use of simulation for education and training. These chapters will certainly interest psychologists, but are also written so that technologists and other practitioners can glean some insight into the important science surrounding learning. Throughout the set, training technologies are explored in more detail. In particular, Volume 2, Sections 1 and 2 include several diverse chapters demonstrating how learning theory can be effectively applied to simulation for training.

The use of simulation for training goes back to the beginning of time. As early as 2500 B.C., ancient Egyptians used figurines to simulate warring factions. The precursors of modern robotic simulations can be traced back to ancient China, from which we have documented reports (circa 200 B.C.) of artisans constructing mechanical automata, elaborate mechanical simulations of people or animals. These ancient "robots" included life-size mechanical humanoids, reportedly capable of movement and speech (Kurzweil, 1990; Needham, 1986). In those early days, these mechanical devices were used to train soldiers in various phases of combat, and military tacticians used war games to develop strategies. Simulation technology as we know it today became viable only in the early twentieth century.

Probably the most significant event was Ed Link's development of the Link Trainer (aka the "Blue Box") for pilot training. He applied for its patent in 1929. Yet, simulation did not play a major role in training until the start of World War II (in 1941), when Navy captain Luis de Florez established the Special Devices Desk at the Bureau of Aeronautics. His organization expanded significantly in the next few years as the value of simulation for training became recognized. Captain de Florez is also credited with the development of the first flight simulation that was driven by an analog computer. Developed in 1943, his simulator, called the operational flight trainer, modeled the PBM-3 aircraft. In the period after World War II, simulators and simulation science grew exponentially based upon the very successful programs initiated during the war.

There are two fundamental components of any modern simulation system. One is a sound mathematical understanding of the object to be simulated. The other is the real time implementation of those models in computational systems. In the late 1940s the primary computational systems were analog. Digital computers were very expensive, very slow, and could not solve equations in real time. It was not until the late 1950s and early 1960s that digital computation became viable. For instance, the first navy simulator to use a commercial digital computer was the Attack Center Trainer at the FBM Facility (New London, Connecticut) in 1959. Thus, it has been only for the past 50 years that simulation has made major advancements.

Even today, it is typical that user requirements for capability exceed the ability of available technology. There are many areas where this is particularly true, including rapid creation of visual simulation from actual terrain environment databases and human behavior representations spanning cognition to social networks. The dramatic increases in digital computer speed and capacity have significantly closed the gap. But there are still requirements that cannot be met; these gaps define the next generation of science and technology research questions.

In the past decade or so, a number of major simulation initiatives have developed, including distributed interactive simulation, advanced medical simulation, and augmented cognition supported simulation. Distributed simulation enables many different units to participate in a joint exercise, regardless of where the units are located. The requirements for individual simulations to engage in such exercises are mandated by Department of Defense standards, that is, high level architecture and distributed interactive simulation. An excellent example of the capabilities that have resulted are the unprecedented number of virtual environment simulations that have transitioned from the Office of Naval Research's Virtual Technologies and Environments (VIRTE) Program to actual military training applications discussed throughout this handbook. The second area of major growth is the field of medical simulation. The development of the human patient simulator clearly heralded this next phase of medical simulation based training, and the field of medical simulation will certainly expand during the next decade. Finally, the other exciting development in recent years is the exploration of augmented cognition, which may eventually enable system users to completely forgo standard computer interfaces and work seamlessly with their equipment through the utilization of neurophysiological sensing.

Now let us address some of the issues that occur during the development process of a simulator. The need for simulation usually begins when a customer experiences problems training operators in the use of certain equipment or procedures; this is particularly true in the military. The need must then be formalized into a requirements document, and naturally, the search for associated funding and development of a budget ensues. The requirements document must then be converted into a specification or a work statement. That then leads to an acquisition process, resulting in a contract. The contractor must then convert that specification into a hardware and software design. This process takes time and is subject to numerous changes in interpretation and direction. The proof of the pudding comes when the final product is evaluated to determine if the simulation meets the customer's needs.

One of the most critical aspects of any modeling and simulation project is to determine its effectiveness and whether it meets the original objectives. This may appear to be a rather straightforward task, but it is actually very complex. First, it is extremely important that checks are conducted at various stages of the development process. During the conceptual stages of a project, formal reviews are normally conducted to ensure that the requirements are properly stated; those same reviews are also conducted at the completion of the work statement or specification. During the actual development process, periodic reviews should be conducted at key stages. When the project is completed, tests should be conducted to determine if the simulation meets the design objectives and stated requirements. The final phase of testing is validation. The purpose of validation is to determine if the simulation meets the customer's needs. Why is this process of testing so important? The entire development process is lengthy, and during that process there is a very high probability that changes will be induced. The only way to manage the overall process is by performing careful inspections at each major phase of the project.

As the organization and content of this handbook make evident, this process has been the fundamental framework for conducting most of today's leading research and development initiatives. Following section to section, the reader is guided through the requirements, development, and evaluation cycle. The reader is then challenged to imagine the state of the possible in the final, Future Directions, section.

In summary, one can see that the future of simulation to support education and training is beyond our comprehension. That does not mean that care must not be taken in the development process. The key issues that must be addressed were cited earlier. There is one fact that one must keep in mind: No simulation is perfect. But through care, keeping the simulation objectives in line with the capabilities of modeling and implementation, success can be achieved. This is demonstrated by the number of simulations that are being used today in innovative settings to improve training for a wide range of applications.

REFERENCES

Kurzweil, R. (1990). *The age of intelligent machines.* Cambridge, MA: MIT Press.
Needham, J. (1986). *Science and civilization in China: Volume 2.* Cambridge, United Kingdom: Cambridge University Press.

ACKNOWLEDGMENTS

These volumes are the product of many contributors working together. Leading the coordination activities were a few key individuals whose efforts made this project a reality:

Associate Editor
Julie Drexler

Technical Writer
Kathleen Bartlett

Editing Assistants
Kimberly Sprouse and Sherry Ogreten

We would also like to thank our Editorial Board and Review Board members, as follows:

Editorial Board

John Anderson, Carnegie Mellon University; Kathleen Bartlett, Florida Institute of Technology; Clint Bowers, University of Central Florida, Institute for Simulation and Training; Gwendolyn Campbell, Naval Air Warfare Center, Training Systems Division; Janis Cannon-Bowers, University of Central Florida, Institute for Simulation and Training; Rudolph Darken, Naval Postgraduate School, The MOVES Institute; Julie Drexler, University of Central Florida, Institute for Simulation and Training; Neal Finkelstein, U.S. Army Research Development & Engineering Command; Bowen Loftin, Texas A&M University at Galveston; Eric Muth, Clemson University, Department of Psychology; Sherry Ogreten, University of Central Florida, Institute for Simulation and Training; Eduardo Salas, University of Central Florida, Institute for Simulation and Training and Department of Psychology; Kimberly Sprouse, University of Central Florida, Institute for Simulation and Training; Kay Stanney, Design Interactive, Inc.; Mary Whitton, University of North Carolina at Chapel Hill, Department of Computer Science

Review Board (by affiliation)

Advanced Brain Monitoring, Inc.: Chris Berka; Alion Science and Tech.: Jeffery Moss; Arizona State University: Nancy Cooke; AuSIM, Inc.: William Chapin; Carlow International, Inc.: Tomas Malone; CHI Systems, Inc.: Wayne Zachary; Clemson University: Pat Raymark, Patrick Rosopa, Fred Switzer, Mary Anne Taylor; Creative Labs, Inc.: Edward Stein; Deakin University: Lemai Nguyen; Defense Acquisition University: Alicia Sanchez; Design Interactive, Inc.: David Jones; Embry-Riddle Aeronautical University: Elizabeth Blickens-derfer, Jason Kring; Human Performance Architects: Richard Arnold; Iowa State University: Chris Harding; Lockheed Martin: Raegan Hoeft; Max Planck Insti-tute: Betty Mohler; Michigan State University: J. Kevin Ford; NASA Langley Research Center: Danette Allen; Naval Air Warfare Center, Training Systems Division: Maureen Bergondy-Wilhelm, Curtis Conkey, Joan Johnston, Phillip Mangos, Carol Paris, James Pharmer, Ronald Wolff; Naval Postgraduate School: Barry Peterson, Perry McDowell, William Becker, Curtis Blais, Anthony Ciavarelli, Amela Sadagic, Mathias Kolsch; Occidental College: Brian Kim; Office of Naval Research: Harold Hawkins, Roy Stripling; Old Dominion Uni-versity: James Bliss; Pearson Knowledge Tech.: Peter Foltz; PhaseSpace, Inc.: Tracy McSherry; Potomac Institute for Policy Studies: Paul Chatelier; Renee Stout, Inc.: Renee Stout; SA Technologies, Inc.: Haydee Cuevas, Jennifer Riley; Sensics, Inc.: Yuval Boger; Texas A&M University: Claudia McDonald; The Boeing Company: Elizabeth Biddle; The University of Iowa: Kenneth Brown; U.S. Air Force Academy: David Wells; U.S. Air Force Research Laboratory: Dee Andrews; U.S. Army Program Executive Office for Simulation, Training, & Instrumentation: Roger Smith; U.S. Army Research Development & Engineer-ing Command: Neal Finkelstein, Timothy Roberts, Robert Sottilare; U.S. Army Research Institute: Steve Goldberg; U.S. Army Research Laboratory: Laurel Allender, Michael Barnes, Troy Kelley; U.S. Army TRADOC Analysis Center–Monterey: Michael Martin; U.S. MARCORSYSCOM Program Manager for Training Systems: Sherrie Jones, William W. Yates; University of Alabama in Huntsville: Mikel Petty; University of Central Florida: Glenda Gunter, Robert Kenny, Rudy McDaniel, Tim Kotnour, Barbara Fritzsche, Florian Jentsch, Kimberly Smith-Jentsch, Aldrin Sweeney, Karol Ross, Daniel Barber, Shawn Burke, Cali Fidopiastis, Brian Goldiez, Glenn Martin, Lee Sciarini, Peter Smith, Jennifer Vogel-Walcutt, Steve Fiore, Charles Hughes; University of Illinois: Tomas Coffin; University of North Carolina: Sharif Razzaque, Andrei State, Jason Coposky, Ray Idaszak; Virginia Tech.: Joseph Gabbard; Xavier Univer-sity: Morrie Mullins

SECTION 1

INTEGRATED TRAINING SYSTEMS

SECTION PERSPECTIVE
Neal Finkelstein

HISTORY

The uses of modeling and simulation for military purposes have come a long way with the advances of information and computer technology. Long gone are those early days in the 1930s with the ANT-18 Basic Instrument Trainers, known to tens of thousands of fledging pilots as Ed Link's Blue Box Trainers. The noisy ANT-18s were guaranteed to offer an individual a flight training course for $85 (Vintage Flying Museum, 2005). The military was soon sold on the promise of being able to provide flight training instruction with instruments alone after seeing a demonstration of the ANT-18 technology under some of the harshest conditions. Shortly thereafter, the Army Air Corps purchased six of Link's trainers for $3,500 apiece, thus beginning Link Aviation Devices, Inc., and creating an ever-changing industrial base for military training with simulation.

Throughout the decades since Link's ANT-18 was discovered by the military, the modeling and simulation industry has come to produce many bright, committed, and innovative men and women working to improve methods for modeled and simulated military training. This section includes chapters authored by experts in the fields of modeling and simulation for the purposes of military training. The authors will address specific training applications that use some of the same concepts once employed in the early ANT-18s, as well as cutting-edge research and development, some of which is still in its infancy awaiting Moore's, Metcalfe's, and Gilder's laws to move the industry a little farther along its path (Pinto, 2002).

Throughout this section, the authors discuss a cross-section of modeling and simulation applications used by the various branches of military service. Some of the applications began from the very outset with specific requirements, goals, and customers, while others were started in the great minds of engineers, scientists, and warfighters in search of new ways to improve upon warfighters' abilities. However, no matter how projects are started, developed, or challenged by funding issues or requirements creep, the bottom line is that these applications are developed with the best intentions to aid warfighters by physically, mentally, and emotionally immersing them in specific environments to prepare them for the next phase of training or the ultimate test on the battlefield.

RECENT PAST

As the book editors and chapter leads began to discuss a section on the applied use of simulation for military applications, I was quickly reminded of how far we have come just in the last few decades. Specifically, I thought of a meeting on June 24, 1996, at the Pentagon, when a Department of the Army Inspector General LTC Elms gave a 90 minute briefing to the Vice Chief of Staff of the Army on modeling and simulation management. During this briefing, the Vice Chief of Staff of the Army directed the U.S. Army staff to conduct a functional area assessment on modeling and simulation management due to the Vice Chief of Staff of the Army's three areas of concern.

a. The army is spending more on modeling and simulation than it can afford.
b. The army does not have total visibility on current modeling and simulation investments, especially operations and maintenance accounts.
c. The army cannot define the value added of its modeling and simulation investments.

Many changes occurred following this briefing, including organizational transitions, the drafting of the Army Investment Plan, and reallocations of funding (Department of the Army, 2005), among many other changes I will not go into within the space of this introduction. However, looking back on that historical meeting, it is important to note how the military, industrial, and academic communities rose to the challenge in answering those questions in the following decade, although some could debate that we still have a long way to go.

From either perspective on that debate, walking through the 400-plus exhibit booths at the annual Interservice/Industry Training, Simulation, and Education Conference (I/ITSEC; www.iitsec.org) in Orlando, Florida, which was visited by over 16,000 visitors from around the globe, reveals that something special has happened in the modeling and simulation community. The I/ITSEC experience and offerings clearly demonstrate the industry's worth, introducing a vast array of new technologies each year and highlighting the drive toward research in various fields, such as embedded simulation; interoperability for live, virtual, and constructive simulations; and holographic visual imaging, realistic environments of the large training centers to the smaller mobile military operations in urban terrain facilities being fielded all over the world to game based/mobile computing platforms and improvised explosive device trainers. That list clearly demonstrates an industry that has shown its value in a very short time.

Those early questions from a Vice Chief of Staff of the Army can be seen as a testament to the advances of the modeling and simulation community and programs. Many times during the early phases of technology growth there is an obligation by leadership not to stifle that growth (for example, the hundreds of unmanned systems companies competing for the Department of Defense marketplace today), and there is a "budget pass" given on funding efficiency in order to further science and that specific growth technology. However, when the big dollars start rolling toward a business area or technology, much more scrutiny is brought to bear. And that increased scrutiny is what the modeling and simulation industry saw in the mid-1990s. There should be no doubt that the big dollars are still rolling into the industry. For example, one estimate in 2005 by the National Training Systems Association suggests that government spending on U.S. military training, training support, and simulation is generally pegged at somewhere around 8 percent of overall departmental budgets, about $35 billion. In 2008, that total would be approximately $40 billion.

Whether or not this funding has fueled a renewed debate on those three questions presented over a decade ago, there is little debate on the potential benefit of modeling and simulation technologies when executed in the truest sense of best value, with correct pedagogy, and tied to acceptable and validated requirements for the military community. Under these conditions these technologies easily demonstrate their abilities to benefit the military, homeland defense, academia, and commercial applications.

Although the potential for the modeling and simulation industry is certainly there, and the military is certainly seizing that potential, it has not been without its challenges and pitfalls. Some of the pitfalls are driven by technological immaturity, others by political immaturity, and still others by the economy of bottom lines. The primary challenges the industry still faces are hardware, software engineering practices, software reuse, standards, interoperability, games management, industry and distribution rights, embedded training, and information protection. The chapters of this section discuss many of these challenges, so, as a means of transition, I will discuss three future trends for the next decade: embedded training, game based simulation, and the move toward Hollywood.

TRENDS IN THE MODELING AND SIMULATION COMMUNITY

Embedded Training

The move toward embedded training has been under way within the U.S. Army since 1983 when the U.S. Army Training and Doctrine Command (TRADOC) made it the preferred way of training the U.S. Army (Witmer & Knerr, 1996). Embedded training has a wide variety of definitions in the army, but one that serves to encompass the complex nature of the term is given by TRADOC Pamphlet 350-37, Training: Objective Force Embedded Training (OFET) Users' Functional Description (Department of the Army, 2003), which states that it is hardware and/or software, integrated into the overall equipment configuration, that supports training, assessment, and control of exercises on the operational equipment, with auxiliary equipment and data sources, as necessary. Embedded training, when activated, starts a training session, or overlays the system's normal operational mode, to enter a training and assessment mode. A case can be made that, to some degree, embedded training has been successful, especially in the U.S. Navy and the U.S. Air Force where many embedded training systems have been deployed. In a sense, embedded training may be a little easier to deploy in those services as opposed to the U.S. Army, in which the distribution and numbers of systems far outweigh the ability to move training down to the "boots on the ground."

A few embedded training programs, such as the multiple integrated laser engagement system, tank weapons gunnery simulation system, precision gunnery system, and some air defense systems have been highly successful. However, the challenges arise in the extreme complexity needed to meet the vision of the leaders for embedding all training for future systems on the systems themselves. For example, embedding an advanced gunnery training system into the future combat systems so that a soldier, a crew, and even a platoon can seamlessly train within their actual battle-tested vehicles while steaming across an ocean, waiting in a motor pool, or deployed in a desert is no easy task. All this must be accomplished while limiting or eliminating the need for any new hardware for the training to be accomplished in those systems. Embedded training remains a large challenge for our industry and one that will continue to be with us for some time. One of the lights somewhere near the end of the tunnel for embedded training is the reduction of operations and maintenance accounts costs and negative training as the actual battle-tested system is being used with concurrent software revisions. As long as the first thought on embedded training is a good pedagogy and solid requirements, embedded training has tremendous potential to save costs, provide solid returns on investments, and show great benefits. This would have made the Vice Chief of Staff of the Army smile.

Games—Individual, Team, Massively Multiplayer, and Mobile

Another trend for the foreseeable future is computer based learning, which has been going on for decades; however, recent advances in personal computer (PC) based graphics and image generation have led to the building of computer games and massively multiplayer (MMP) environments to train the military warfighter. The first attempt to use a computer game for the military was when Lt. Scott Barnett, a project officer, in Quantico, Virginia, working for the Marine Corps Modeling and Simulation Management Office, obtained a copy of the commercial *Doom,* released in 1993 (http://www.tec.army.mil/TD/tvd/survey/Marine _Doom.html). General Charles C. Krulak (1997), Commandant of the U.S. Marine Corps, issued a directive (Marine Corps Order 1500.55) to use war games for improving "Military Thinking and Decision Making Exercises." Moreover, he entrusted the Marine Combat Development Command with the tasks of developing, exploiting, and approving computer based war games to train U.S.

Marines for "decision making skills, particularly when live training time and opportunities were limited" (p. 1).

In the U.S. Army, the first army game was developed by the U.S. Army Simulation Training and Instrumentation Command. The game was called *Battle Command 2010* with the stated goal of being able to supplement lecture training by supplying a more engaging and stimulating system in order to provide warfighters with a better ability to retain information provided to them (Stottler, Jensen, Pike, & Bingham, 2002). With the hope of low cost solutions and relatively quick development schedules, games seem to be an avenue for the military to supplement live training and provide mission rehearsal and after action capability. After the success of *America's Army* or Army Game Project (http://www.americasarmy.com/), which was originally developed as a tactical multiplayer first-person shooter game for the public relations initiative to help with U.S. Army recruitment, gaming has now drifted into many aspects of the acquisition process.

No matter what generation is involved, a case can be made that humans like playing games. Games are here to stay in the military because warfighters want them and use them. The amount of use these games get is shown by the many hours warfighters will spend playing or training themselves on a game even during their free time both individually and with their teammates (Solberg, 2006). No matter if they are deployed, on the move, or at a home station, the games are capable of getting both attention and time from the warfighters. Warfighters freely choosing to use an approved computer based game for their personal downtime seems to be an ultimate achievement for the creators, trainers, and investors in military training games. Currently, many studies are under way to investigate the effectiveness of these games for schools, businesses, and government agencies. The outcome of these studies is wide raging; however, once you discount the obligatory statement noting that individual differences may play as large a role as any other factor, you get down to what the scientific community has concluded over and over again with games. Games bring to training certain elements: competition, engagement, repetition, fun, and individual pace, all of which are key components to memory and retention (Farr, 1986).

However, the many challenges of game based initiatives are discussed in this section. Some of these challenges include the digital divide (not everyone plays games) and the fact that some of these games are not built by folks who understand the cultural differences of the community they are trying to reach (Rosser, 2007). Other challenges include those faced by the larger-scale simulations, such as data rights, corporate business models, standards, interoperability, requirements creep, and most importantly, building a well-constructed game for good pedagogy, based on solid requirements that meet the goals set out for learning provided by dedicated trainers. One thing is for certain, based on the amount of games and game environments being built for industry, academia, and the military, games are here to stay, whether built in a MMP environment, computer based PC, or a mobile computing platform, such as a cell phone or iPod.

Hollywood

A trend that had been ramping up in the late 1990s and early twenty-first century, but which seems to be slowing now, was the U.S. Army move to solicit the help of Hollywood as a research and development agency for military training needs. This partnering plan served as the "big idea" for the army to further develop the future of military training and infuse the military and industrial base with new ideas (Ferren, 1999). As such, Hollywood seemed like an idea worth pursuing. Hollywood certainly can make immersive movies and excellent video games, and rides at Disneyland and Sea World capture the science of storytelling and immersion, making a two minute ride seem like an adventure.

Hollywood and the Pentagon have a long history of making movies together; in fact, throughout much of the last century the military has frequently needed Hollywood just as Hollywood has frequently needed the military. This mutually beneficial partnering can be seen as far back as the early days of silent films (Roberts, 1997). Hollywood producers get what they want—access to billions of dollars' worth of military hardware and equipment, such as tanks, jet fighters, nuclear submarines, and aircraft carriers, while the military gets what it wants —films that portray the military in a positive light, as well as films that help the services recruit new soldiers. This is well documented in the official U.S. Army publication *A Producer's Guide to U.S. Army Cooperation with the Entertainment Industry,* in which Lawrence (2005) states film productions seeking the army's assistance "should help Armed Forces recruiting and retention programs." This continues today as the military budgets are used for NASCAR, motorcycle racing, extreme events, and thousands of hours of commercial time during television shows or advertising during the World Series and the Super Bowl.

However, at times that partnership can become troublesome. This situation is what Hollywood journalist David L. Robb (2004) revealed in his book *Operation Hollywood,* quoting filmmaker Oliver Stone after he refused military assistance for his Vietnam War films *Platoon* and *Born on the Fourth of July,* "They make prostitutes of us all because they want us to sell out to their point of view" (p. 25).

One of the main reasons the army looked to Hollywood during the quest for immersive simulation technologies was to help develop an immersive virtual environment that would capture the spirit of the battlefield and immerse soldiers in a system best described like the holodeck from *Star Trek* (Pollack, 1999). This system was to be so real that soldiers would unquestionably believe they had been physically, mentally, and emotionally deployed to the actual battle. In 1999, when E! Entertainment ran a piece just over three minutes in length to promote the $45 million army investment into the University of Southern California Institute for Creative Technology stating at the promotion's end that the army could see this holodeck within the next two years, many legitimate acquisition professionals knew that this was an overly optimistic timeline. Perhaps the reason for the optimism was that Hollywood would easily be able to develop inventive immersive technologies with the artistry and creative genius of the Hollywood mindset, but the military and the entertainment complex are two very different communities, and it is going to take a lot longer to get to know each other with regard to the military's requirements based environment. Additionally, the business model of Hollywood dictates a small percentage of successes within a large number of trials. How many films must Hollywood make to achieve that one big blockbuster? If we are willing to give Hollywood the reins to creatively come up with a solution, then so be it. But get ready for many failures in order to achieve a few very big blockbuster solutions. One way or the other, this relationship is likely to be a rocky one where both parties frequently do not see eye to eye. As Bette Davis, the great American actress (1908–1989), once said, "Hollywood always wanted me to be pretty. But I fought for realism." Hollywood has hired more and more ex-military people to help in these efforts, and there is tremendous potential for the military/entertainment complex to provide solutions that are just as good as any solutions the military/industrial complex can provide. This should prove to be an interesting relationship to watch in the coming years.

CONCLUSION

While the modeling and simulation community has come a long way since the Pentagon briefing in 1996, many challenges remain. By minimizing the point solutions of the past and embracing new technologies, the road ahead is well paved. With such concepts as blended simulation, sometimes referred to as blended learning (BL), continual progress will be made (Alvarez, 2005). Concepts like BL support opportunities to harness the best of face-to-face interaction with live, virtual, constructive, linked, unlinked, embedded, and mobile technologies to deliver the advantages of all forms of learning together when appropriate. What is unique today about BL is that never before have trainers had so much overlap in their abilities to bring training to the warfighter. With our newfound technology and the ability to move it on the digital highway comes the responsibility of spending as much time as possible on the front end of training requirements and the back end with scientific training effectiveness evaluations ensuring we get the training to the warfighter on time and on budget while meeting the performance needed for knowledge and skill transfer. For this reason, we have chosen a wide variety of experts in research, development, and acquisition of military applications in order for the readers of this section to be exposed to simulation work that may provide a way for more warfighters to be trained in a more effective manner than ever before possible.

REFERENCES

Alvarez, S. (2005). Blended learning solutions. In B. Hoffman (Ed.), *Encyclopedia of educational technology*. Retrieved November 6, 2007, from http://coe.sdsu.edu/eet/articles/blendedlearning/start.htm

Department of the Army. (2003, June 1). *Training: Objective Force Embedded Training (OFET) users' functional description* (TRADOC Pamphlet No. 350-37). Fort Monroe, VA: Author, Headquarters.

Department of the Army. (2005, February 1). *Management: Management of Army models and simulations* (Army Regulation No. AR-5-11). Washington, DC: Author, Headquarters.

Elms, P. (1996, June 24). *DAIG briefing to the VCSA on modeling and simulation management.* Alexandria, VA: Pentagon.

Farr, M. (1986). *The long-term retention of knowledge and skills: A cognitive and instructional perspective* (IDA Memorandum Rep. No. M-205). Alexandria, VA: Institute for Defense Analyses.

Ferren, B. (1999, May–June). Some brief observations on the future of Army simulation. *Army Research, Development and Acquisition Magazine.*

Krulak, C.C. (1997, April). *Military thinking and decision making exercises* (Marine Corps Order 1500.55). Washington, DC: U.S. Marine Corps Headquarters, Department of the Navy.

Lawrence, J. S. (2005). Operation Hollywood: How the Pentagon shapes and censors the movies [Book review]. *Journal of American Culture, 28*(3), 329–331.

National Training Systems Association. (2005). *Training and simulation industry market survey 1.0.* Arlington, VA: Author.

Pinto, J. (2002). *The 3 technology laws.* San Diego, CA: Automation.com.

Pollack, A. (1999, August 18). Pentagon looks for high-tech help from film. *New York Times.* Available online: http://www.amso.army.mil/resources/smart/add-nfo/articles/ict.htm

Robb, D. L. (2004). *Operation Hollywood: How the Pentagon shapes and censors the movies.* Amherst, NY: Prometheus Books.

Roberts, R. (1997). Sailing on the Silver Screen: Hollywood and the U.S. Navy. *The American Historical Review, 102*(4), 1246.

Rosser, J. B. (2007, October). *We have to operate on you, but let's play games first!* Keynote Address Follow-Up presented at the Learning2007 Conference, Orlando, FL.

Solberg, J. (2006, August). *Researching serious games: Asking the right questions.* Paper presented at the Intelligent Tutoring in Serious Games Workshop, Marina del Rey, CA.

Stottler, R. H., Jensen, R., Pike, B., & Bingham, R. (2002, December). *Adding intelligent tutoring system to an existing training simulation.* Paper presented at the 2002 Interservice/Industry Training, Simulation & Education Conference, Orlando, FL.

Vintage Flying Museum. (2005, September). Fort Worth, TX: Meacham International Airport [http://www.vintageflyingmuseum.org/]

Witmer, B. G., & Knerr, B. W. (1996). *A guide for early embedded training decisions—Second edition* (Research Product No. 96-06). Alexandria, VA: U.S. Army Research Institute for Behavioral and Social Sciences.

Part I: Systems Engineering and Human-Systems Integration

Chapter 1

SYSTEMS ENGINEERING APPROACH FOR RESEARCH TO IMPROVE TECHNOLOGY TRANSITION

Denise Nicholson and Stephanie Lackey

The scientific community is typically trained within conventional sciences, such as physics, psychology, and computer sciences, based on the scientific method for research. However, in this highly competitive era, the need to deliver ready-to-be-applied or transitioned products from research, in addition to publishing conclusions, is ever increasing.

This emphasis on deliverable products creates a need for project teams to be multidisciplinary and to bring a more rigorous project management technique to research projects, such as systems engineering procedures typically utilized for the development of complex systems, for example, automobiles and planes. This chapter addresses a novel approach adapted from a more complex environment to merge with the scientific method. Following this approach will result in research that can answer both of the following questions: "What is the science?" and "What is the deliverable product?"

SCIENCE AND TECHNOLOGY FUNDAMENTALS

Advancement of scientific methodologies and tools through research and development is intended to benefit follow-on enterprise elements. Research and development enables the introduction of products, systems, and capabilities in order to "maximize value realized and minimize time until realization" (Bodner & Rouse, 2007). Efforts typically correlate to three phases: basic research, applied research, and advanced technology development (Department of Defense, 2006).

Basic research focuses on the development of theoretical underpinnings to future technological advancement through systematic investigation (Department of Defense, 2006). These research activities do not focus on specific applications, hardware, or products. However, success at this level may transfer results to subsequent research in applied projects. One example would be an experiment designed to study cognitive processes and memory using college student participants who complete basic tasks, such as recalling a list of numbers, words, or objects, within a controlled laboratory setting. It is typically challenging to interpret the impact of the results to real world applications.

Applied research aims to address specified needs through targeted development and advancement of an existing knowledge base. Results of these efforts may be presented as designs, systems, methods, or prototype devices in response to general mission requirements. Frequently, products provide insight into initial technical feasibility or potential technical solutions to general military needs (Department of Defense, 2006). In applied research, the above study would change by asking participants to perform a task similar to an operational task within a laboratory setting, such as simulated driving, operating a control panel, or making decisions based on information presented on a display. The results are easier to translate into use since real requirements are taken into account when the original problem is defined.

Advanced technology development concentrates on subsystem and component development intended for integration into field experimentation or simulation efforts (Department of Defense, 2006). Technology demonstrations provide opportunities for product exposition, initial testing outside of a laboratory environment, and technology readiness reviews. Success at this level indicates that a technology should be made available for transition. An example would be the study of interface designs to improve cognitive processes during a real operational task. Experiments can include actual operators as participants within operational environments where the system will be used. In this case it is imperative that the end-use environment be well understood; otherwise, the results could go unused.

Current trends in research indicate increased demand for products earlier in the traditional research cycle. The increased demand for insertion of innovative products often correlates to increased technical, cost, and schedule risks. Systems engineering (SE) principles and practices offer opportunities to address the technology transition challenges faced by such endeavors.

SYSTEMS ENGINEERING FUNDAMENTALS

The genesis of modern system engineering practices emerged from the weapons race between the United States and the Soviet Union following World War II. The complexity of post–World War II military projects increased exponentially compared to their predecessors (Hallam, 2001). Prior to World War II, military departments would rely on a "prime contractor" with expertise in the field required (for example, aircraft) to manage subcontractors and develop military systems. However, the Cold War motivated the integration of aircraft characteristics into advanced weapon systems. The technical advancement of weapon system technology resulted in significantly increased complexity and required an alternative development approach: a systems engineering approach. The Atlas Intercontinental Ballistic Missile Program served as the flagship project upon which systems engineering principles were founded (Hughes, 1998). This project team is credited with pioneering the development of quantitative methods that form the basis of analytic tools and decision aids still prevalent in the twenty-first century (Hallam, 2001; Hughes, 1998).

From an academic perspective, the International Council on Systems Engineering (2004) summarizes systems engineering as an "interdisciplinary approach and means to enable the realization of successful systems." SE requires balancing several related developmental components: operations, performance, testing, manufacturing, cost and schedule, training and support, and disposal. In addition, the Council also emphasizes the importance of early identification of customer needs and functional requirements, and documentation of those requirements before system design and validation occur (International Council on Systems Engineering). Thus, the underlying philosophy of systems engineering focuses on what system entities do prior to determining what the entities are (Badiru, 2006).

SYSTEMS ENGINEERING APPROACHES

Since systems engineering is not dictated by physical properties leading to strict mathematical relationships, an abundant variety of approaches exists. Applications of systems engineering principles often differ depending on the nature of each specific project. However, Bahil and Dean (2007) specify seven tasks inherent to any systems engineering approach:

1. State the problem.
2. Investigate alternatives.
3. Model the system.
4. Integrate.
5. Launch the system.
6. Assess performance.
7. Reevaluate.

This process is known by the acronym SIMILAR (Bahil & Gissing, 1998). Commonly used systems engineering approaches that incorporate SIMILAR components include the waterfall method, the spiral approach, and the "Vee" model.

The traditional waterfall method is composed of successive steps leading from problem formulation to system testing (Royce, 1970). Each of the typical waterfall phases is followed by a review of progress and documentation in order to determine whether the project is prepared to proceed to the next phase. This approach represents a structured engineering methodology aimed at designing and constructing large, complex systems.

The waterfall method is easy to understand and well recognized. However, the waterfall approach fails to allow for adequate executive control, nor does it accommodate highly complex systems (Wideman, 2003). Furthermore, the waterfall approach is time consuming and costly, and prototyping is not accommodated (Lackey, Harris, Malone, & Nicholson, 2007).

Spiral approaches also include a series of steps to be completed in succession, but multiple iterations of the process are planned. For example, project objectives, requirements, and constraints provide input for requirements analysis. Lower level functions are then decomposed during functional analysis. Next, the design phase identifies and specifies the elements required to produce system components meeting the specified requirements. Components are then developed based upon the established design. Feedback loops provide opportunities to revisit previous phases if necessary. Each iteration constitutes a spiral and is followed by a progress and documentation review by an oversight function. Oversight typically takes the form of management responsible for trade-off analyses, decision support, scheduling, and integration of technical disciplines (Lackey et al., 2007). Multiple spirals are employed until completion of the final product.

The spiral approach provides greater flexibility than the waterfall approach. Each phase is preceded by a requirements phase. Amended requirements and prototyping may be incorporated if required (Lackey et al., 2007). Thus, the spiral approach facilitates faster development, but still requires a significant time investment. Additional advantages of the spiral method include close interaction with the user, iterative requirement refinements, and the cyclic development that continues until the product is accepted. A key disadvantage also exists. Executive control is challenged by a lack of schedule and budget accountability on the part

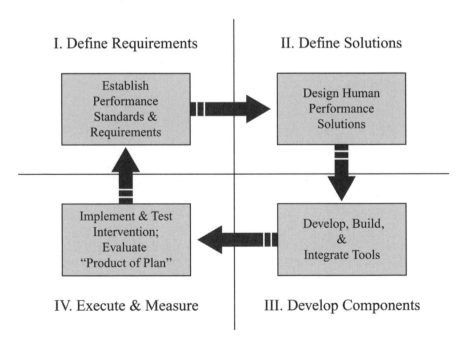

Figure 1.1.   The Human Performance Systems Model: Integrating the Scientific Method and Systems Engineering

of developers. Without disciplined implementation of the spiral method, indefinite additions contribute to cost and schedule overruns (Wideman, 2003).

The Vee model represents a linear approach that can be applied iteratively. Decomposition and definition of tasks comprise the downward slope of the "V." Product requirements, design specifications, and test plans are documented. Fabrication and assembly follow. Once built, component verification and validation lead to system integration, testing, and demonstration on the upward slope of the V. Positive attributes to the approach involve well-documented requirements and designs, which facilitate auditing purposes. However, documentation processes can be cumbersome, and stakeholder accountability may be questionable.

An alternative approach that incorporates elements from the previously described approaches is known as the human performance systems model (HPSM). Establishment of the HPSM reflects the U.S. Navy's desire to apply system processes to address human performance issues. However, recent research has demonstrated the applicability of HPSM to technical research, development, and engineering. The HPSM (see Figure 1.1) is comprised of four iterative phases: (1) define requirements, (2) define solutions, (3) develop components, and (4) execute and measure. Each of the HPSM phases is presented in detail in the figure.

Table 1.1 summarizes the steps for each of the systems engineering approaches.

HPSM: INTEGRATING THE SCIENTIFIC METHOD AND SYSTEMS ENGINEERING

Although the genesis of the HPSM resides within human systems development, the desire for products within the uncertainty of research endeavors makes the HPSM a viable alternative to traditional systems engineering approaches. This model has demonstrated applicability to prototype and research efforts due to its organized, yet flexible, structure (Lackey et al., 2007). The growing complexity of technology development amplifies the need to carefully plan and execute highly advanced research. Sauser (2007) stresses the importance of identifying appropriate systems engineering approaches to facilitate project success. Both positive and negative consequences must be considered when choosing a systems engineering method. The categorization of work to be performed plays an important role in systems engineering method selection (Bodner & Rouse, 2007; Saucer, 2007; Maier, 2006).

HPSM Phase I—Define Requirements

Requirements established through knowledge acquisition are documented and defined based upon the theories and techniques previously covered in Volume 1, Sections 1 and 2 of this handbook. Establishment of baseline or standard performance criteria is the focus. Statement of the problem occurs here as in the SIMILAR method. Constraints and derived requirements are delineated in a

Table 1.1. Systems Engineering Approaches

Step	Scientific Method	SIMILAR	Waterfall	Spiral	Vee	HPSM
1	State the problem	State the problem	Define requirements	Analyze system functionality and define requirements	Define requirements	Specify knowledge acquisition and requirements
2	Make observations	Investigate alternatives	Specify design	Specify design	Specify design and verification plan	Define solutions and specify design
3	Form a hypothesis	Model the system	Build	Develop components	Fabricate and assemble	Model system and develop components
4	Perform experiment	Integrate	Test	Test performance	Verify and validate components	Conduct experimentation and measure performance
5	Draw and publish conclusions	Launch the system	Deploy	Provide feedback	Integrate and test	Deliver results: prototype and reports
6		Assess performance		Iterate	Iterate	Iterate
7		Reevaluate				

similar manner to the spiral method. Like the Vee model, input for test plans and evaluation products are also crafted in this phase.

In HPSM Phase I, researchers focus on functionality, scope, and performance criteria. These aspects are defined to the level of detail commiserate with the level of effort allowable. Due to the typically smaller investment in research, as opposed to full cycle acquisition (executed under the waterfall, spiral, or Vee methods), the type of freedom provided by the HPSM is beneficial. Although the formality of reporting may be reduced, traceability may be maintained.

HPSM Phase II—Define Solutions

The second phase presents solutions to the problems and requirements defined in Phase I of the HPSM. (For VE training and education systems, solutions can include the methodologies and technologies described in Volumes 1 and 2 of this handbook.) Clear parallels exist between the activities of this phase of the model and SIMILAR's "investigation of alternatives." This phase comprises typical design tasking found in the waterfall, spiral, and Vee models.

The design phase in research benefits from the simplicity of the HPSM. Large-scale programs warrant formal design review processes. However, research projects suffer from programmatic overload when required to function as full acquisition programs. For example, a preliminary design review for an acquisition program requires the same level of programmatic effort (for example, completion of formal and labor-intensive entrance and exit criteria checklists) as a critical design review. Additionally, the time frame between preliminary design review and critical design review may be several months or more depending on the level of system complexity. Within research the appropriate level of effort for a preliminary design review may be peer review, and the entrance and exit criteria may not require intervention by program management. Moreover, the time between preliminary design review and critical design review may be a period of weeks, rather than months. The flexibility of the HPSM permits the research team to tailor systems engineering processes to meet each project's unique needs without sacrificing quality.

HPSM Phase III—Develop Components

Next, Phase III focuses on component development that typifies the other systems engineering methods discussed. (Many of the chapters in this section of the handbook describe prototypes developed via this process.) The team will finalize the design based on feedback from the design review. In relation to the scientific method, the developers "hypothesize" that the design will satisfy the requirements. Any requirements that cannot be satisfied with readily available approaches are documented as new hypotheses for further scientific exploration.

Many times the development is designed to be implemented in phases or builds. Each build is named using a version nomenclature, that is, v1.0, 1.5, 2.0, and so forth. Depending upon the final products to be developed and the iteration of the HPSM, each version of the development may focus on an initial or upgraded model, process, software, hardware, or the final integration of these components into a system.

HPSM Phase IV—Execute and Measure

Finally, the product's performance is measured through experimentation, test, and measurement against stated requirements. (See Volume 3, Section 2 for chapters related to this step of the process.) This phase shares similarities with the spiral method's verification loop and the Vee model's integration and verification branch. Like the spiral and Vee methods, the output of the execution and measurement phase provides input for the first phase of the next HPSM iteration. Multiple iterations, or spirals, of the HPSM are conducted based upon predetermined evaluation criteria (Human Performance Center, 2003). The results of experimentation verify that the design satisfies the requirements and hypothesis. Unfavorable results can be used as requirements for design improvements in the next cycle. In the meantime, the current version could be transitioned to the user as a prototype for initial implementation while the components are being upgraded. This approach is routinely seen in the software development field where initial "Beta" versions are released to users, and their feedback is a valuable step toward improving the final delivery.

HPSM Iteration

By iterating the HPSM execution, some of the spiral method advantages are available. However, clear metrics defined in Phase I alleviate the endless spiral syndrome addressed by the Vee model. Iterative applications of the HPSM allow the user to tailor systems engineering steps to an appropriate level. These features present advantages for smaller engineering efforts, and recent research findings (Lackey et al., 2007) indicate that this approach also reduces technical, cost, and schedule risks during prototype development.

SUMMARY AND CONCLUSIONS

The human performance systems model offers features similar to those found in the SIMILAR, waterfall, spiral, and "Vee" approaches, without the associated cost and schedule overhead. By enlarging the goal of the traditional scientific method to include the development of deliverables, HPSM provides a systems engineering alternative that can be customized to meet the needs of individual projects. The flexibility of the model lends itself to efforts that do not require exhaustive documentation and review cycles and/or efforts that may not be suited to complex, large-scale engineering efforts in need of detailed auditing.

This handbook has been organized to help facilitate the human performance systems model systems engineering process. Applying the model to research allows project managers to plan and track milestones and to align deliverables to transition program requirements. HPSM also permits research to proceed without unwarranted programmatic overhead. Thus, this process enables researchers to balance tasking to most effectively achieve mission objectives. As described in following chapters, research conducted via this process has resulted in successful programs with substantial contributions to both the scientific literature and delivered products that have transitioned to satisfied end users.

REFERENCES

Badiru, A. B. (2006). *Handbook of industrial and systems engineering*. Boca Raton, FL: CRC Press.

Bahil, A. T., & Dean, F. F. (2007). What is systems engineering? A consensus of senior systems engineers. *Systems Engineering*. Retrieved June 24, 2007, from http://www.sie.arizona.edu/sysengr/whatis/whatis.html

Bahil, A. T., & Gissing, B. (1998). Re-evaluating systems engineering concepts using systems thinking. *IEEE Transactions on Systems, Man, and Cybernetics—Part C: Applications and Reviews, 28*(4), 516–527.

Bodner, D. A., & Rouse, W. B. (2007). Understanding R&D value creation with organizational simulation. *Systems Engineering, 10*(1), 64–82.

Department of Defense (2006). *DoD financial management regulation 7000.14-R 2B budget formulation and presentation*. Retrieved October 26, 2007, from http://www.defenselink.mil/comptroller/fmr/

Hallam, C. R. A. (2001). An overview of systems engineering—the art of managing complexity. *Massachusetts Institute of Technology Research Seminar in Engineering Systems*. Retrieved June 19, 2007, from http://web.mit.edu/esd.83/www/notebook/NewNotebook.htm

Hughes, T. P. (1998). *Rescuing Prometheus*. New York: Pantheon Books.

Human Performance Center. (2003). *Human performance system model*. Retrieved November 18, 2005, from https://www.spiderhpc.navy.mil

International Council on Systems Engineering (2004). *What is systems engineering?* Retrieved June 22, 2007, from http://www.incose.org/practice/whatissystemseng.aspx

Lackey, S. J., Harris, J. T., Malone, L. C., & Nicholson, D. M. (2007). Blending systems engineering principles and simulation-based design techniques to facilitate military prototype development. *Proceedings of the 2007 Winter Simulation Conference* (pp. 1403–1409). New York: Institute of Electrical and Electronics Engineers.

Maier, M. W. (2006). System and software architecture reconciliation. *IEEE Transactions on Systems Engineering, 9*(2), 146–152.

Royce, W. W. (1970, August). Managing the development of large software systems. *Proceedings of IEEE WESCON, 26,* 1–9.

Sauser, B. (2007). Toward mission assurance: A framework for systems engineering management. *IEEE Transactions on Systems Engineering, 9*(3), 213–227.

Wideman, M. (2003). *Software development and linearity (or, why some project management methodologies don't work): Part 1*. Retrieved August 18, 2007, from http://www.maxwideman.com/papers/linearity/linearity1.pdf

Chapter 2

HUMAN-SYSTEMS INTEGRATION FOR NAVAL TRAINING SYSTEMS

Katrina Ricci, John Owen, James Pharmer, and Dennis Vincenzi

For the past several decades, there has been a surge of activity associated with human-systems integration (HSI). Affordability and human performance factors have forced the U.S. Navy, and the Department of Defense [DoD] at large, to reconsider its way of doing business. Within the U.S. Navy, recent HSI activity is seen in the context of policy and organizational changes, research and development activities, and educational initiatives (Pharmer, 2006).

Problems of today's acquisition processes echo issues raised in the 1970s (Smootz, 2003). Such acquisition programs as Aquila, a remotely piloted vehicle that was to capitalize on emerging technology, illustrated the need for well-defined HSI policies and procedures. This program started in 1979 and was originally estimated to cost $123 million for a 43 month development effort, followed by planned expenditures of $440 million for procurement of 780 air vehicles (U.S. General Accounting Office, 1997). By the time the army abandoned the program in 1987 due to cost, schedule, and technical difficulties, Aquila had cost over $1 billion (Smootz, 2003; U.S. General Accounting Office, 1997), and future procurement costs were expected to be an additional $1.1 billion for 376 aircraft (U.S. General Accounting Office). Yet even after lessons learned from acquisition programs such as Aquila, and considerable investment in initiatives, policies, and processes (for example, MANPRINT [manpower and personnel integration], HARDMAN [hardware versus manpower], and the DoD 5000 series), cost overruns, performance failures, and delivery delays continue to occur.

From a training perspective, failures in the acquisition and design process inevitably lead to problems for the training community (Office of the Chief of Naval Operations, 2001). Further, advances in weapon system technology compel the training community to continuously examine the knowledge, skills, and abilities and corresponding training technologies and methodologies needed for today's navy. Among those technologies is the application of virtual environments, affording not only an immersive training environment, but a tool to examine human performance in the systems engineering process as well.

In order to understand current issues in HSI and training, it is useful to examine the history associated with HSI, the implementation strategies applied to the navy and DoD acquisition programs, and the changes that have occurred that require further transformation in the manner with which we design, develop, deliver, and maintain total system performance.

Following a brief description of HSI, this chapter will detail the rise of HSI in the Department of Defense and the U.S. Navy. Advances in the processes and tools enhancing HSI and training will be discussed, particularly in the context of recent research and acquisition programs where advanced technology for both design and training applications are being used. Finally, this chapter will examine future issues and emerging challenges for HSI and training.

HSI DEFINED

While a number of definitions exist for describing HSI (for example, Booher, 2003; Defense Systems Management College, 2001; U.S. Department of Defense, 1999), the basic underlying elements are that it is a *process* that includes the *human component* in the context of *systems engineering*. It is a continuous, cyclical, and ever-evolving process over the course of system design that integrates human centric disciplines across the entire lifecycle of the system. These disciplines, or HSI domains, include manpower, personnel, training, human factors engineering, survivability, habitability, safety, and occupational health.

Integration processes occur at several levels: between the domains themselves, within the systems engineering process, and within the acquisition strategy. First, individual HSI domains must recognize implications of design options and decisions in other domains. The manpower, personnel, and training (MPT) and human factors engineering domains and the interactions inherent to those domains embody the importance of HSI processes. A push to reduce manpower has enormous implications for proper human factors engineering, the personnel required to man the system, and the training necessary to ensure successful performance. Failure to recognize this interaction can lead to subsequent system inadequacies.

A second interaction must take place between the HSI domains and the systems engineering process, which consists of the iterative execution of the activities to analyze and decompose system requirements, to perform functional analyses and allocations, and to synthesize these requirements and functions into a product baseline. At each of these phases in the process, decisions are made that will have an impact on who the end user will be (personnel), how many users will be required (manpower), the characteristics the system must have to support the user (human factors engineering), and how that user must gain the competencies (training) to safely (safety) and effectively operate and maintain the designed system. Without direct involvement in the systems engineering process of individuals with expertise in the HSI domains, these critical decisions impacting the end user are made with too much emphasis on hardware and software considerations and not enough emphasis on the end user.

To some degree, this past approach was less problematic than it is today, owing to the resourcefulness and resilience of end users. However, the navy has recognized that manning costs are a very large component of the total cost of ownership of a new system, and this knowledge has driven decisions to reduce the number of operators and maintainers of increasingly complex systems. In this environment of doing more with less, the inclusion of human considerations on equal par with hardware and software considerations in the systems engineering process has become much more important.

While human considerations need to be integrated into the systems engineering processes, a need also exists to integrate these considerations within the acquisition strategy and management processes. While systems engineering focuses on development of the end product, the acquisition strategy and management processes ensure that the product is developed to meet the performance capability requirements within cost and schedule constraints. Again, each of the HSI domains plays a role in trade-offs among cost, schedule, and performance that characterize the acquisition of a new system. Without advocacy for the end user in the process, decisions would be made on the basis of reducing cost or maintaining the acquisition schedule.

Of particular concern is the fact that huge advances in technology over the last several decades have made it possible to automate many tasks that, in the past, could be performed only by a human operator. On the surface, the widespread application of automation to a program would appear to resolve a number of the concerns of an acquisition program, including the cost of manpower and meeting performance goals. However, with the application of automation, the role of the end user moves from direct manipulator and maintainer of the system to a supervisor of, perhaps, multiple complex systems. This has implications on the knowledge, skills, and abilities required of the users (personnel and training), as well as the characteristics the system itself must have to support situational awareness and maintain manageable cognitive workload (human factors engineering) and protect against errors (safety). Thus, what may have appeared to be a simple solution creates a number of human-related challenges that can be addressed only by the inclusion of expertise on these issues into the daily trade-offs between cost, schedule, and performance that characterize the acquisition management process.

Unfortunately, one challenge for those who advocate human systems integration is that many of the benefits are not readily seen in the early phases of exploring and refining the developing technology when the most important decisions about the design of a system are being made. In fact, including human considerations can increase the cost of developing the system in the short term. However, minor changes made early in the system development process are likely to cost substantially less than modifications made after fielding a system. The following section provides a brief history of the roots of this cultural change within the military.

HSI: A BRIEF HISTORY

Imagine this scenario: a major military acquisition is running behind schedule. Costs are skyrocketing, and from what has been tested to date, system performance is significantly lower than promised. This certainly sounds like a scenario happening today. However, this story has been recurring for decades. Over 30 years ago, as the Vietnam era drew to a close and the ranks of the armed forces decreased, the U.S. military began a force modernization program. Technology insertion was seen as a secure undertaking that promised to increase capability and readiness and to help regain a powerful force. Unfortunately, new weapons systems incorporating advanced technologies often proved difficult to operate, maintain, and support. New systems were delivering far poorer performance than expected, forcing a demand for higher levels of manpower and for more highly educated personnel.

A passage from a 1981 General Accounting Office (GAO) report describes just a few of the problems the Department of Defense was experiencing:

A tank hatch that a soldier, clothed for winter, cannot fit through; a major shipboard fire control system that cannot be adequately supported; aircraft test equipment that causes more problems than it solves; and a handheld missile that when fired startles the person that fires it, resulting in misses, are some examples of the problems with currently fielded weapon systems.

(p. 4)

The purpose of the 1981 GAO report was to identify some of the more prominent causes of problems with acquiring and fielding major weapon systems and to recommend some meaningful actions to reduce problems with deployed systems in the future. The report focused on a concept termed "ownership considerations"—the factors other than cost, performance, and schedule that influence the effectiveness of a weapon system. These considerations today are termed "ilities" and include aspects of the acquired system that manifest themselves after the system is delivered. Maintainability, survivability, interoperability, and transportability are just a few of the factors that, when not considered in the system design process, can force huge cost surges during the lifecycle of the system.

Policy, Practices, and Technology

The acquisition issues of the post-Vietnam era stirred a number of policy and process changes designed to avoid the types of pitfalls experienced within system acquisition. More recently, research efforts have taken aim at underlying technology advances that on one side allow trade-offs between manpower and design (for example, automation to reduce manning), yet foster the development of personnel to meet new knowledge, skill, and ability targets (for example, advanced training technologies).

The Department of Defense Directive (DoDD) 5000.1 and the accompanying instruction (DoDI 5000.2) were first released in 1971 and 1975, respectively. Both were seen as a mechanism for effectively managing defense acquisition and controlling cost growth. This first DoDD 5000.1 was relatively small—only seven pages in length, but contained the cornerstone for future releases of the directive: centralizing policy, decentralizing execution, and streamlining organizations (Ferrarra, 1996). Key components and continuing themes in DoDD 5000.1 included the need for acquisition workforce competency and clear and logical requirements definition. Most recently, the May 2003 issuance of DoDD 5000.1 specifies that the acquisition program manager "shall apply human systems integration to optimize total system performance" (Ferrara).

In 1985, long before the DoDD requirement to apply HSI, the U.S. Navy introduced HARDMAN as a process to inject manpower and personnel considerations into system design for all major acquisition categories. The process utilized a baseline comparison system to project MPT requirements. In doing so, program managers could make decisions on alternate designs in order to avoid acquisitions that could prove costly to operate and maintain. However, the HARDMAN methodologies themselves were cost and labor intensive and, thus, not always totally embraced. In reaction to this, the navy produced the training planning process methodology (TRPPM). The TRPPM prescribed a much more tailored approach to MPT analysis, allowing that smaller systems—systems not as likely to produce huge MPT dilemmas—required less analysis.

In a similar effort, the army introduced the MANPRINT program in 1982. While the HARDMAN and TRPPM processes dealt specifically with analysis of MPT, MANPRINT is divided into seven domains: manpower, personnel capabilities, training, human factors engineering, system safety, health hazards, and soldier survivability. Although each domain is called out as a separate entity, in practice there is considerable overlap, and, in fact, the success of MANPRINT relies on the interaction of the seven domains. Throughout the acquisition process, the backbone of MANPRINT relies on the constant communication, interaction, and coordination between the MANPRINT domains, the program manager, and program integrated product teams (IPTs).

Advanced Technology and Training Considerations

While policy and guidance provide a strong supporting plank for sound HSI practices, ever-evolving technology must also be addressed. As Foushee (1990) argues, improved design and increased automation must be met with corresponding changes in training. Where previously emphasis was placed on individual knowledge and skills, an added emphasis must be placed on the interactions of a team and the judgment and decision-making skills of its members.

Several research efforts have helped define and optimize this emerging training challenge. The tactical decision making under stress research program was specifically designed to meet the increasingly complex decision-making environment through both decision support tools and training technologies. The events of July 3, 1988, in which the USS *Vincennes* mistakenly shot down an Iranian airbus killing all 290 passengers and crew onboard, dramatically emphasized the relationship between advance technology and training considerations. Under a perceived threat and operating in a hostile environment, the crew members of the *Vincennes,* a guided missile cruiser (CG-49) with the U.S. Navy's most sophisticated battle-management system, mistook the Iranian airbus for a probable hostile F-14 (Collyer & Malecki, 1998). Subsequent research provided invaluable knowledge on the human decision-making process (for example, Zsambok & Klein, 1997), the development of critical thinking skills (for example, Cohen, Freeman, Wolf, & Militello, 1995), the behaviors that characterize high performing teams (for example, Smith-Jentsch, Johnston, & Payne, 1998), and the training and performance measurement strategies that produce and identify successful team performance (for example, Smith-Jentsch, Payne, & Johnston, 1996; Cannon-Bowers & Salas, 1998).

Additional research has examined the use of technology within the context of team training. Team training in a complex environment, such as a shipboard combat information center, can require a large number of trainers in order to observe and record performance and provide feedback in a timely manner. Embedded training, defined as simulation based training seamlessly integrated into the operational setting (Lyons & McDonald, 2001), provides an opportunity to practice and to receive feedback on critical job skills within the context of the job. Further, advance technologies minimize the number of instructors needed in order to conduct training (Lyons & McDonald). The application of automated performance recording capabilities, the tracking of trainee performance as compared to experts, and the automation of the feedback process provided the opportunity to practice and learn individual and team skills without the traditionally manpower intensive context.

HSI and Training Systems

Training is not only a domain of HSI—there is also an element of HSI in the design of training systems. As one example, the LPD-17 San Antonio class, the navy's newest class of amphibious assault ships, embraced training as an important design consideration. As the functional replacement for the 41 ships of the LST-1179, LKA-113, LSD-36, and LPD-4 classes built in the 1960s, there was little doubt that ships of the San Antonio class would be called upon to support a wide range of missions. With that in mind, the LPD-17 class training required flexibility and adaptability to changing missions and demands (Phillips et al., 1997). Optimized manpower levels also underscored the need for organic training capabilities—covering introductory, or familiarization training, all the way to total crew team training. Further, it was recognized that the addition of automation that allowed for a reduction in crew size would require additional training resources, as the expertise required to work with new technology would require additional practice and exposure to a variety of environmental cues.

As with the introduction of any capability, training considerations are required early in the design process. Requirements from the LPD-17 required operational capabilities document included provisions to provide fleet training services and to maintain readiness by providing for training of one's own unit's personnel, including onboard medical personnel and the embarked U.S. Marine Corps. Thus, critical training components, such as dedicated training compartments, embedded training systems, and an onboard training management system, can be traced to early program documentation. Further, the development of these capabilities necessitated the formal requirements documentation, software/hardware requirements reviews, and tests and evaluations as would be required for any capability.

The development of training resources for the LPD-17 took into account the emerging technology by providing an electronic resource center and an advanced electronic classroom, as well as supporting access to electronic training through any of the over 300 shipwide area network drops located throughout the ship. Design considerations were also impacted by the need to support a virtual environment training capability for the Marine Corps.

Finally, the LPD-17 training program worked along side the manpower and personnel IPTs in defining the personnel, knowledge, and skills necessary to man the five-crew-member training department—a new department onboard the LPD-17 whose tasks represented jobs usually allocated as collateral duties onboard other navy ship classes.

As a part of a larger ship design team, over 100 workshops were conducted that made use of three-dimensional modeling capabilities that allowed the visualization of sailors and marines moving through the food service lines, stretcher movement through triage, fully equipped marines moving through air locks and stairs, forklifts operating in storage areas, and even trainees entering and engaging in learning in the advanced electronic classroom. These workshops produced feedback and design recommendations that significantly enhanced the ship's design. Individuals were able to visualize design problems both small (for example, a phone located too far from a workstation) and large (for example, a welding shop located too close to a fuel compartment) that could be easily remedied early in the engineering process, but would be costly if not impossible later on.

THE FUTURE OF HSI AND NAVY TRAINING

Certainly, the challenges of integrating human considerations into system design are still visible in Department of Defense acquisition. However, there have been a number of innovations in both processes and technologies that are paving the way for how we conduct human systems integration. Although technology can afford newer tools and processes to support the infusion of human considerations into system design, changes will always continue in the technology associated with modern weapons systems. As these changes occur, the training community—as well as other HSI domains—must also evolve.

One emerging challenge for the HSI community is the use of unmanned vehicles, prominently as aviation assets, but as surface and subsurface platforms, as well. The use and demand for autonomous vehicles (AVs) has risen dramatically since the onset of the Global War on Terrorism and military operations in both Iraq and Afghanistan. In fact, a recent National Research Council study recommended that the navy accelerate the introduction of existing AVs and pursue new AV concepts and technologies (U.S. Department of Defense, 2005).

As the surge of activity for a relatively new military asset continues, ironically, there is a concurrent debate related to the manpower, personnel, training, and human factors issues associated with an "unmanned" system. With the mishap rate for unmanned aircraft systems much higher than that of the manned flight community (Tvaryanas, Thompson, & Constable, 2006), a growing body of empirical data strongly suggest the need for immediate research and development to address fundamental human performance areas associated with the current inventory of AVs. Such questions as how many vehicles a single operator or control station can operate, whether or not the operator should be a certified aviator, and how much training and what type of training is necessary, are all unique challenges for the HSI community.

More and more, advanced training technologies are providing the tools to meet the training demands of today's navy. Embedded training capabilities, networked personal computer based simulation systems, and distant learning technologies provide contextual opportunities to practice and hone both individual and team skills. Further, immersive interactive training applications provide not only training capabilities, but offer opportunities to study design consideration and their impact on human performance.

REFERENCES

Booher, H. (2003). *Handbook of human systems integration.* Hoboken, NJ: Wiley & Sons.

Cannon-Bowers, J. A., & Salas, E. (1998). Individual and team decision making. In J. A. Cannon-Bowers & E. Salas (Eds.), *Making decisions under stress* (pp. 17–38). Washington, DC: American Psychological Association.

Cohen, M. S., Freeman, J. T., Wolf, S. P., & Militello, L. (1995). *Training metacognitive skills in naval combat decision making* (Tech. Rep. No. 95-4). Arlington, VA: Cognitive Technologies.

Collyer, S. C., & Malecki, G. S. (1998). Tactical decision making under stress: History and overview. In J. A. Cannon-Bowers & E. Salas (Eds.), *Making decisions under stress* (pp. 3–15). Washington, DC: American Psychological Association.

Defense Systems Management College. (2001, January). *Defense acquisition acronyms and terms* (10th ed.). Washington, DC: U.S. Government Printing Office.

Ferrara, J. (1996, Fall). DOD's 5000 documents: Evolution and change in defense acquisition policy. *Acquisition Review Quarterly,* 109–130.

Foushee, C. H. (1990). Preparing for the unexpected: A psychologist's case for improved training. *Flight Safety Digest* [Electronic version]. Retrieved July 28, 2007, from, http://www.mtc.gob.pe/portal/transportes/aereo/aeronauticacivil/alar_tool_kit/pdf/fsd_mar90.pdf

Lyons, D. M., & McDonald, D. P. (2001). Advanced embedded training with real-time simulation for Navy surface combatant tactical teams. In M. Smith & G. Salvendy (Eds.), *Systems, social and internationalization design aspects of human computer interaction* (Vol. 2, pp. 859–863). Hillsdale, NJ: Lawrence Erlbaum.

Office of the Chief of Naval Operations. (2001, August). *Revolution in training executive review of Navy training.* Washington, DC: Author.

Pharmer, J. (2006). The challenges and opportunities of implementing human system integration into the navy acquisition process. *Defense Acquisition Review Journal, 14*(1), 279–291.

Phillips, D., Sujansky, J., Hontz, E. T., Cannon-Bowers, J. A., Salas, E., & Villalonga, J. (1997). Innovative strategies and methods for total ship training on LPD-17. *Proceedings of the 19th Annual Interservice/Industry Training, Simulation and Education Conference* (CD-ROM). Arlington, VA: National Training Systems Association.

Smith-Jentsch, K. A., Johnston, J. H., & Payne, S. C. (1998). Measuring team-related expertise in complex environments. In J. A. Cannon-Bowers & E. Salas (Eds.), *Making decisions under stress* (pp. 61–87). Washington, DC: American Psychological Association.

Smith-Jentsch, K. A., Payne, S. C., & Johnston, J. H. (1996, April). *Guided team self-correction: A methodology for enhancing experiential team training.* Paper presented at the 11th annual conference of the Society for Industrial and Organizational Psychology, San Diego, CA.

Smootz, E. R. (2003). Human systems integration and systems acquisition interfaces. In H. R. Booher (Ed.), *Handbook of human systems integration* (pp. 101–119). Hoboken, NJ: Wiley & Sons.

Tvaryanas, A. P., Thompson, W. T., & Constable, S. H. (2006). Human factors in remotely piloted aircraft operations: HFCAS analysis of 221 mishaps over 10 years. *Aviation, Space, and Environmental Medicine, 77,* 724–732.

U.S. Army. (2005). *MANPRINT handbook.* Retrieved July 26, 2007, from http://www.manprint.army.mil/manprint/mp-ref-works.asp

U.S. Department of Defense. (1999, May). *Department of Defense handbook: Human engineering program process and procedures* (MIL-HDBK-46855A). Retrieved April 25, 2008, from http://hfetag.dtic.mil/docs-hfs/mil-hdbk-46855a.pdf

U.S. Department of Defense, Office of the Secretary of Defense. (2005). *Unmanned aircraft system roadmap 2005–2030.* Retrieved March 2, 2006, from http://www.acq.osd.mil/usd/uav_roadmap.pdf

U.S. General Accounting Office. (1981). *Effectiveness of U.S. forces can be increased through improved weapon system design* (Report No. PSAD-81-17). Washington, DC: General Accounting Office.

U.S. General Accounting Office. (1997, April). *Unmanned aerial vehicles: DOD's acquisition efforts* (Report No. T-NSIAD-97-138). Washington, DC: Author.

Zsambok, C., & Klein, G. (1997). *Naturalistic decision making—Where are we now?* Mahwah, NJ: Erlbaum.

Chapter 3

VIRTUAL ENVIRONMENTS AND UNMANNED SYSTEMS: HUMAN-SYSTEMS INTEGRATION ISSUES

John Barnett

People work with unmanned vehicles (UVs) and similar systems in a considerable number of applications. The employment of these systems has increased significantly in recent years, and indications are that the number of applications for UVs will increase in the future. Unmanned systems may include teleoperated vehicles or fixed systems where the operator controls the system's actions, or they may be semi-autonomous vehicles that perform certain functions on their own under the operator's direction. UVs are used primarily where it is dangerous or costly for humans to work. For example, they may explore other planets (Jet Propulsion Laboratory, 2007), neutralize improvised explosive devices (Lawlor, 2005), or operate in dangerous industrial environments (Brumson, 2007).

UVs are relevant to virtual environments (VEs) because there are a number of advantages to using VEs to research unmanned systems (Evans, Hoeft, Jentsch, Rehfeld, & Curtis, 2006) and also to train users to operate them. Unmanned systems are controlled primarily by software. Similarly, objects operating in VEs are controlled by software. Although the lines of code may be different, the core architecture is the same. This similarity means that, from a human operator's perspective, objects modeled in VEs will tend to act similar to robotic systems in the real world. Thus, the behavior of virtual unmanned systems can be used as an analog to real unmanned systems.

This can be important because humans and UVs sometimes do not work together well. When people give directions to semi-autonomous systems, they can be unpleasantly surprised by the way the system carries out their commands, a phenomenon known as "automation surprise" (Sarter & Woods, 1997). This disconnect occurs because the way people process information is significantly different from the way software processes information, which means that human-system integration can be a challenge, both in the real world, and, correspondingly, in VEs.

Unmanned systems are a form of automation; thus to understand how VEs can facilitate their use, it is important to understand the benefits and challenges of automation. Understanding some of the fundamental differences between how humans and automation operate might illustrate why communication failures and surprises occur and how they could possibly be avoided.

THE PROMISE AND CHALLENGES OF AUTOMATION

Considerable research has been done in the last several decades about how people work with automation. This research is relevant to UVs since many of the issues involving people working with automation apply to people operating UVs. Automation was originally seen as a means of reducing (human) errors and also reducing human workload (Billings, 1997; Bowers, Oser, Salas, & Cannon-Bowers, 1996). What research has found is that, although automation has significant benefits, there are also new challenges associated with human-automation interaction (Billings, 1997), including workload, operator trust, automation failures, and human/automation interface problems.

Workload/Boredom Continuum

Automation often does not reduce human workload as much as change it from performing functions to monitoring the automation. When it does reduce workload, it can come at the wrong time, so that the operator experiences boredom with its consequent loss of vigilance (Billings, 1997). Since vigilance is necessary for monitoring, if the automation malfunctions, the operator will be less likely to notice it immediately.

Dealing with Automation Failure

If the automation does malfunction, it is frequently difficult for the operator to recover from the failure (Sarter & Woods, 1997). Often this is because the operator may not be aware of what the automation was doing at the time of the failure (known as being "out of the loop"), may not notice the failure right away, and may have difficulty reverting to manual means of performing the function.

Trust and the Complacency/Distrust Continuum

The way people establish a level of trust in semi-autonomous and automated systems generally mirrors how they trust other people (Muir, 1994). However, there is one major exception: people who have little experience with a system tend to put too much trust in it. They tend to have overconfidence in the ability of the system to perform its task, a phenomenon known as automation bias (Mosier, Skitka, Heers, & Burdick, 1998), and are surprised when the automation does not perform as expected.

Research shows that once people experience automation failure, they tend to lose confidence in the system. Their confidence is slowly restored as they see the system perform correctly, but it never reaches the pre-failure level (Lee & Moray, 1994; Eidelkind & Papantonopoulos, 1997).

Human-Automation Interface Problems

Human-automation teams are significantly different from human-human teams, and if this fact is not considered in the design of the human-automation interface, problems ensue. Often automated systems will have too many ways to do the same thing (known as mode proliferation), which tends to confuse human operators. This is especially true when feedback about what the automation is doing is not available to the operator, resulting in a lack of mode awareness.

A major reason that human-automation teams are significantly different from human-human teams is that people and software process information in very different ways. Some of the major differences in human and machine information processing involve what each "knows" and how that information is "known."

INFORMATION PROCESSING: SOFTWARE VERSUS "BIOWARE"

At the most fundamental level, people and software process information differently. Software processes information as a series of logic trees and finite algorithms. Digits and states are discreet, and a bit is either one or zero. On the other hand, people process information as probabilistic networks. A nerve impulse does not necessarily trigger a subsequent nerve; it increases (or decreases) the probability the subsequent nerve will fire. This fundamental difference means that software will tend to react the same way to the same stimulus, whereas people may react differently to similar stimuli presented at different times.

Another major difference between people and automated systems is that people perceive things by matching patterns, but automated systems normally require exact matches to perceive something. This means that people can recognize something if there are pieces missing or obscured, but semi-autonomous systems find it much more difficult.

One difference between people and automation that has probably the greatest impact on misunderstandings is that people share a vast amount of implicit knowledge about the world, but automated systems do not share this same knowledge. Because this knowledge is shared by nearly all humans, it is commonly understood and therefore unstated. For example, if one person asks another to go into a dark room and find an item, it is not necessary to tell him or her to turn on the light. Humans share the implicit knowledge that providing enough light to see is a preliminary step to conducting a visual search. Semi-autonomous systems do not share this implicit knowledge the way people do, and if they are not given explicit instructions, they do not perform as expected.

One example from a computer game may illustrate this point. In one game, the player leads a team of commandos with the mission of rescuing "hostages" held by "terrorists" in an urban environment. Aside from the human player, all of the other entities are computer-generated avatars. In one instance, the human player commanded an avatar teammate to throw a stun grenade in a room where it was suspected hostages were being held by terrorists. The avatar dutifully selected a stun grenade, pulled the pin, and threw it at the closed door. The grenade bounced off the door and stunned the team. Not surprisingly, the human player was startled at this turn of events. The player never thought to open the door, because a human teammate would have understood that as a preliminary step to throwing something into the room.

The reason people are surprised at the lack of implicit knowledge in automated systems is twofold. First, people's implicit knowledge of the world is so deeply ingrained it operates automatically (that is, without conscious thought). Therefore, it rarely occurs to them that the automated system would not have the same world knowledge. Second, people tend to interact socially with automated systems (Sundar, 2004) and frequently expect them to behave as people do (Bergeron & Hinton, 1985; Muir, 1994). When automation violates these expectations, people are caught by surprise. Fortunately, even with such fundamental differences, human-automation teams can work well together if a conscious effort is made to integrate people and automation.

IMPROVING HUMAN-SYSTEM INTERACTION

Given that there are sometimes some surprising disconnects between humans and UVs, there are two basic approaches to reducing these difficulties. The first is to provide comprehensive training to UV operators so that they understand how the UVs function at the most basic level and also to provide extensive practice so that the operators will overcome their natural tendency to think of the them in human terms. Although feasible, this approach is a time consuming and expensive process. It would require retraining behaviors that people have developed over decades of working with other humans. Such training would be time consuming and could easily break down under stress.

The second approach would be to program UVs to function more like operators would expect and to provide feedback to the operators so that they can better predict the system's actions. The advantage of this method is that, unlike people, changing the software for a UV changes its actions permanently. Software based systems do not need to be trained and retrained. The complicated part of this approach is that it would require user testing to identify where human-automation conflicts occur so that they could be addressed. However, once the incompatibilities are identified and addressed, the system would not only be easier for the operators to use, but could serve as a model for future user-friendly systems. Conversely, training operators to work with the system, as in the first approach, means each new operator would not only require initial training, but periodic retraining as well.

Programming semi-autonomous systems to act more like people expect them to is especially important in VEs when the system is an avatar modeling a human.

Obviously, avatars that model humans should be expected to act like humans as much as possible. If a person in a virtual environment can interact with an avatar in a natural way, it tends to increase his or her sense of presence in the VE (Schroeder, 2006).

From a practical standpoint, it may not be possible to make UVs act exactly like humans. There may be cases in which the technical challenges make it impractical to change the actions of the unmanned system. Therefore, the best technique may be a melding of the two approaches; that is, modify the software to reflect human expectations where possible, but, when necessary, train operators to understand system limitations so as to mitigate violations of user expectations and thus minimize surprises.

USING VES TO IMPROVE HUMAN-SYSTEM INTEGRATION

A VE may be the ideal environment to help improve the fit between people and semi-autonomous systems. Since UVs act similarly in a VE as they would in the real world, a VE can be used to assess the quality of the interaction between UVs and people under a variety of controlled simulated conditions. A VE can be used to identify those situations where misunderstandings occur between people and the unmanned systems. It can also be used to develop more human-friendly software.

The virtual world may be the best environment to test human-system fit when it is impractical or dangerous to test in the real world. For example, testing UVs that carry weapons such as unmanned combat air vehicles or unmanned ground combat vehicles requires considerable space to maintain safety. Virtual unmanned combat vehicles would not require the same large weapons ranges or have the same safety concerns as their real counterparts.

Testing human-system integration would include not only testing the software routines, but the physical interfaces as well. Again, a VE would be a good environment to test the interface since it is generally easier and less costly to make software changes versus building hardware interfaces. Some interface testing of this type is already being accomplished (Neumann, 2006).

Obviously, VE would be good for training operators of unmanned systems. VE is essentially a training environment. It could be used to train people how to operate the interfaces, as well as introducing operators to any automation quirks of the system.

CONCLUSION

Semi-autonomous UVs and similar robotic systems are likely to become commonplace in the future. For this to happen, the human operators and the automated systems must learn to work well together. The VE promises to be the ideal environment for designing, testing, and improving the human-systems interface, as well as training the operator to take advantage of the benefits of the unmanned system. Both VEs and unmanned systems are evolving technologies that promise to enhance significantly how people work with future technology.

REFERENCES

Bergeron, H. P., & Hinton, D. A. (1985). Aircraft automation: The problem of the pilot interface. *Aviation, Space, and Environmental Medicine 56*(2), 144–148.

Billings, C. E. (1997). *Aviation automation: The search for a human-centered approach.* Mahwah, NJ: Lawrence Erlbaum.

Bowers, C. A., Oser, R. L., Salas, E., & Cannon-Bowers, J. A. (1996). Team performance in automated systems. In R. Parasuraman & M. Mouloua (Eds.), *Automation and human performance: Theory and applications* (pp. 243–263). Mahwah, NJ: Lawrence Erlbaum.

Brumson, B. (2007). Chemical and hazardous material handling robots. Robotics online. Retrieved May 17, 2007, from http://www.roboticsonline.com/ public/ articles/ articlesdetails.cfm?id=2745

Eidelkind, M. A., & Papantonopoulos, S. A. (1997). Operator trust and task delegation: Strategies in semi-autonomous agent system. In M. Mouloua & J. M. Koonce (Eds.), *Human automation interaction: Research and practice* (pp. 46–52). Mahwah, NJ: Lawrence Erlbaum.

Evans III, A. W., Hoeft, R. M., Jentsch, F., Rehfeld, S. A., & Curtis, M. T. (2006). Exploring human-robot interaction: Emerging methodologies and environments. In N. J. Cooke, H. Pringle, H. Pedersen, & O. Connor (Eds.), *Human factors of remotely piloted vehicles* (pp. 345–358). Amsterdam: Elsevier.

Jet Propulsion Laboratory. (2007). Mars exploration Rover mission. Retrieved May 17, 2007, from http://origin.mars5.jpl.nasa.gov/overview

Lawlor, M. (2005). Robots take the heat. *Signal Magazine.* Retrieved May 17, 2007, from http://www.afcea.org/signal/articles/anmviewer.asp?a=692

Lee, J. D., & Moray, N. (1994). Trust, self-confidence and operators' adaptation to automation. *International Journal of Human-Computer Studies, 40,* 153–184.

Mosier, K. L., Skitka, L. J., Heers, S., & Burdick, M. (1998). Automation bias: Decision making and performance in high-tech cockpits. *International Journal of Aviation Psychology, 8*(1), 47–63.

Muir, B. M. (1994). Trust in automation: Part I. Theoretical issues in the study of trust and human intervention in automated systems. *Ergonomics, 37*(11), 1905–1922.

Neumann, J. L. (2006). Effect of operator control configuration on uninhabited aerial system trainability. *Dissertation Abstracts International B, 67*(11). (UMI No. AAT 3242458).

Sarter, N. B., & Woods, D. D. (1997). Team play with a powerful and independent agent: Operational experiences and surprises on the Airbus A-320. *Human Factors, 39*(4), 553–569.

Schroeder, R. (2006). Being there together and the future of connected presence. *Presence, 15*(4), 438–454.

Sundar, S. S. (2004). Loyalty to computer terminals: Is it anthropomorphism or consistency? *Behavior & Information Technology, 23*(2), 107–118.

Part II: Defense Training Examples

Chapter 4

U.S. MARINE CORPS DEPLOYABLE VIRTUAL TRAINING ENVIRONMENT

Pete Muller, Richard Schaffer, and James McDonough

Deployable Virtual Training Environment (DVTE) is an evolving U.S. Marine Corps program. It is not a monolithic system, but rather a framework to deliver individual and small-team training and mission rehearsal simulations on networked laptops. DVTE has changed significantly over the years, and it has absorbed a number of personal computer (PC) based training systems. Before we discuss the DVTE program, we will address the evolution of distributed simulation technologies in the Department of Defense (DoD) and commercial PC games because they each play a critical role in the development of DVTE.

DEPARTMENT OF DEFENSE DISTRIBUTED SIMULATION

In the early 1990s, the Defense Advanced Research Projects Agency (DARPA), under the leadership of Jack Thorpe, began a visionary program known as simulation network (SIMNET). At the time, the U.S. Air Force had very expensive aircraft simulators that were unable to interoperate. Thorpe envisioned a shared virtual environment where aircrews could train together even if they were not collocated. Ironically, the SIMNET program found a strong supporter not in the air force, but in the U.S. Army, and tank training became a major focus of the program (Cosby, 1999). This eventually led to millions of dollars of army investment, ultimately resulting in the close combat tactical trainer (CCTT) for the M1 Abrams tank. While the army developed the CCTT, Modular Semi-Automated Forces (ModSAF) was developed with DARPA funding to further improve the quality of distributed simulation. As the name implies, it provided a modular architecture upon which researchers could expand. In 1995, DARPA began a three year advanced concept technology demonstration known as synthetic theater of war (STOW) with U.S. Atlantic Command (now Joint Forces Command [JFCOM]) that had very aggressive goals, including developing semi-automated forces for each of the services, developing high resolution terrain, and developing realistic environmental effects, such as weather and smoke (Lenoir & Lowood, 2005; Feldmann & Muller, 1997).

To meet the challenges, DARPA had each service develop its own semi-automated force (SAF) based on ModSAF. Concurrently, the synthetic environment within ModSAF was significantly improved, adding such physically correct features as wind, rain, clouds, fog, smoke, and deformable terrain (Lukes, 1997). Marine Corps Semi-Automated Forces development was led by Naval Research and Development (now known as SPAWAR [Space and Naval Warfare] Systems Center, San Diego, California). Most of the ModSAF development up until that point focused on platforms, such as aircraft and armored vehicles. Hughes Research Laboratories led the development of simulated infantry or individual combatants (Howard, Hoff, & Tseng, 1995). After an intense development period, all of the service SAFs were integrated to form Joint Semi-Automated Forces (JSAF). JSAF became one of the first simulations to use the high level architecture (HLA), which is a DoD-mandated software architecture designed specifically for distributed simulation applications. JSAF continues to evolve and is used today by JFCOM for joint experimentation, the navy for fleet battle experiments, and the Marine Corps for DVTE.

COMMERCIAL VIDEO GAMES IN THE U.S. MARINE CORPS

The U.S. Marine Corps (USMC) has a tradition of doing a lot with limited resources. When PC games became popular in the mid-1990s, innovative marines at the Marine Corps Modeling and Simulation Office in Quantico, Virginia, experimented with them to see how they could be used for training. They discovered that the popular first-person shooter game *Doom,* by id Software, could be modified to look more like a marine training tool. The "space marine" was made to look like an actual marine; realistic-looking weapons replaced the futuristic weapons, and the demons were made to look like conventional opposing forces. When the commercial version of *Doom II* was released in 1996, these modifications, or mods, were compiled and put on a USMC Web server. The commandant of the Marine Corps, Gen. Charles Krulak, approved the use of certain PC based games to be used during duty hours on government computers for marines "to exercise and develop their decision making abilities" (Krulak, 1997). In April 1997, *Marine Doom* even made the cover of *Wired* magazine (Riddell, 1997). Despite the early promise, modification of commercial first-person shooter games had many limitations that hindered their ability to be effective training tools. *Doom* was designed to be entertaining, not tactically accurate. Even though the character looked like a marine, it still acted like a *Doom* space marine. The marines continued to experiment with commercial off-the-shelf (COTS) games, but they did not become integrated into any formal program of instruction. The early efforts with mods showed that there was strong potential for PC based training simulations to supplement or replace the more traditional Silicon Graphics or Sun workstations based simulations of the time.

In 1997, MAK Technologies won a naval phase one Small Business Innovative Research (SBIR) for training amphibious forces with video games. This SBIR evolved into Marine Air-Ground Task Force XXI (MAGTF XXI), one of the first tactical decision simulations (TDSs) that was not a modified game. It used gaming technology, but was designed from the outset for training in a USMC environment (Lenoir, 2000). MAGTF XXI has been continuously improved, including the addition of a capability to stimulate real C4I (command, control, communications, computers, and intelligence) systems.

Throughout the late 1990s and into the next century, COTS games continuously improved, particularly those portraying small unit infantry operations. Marines continued to experiment with games such as *Medal of Honor: Allied Forces* and *Rogue Spear*. In addition, the Office of Naval Research (ONR) began research into modifying COTS games to make them into more effective training tools. This led to such TDS as *Close Combat Marines* and *Tactical Operations Marine Corps* (*TacOpsMC*).

While games were showing their utility for limited training tasks, there was still a need for an integrated system to provide more robust team training. Congress included language and funding in the Defense Authorization Act for Fiscal Year (FY) 2001 that jump-started USMC modeling and simulation. This is known as a "congressional plus-up."

SHIPBOARD SIMULATORS FOR MARINE CORPS OPERATIONS

The budget request included no funding for analysis of shipboard Marine Corps operational simulator technology.

The committee is aware of advances made in training simulation technology and the potential that training and rehearsal planning simulators have in supporting Marines deployed at sea. It is clear that technology exists to provide shipboard simulators for many of the expeditionary missions embarked Marines will have to execute. As these simulators will allow Marines an opportunity to train to the fullest extent possible while in transit, the committee believes it is time to explore the availability and applicability of both existing and new training simulators to meet Marine Corps requirements.

(National Defense Authorization Act, 2000, p. 177)

REQUIREMENTS SPECIFICATION FOR DVTE

To meet the congressional mandate, the Technical Division of the USMC Training and Education Command, led by Dr. Mike Bailey, began an aggressive program to define the future of Marine Corps simulation and developed a plan that was presented to Congress in 2001. The report said the "training goals are to maintain and expand proficiency in individual skills, improve decision-making, and enhance teamwork for both Marine teams and Navy-Marine teams" (Technical Director, Training and Education Command, 2001, p. 2). Because of space limitations aboard ships, the report recommended the use of laptops and proposed that each laptop should be able to run multiple training applications (Technical Director, Training and Education Command).

While formal requirements were drafted, Dr. Bailey's team began putting together a prototype configuration that would be tested by marines. In addition to a number of COTS games, the team assembled a JSAF based federation. They coined the term user scrutiny event (USE) to describe events that put marines in front of the systems. The idea was not to demonstrate a particular capability, but to put as many different capabilities as possible in front of marines to solicit their feedback to support a future program.

To meet the rapid demonstration objectives, requirements were fairly informal. Each simulator had to be interoperable with JSAF via the HLA and the MAGTF federation object model. Each component simulation had to operate on the same configuration Dell laptop and Microsoft joystick. All of the vehicles had to use the same basic keyboard and joystick commands, and all of the software had to reside on the hard drive. Each laptop had to be able to be rebooted to run any simulator, and each simulator had to have a virtual representation of the same piece of terrain.

DEVELOPMENT OF SYSTEM PROTOTYPES

There have always been two main components to the DVTE program, the Combined Arms Network and the Infantry Tool Kit.

Infantry Tool Kit

The Infantry Tool Kit (ITK) is a collection of COTS and government off-the-shelf (GOTS) simulations that are able to run on the DVTE host computers. The individual applications are not interoperable with each other, but some can be networked to the same simulation on another laptop. The most visible portion of the Infantry Tool Kit is the first-person shooter (FPS) application. Coalescent Technologies Corporation had the exclusive license to develop *Operation Flashpoint,* a popular FPS application for the Department of Defense military training market and called it *Virtual Battlefield System 1* (VBS-1). The improved graphics, realism, and networking made it much more suitable than *Operation Flashpoint* for infantry team training. VBS-1 became the backbone of the USMC DVTE Infantry Tool Kit.

Combined Arms Network

By using JSAF as the virtual environment, the developers were able to rapidly build simulators for a large portion of the marine expeditionary unit. The Combined Arms Network (CAN) was intended to be the interoperable suite of laptop simulators envisioned in the congressional mandate. Raydon developed vehicle simulations of the amphibious assault vehicle, the M1 tank, and the light armored vehicle. Naval Air Systems Command manned the flight simulator at Patuxent River, Maryland, and developed the air simulations, including the AH-1 Cobra helicopter and the AV-8 B Harrier. FATS, Inc. built the forward observer trainer, based on its Indoor Simulated Marksmanship Trainer system. A common viewer was initially provided by the naval visualization program, a GOTS product written and maintained by the Naval Surface Warfare Center—Coastal Systems Station, Panama City, Florida.

OPERATIONAL OR USER CONSIDERATIONS

Although marines quickly learned how to operate the DVTE system, it required several contractors to set up and maintain the system. Early experience with the system made it clear that future systems should be much easier to operate and maintain so that contractor support would not be needed. DVTE had a series of two USEs. USE 1 took place in December 2001 at Camp Lejeune, North Carolina, and focused on marines' evaluations of the basic functional pieces in a company-sized combined arms scenario. USE 2 was in July 2002 and was conducted aboard LHD-7, the USS *Iwo Jima,* pierside in Norfolk, Virginia.

Technical Performance

USE 1 validated that networked laptops and their graphics cards could be used as the basis for a shipboard virtual environment based training system. Although many people were skeptical of the performance of laptops, they were very capable of running the simulations in real time. One of the major lessons of USE 1 was the need for a single consistent "view" of the world. Although each of the DVTE systems played in the same "box," each one used its own unique database, and each one presented its unique database using its own individual image generator (IG). In addition to making configuration management difficult, the use of different world views compromised the interoperability of the simulations. After USE 1, DVTE moved to a common database and a common IG, the AAcuity PC-IG developed by SDS International. Another major limitation to the DVTE was that each laptop had to be started manually. This was replaced by an application that started the entire federation from a single control station for USE 2 (Zeswitz, 2001; Bailey & Guckenberger, 2002).

DVTE (Current Generation) Performance Standards

The original DVTE program successfully demonstrated a number of concepts, but the technology did not become part of a program of record. Fortunately, the Office of Naval Research had a research program called Virtual Technologies and Environments (VIRTE) Demo 3: Multi-Platform Operational Team Training Immersive Virtual Environment (MOT[2]IVE) that carried forward the spirit of DVTE. The objective of this multiyear (FY2005 to FY2007) demonstration was to develop reconfigurable, deployable prototype training systems and to demonstrate effective team training in virtual environments across dynamic, heterogeneous, networked, interoperable systems.

System Design

The VIRTE program had a significant interest in using the testbed as a tool to conduct research in how marines learn. One of the limitations that the USMC had with the original DVTE prototypes was that each of the developers produced proprietary components. This led to a complex, difficult-to-maintain system that required expensive licenses. While this is an acceptable situation for a demonstration program, it is not for a research system in which the code must be made freely available to a large number of research teams. One of the first tasks of the MOT²IVE team was to develop GOTS equivalents of all of the proprietary platforms in the original DVTE. Rather than develop each component in a vacuum, the new generation of DVTE was designed as a completely integrated system from the start. Another major change was to replace the proprietary AAcuity IG with an open source IG, DELTA3D. As discussed previously, JSAF, a GOTS product, continued to provide the backbone of the simulation infrastructure, and the DoD HLA that is used by JSAF allowed the simulations to share data.

DEVELOPMENT OF SYSTEM PROTOTYPES

Infantry Tool Kit

COTS simulations have evolved significantly since the original DVTE prototype was tested. VBS2 in the ITK will be replacing VBS-1, and the Marine Corps secured an enterprise license directly from Bohemia Interactive. This next generation FPS application contains many of the improvements that have been requested from the U.S. Army, Marine Corps, and Allied Forces to VBS-1 to include the importation of real world terrain and an HLA networking capability (Marine Corps Systems Command, 2006). A DARPA-sponsored simulation called Tactical Iraqi is in the ITK for language and culture training. The ONR-developed MAGTF XXI, *Close Combat Marines*, and *TacOpsMC* are a part of the ITK. The army-developed Recognition of Combat Vehicle series, including improvised explosive device and suicide bomber, are also included.

Combined Arms Network

The VIRTE team quickly built a testbed federation that could be freely shared with researchers and developers. The team used a spiral development model, with four major integration events and testbed releases. To support Fire Support Team training experiments, the team expanded and improved the forward observer personal computer simulation, originally developed by students at the Naval Post Graduate School.

Due to the critical need to train joint terminal air controllers (JTAC), the VIRTE team added a commercial head-mounted display and inertial tracker to the DVTE suite and demonstrated the utility of the CAN to train JTAC skills. The marines have subsequently purchased over 40 JTAC trainers based on this prototype to be fielded in order to meet an urgent need for JTAC training across the Marine Corps.

Programmatics

As we discussed earlier, DVTE got its start as a result of a Congressional Plus-up. Although formally established as a Program of Record in April 2004,

DVTE's funding profile did not have the required funding necessary to go beyond the initial prototype built by the Congressional Plus-up. In recent years, Congress has appropriated "supplemental funding" to support the Global War on Terrorism. The USMC took advantage of this funding and began fielding DVTE suites of computers with the ITK in FY2007.

DEMONSTRATIONS AND TRANSITIONS

One of the challenges for the DVTE program has been educating marines on the available capabilities. In addition to demonstrations at trade shows, such as Interservice/Industry Training, Simulation, and Education Conference and Modern Day Marine, there has been a concerted effort to educate marine leadership. One thrust is supporting the deployment through the formal simulation centers located with each marine expeditionary force. In addition, there have been a number of experiments at marine formal schools, such as The Basic School, the Infantry Officer Course, and the Expeditionary Warfare School. Suggested improvements from marines and experimenters were prioritized and added to the development program. This exposure helps to educate young officers about capabilities that will be available to them when they reach the fleet.

CONCLUSION

DVTE continues to evolve and absorbs new capabilities from the science and technology and commercial communities to meet the critical training needs of the warfighter. As marines become more familiar with using simulations, we expect DVTE to become an integral part of USMC training and education.

REFERENCES

Bailey, M., & Guckenberger, D. (2002). Advanced distributed simulation efficiencies & tradeoffs: DVTE, DMT, and BFTT experiences. *Proceedings of the Interservice/Industry Training, Simulation and Education Conference* [CD-ROM]. Arlington, VA: National Training Systems Association.

Cosby, N. L. (1999). SIMNET—An insider's perspective. *Simulation Technology, 2*(1). Retrieved April 8, 2008, from http://www.sisostds.org/webletter/siso/iss_39/art_202.htm

Feldmann, P., & Muller, P. (1997). DARPA STOW synthetic forces. *Proceedings of the 19th Interservice/Industry Training, Simulation and Education Conference* (pp. 461–471). Arlington, VA: National Training Systems Association.

Howard, M., Hoff, B., & Tseng, D. (1995). Individual combatant development in Mod-SAF. *Proceedings of the Fifth Conference on Computer Generated Forces and Behavior Representation* (pp. 479–486). Orlando, FL: UCF Institute for Simulation and Training.

Krulak, C. C. (1997, April). *Military thinking and decision making exercises* (Marine Corps Order 1500.55). Washington, DC: US Marine Corps Headquarters, Department of the Navy.

Lenoir, T. (2000). All but war is simulation: The military-entertainment complex. *Configurations, 8*(3), 289–335.

Lenoir, T., & Lowood, H. (2005). Theaters of war: The military-entertainment complex. In H. Schramm, L. Schwarte, & J. Lazardzig (Eds.), *Collection, laboratory, theater: Scenes of knowledge in the 17th Century* (pp. 427–456). Berlin: Walter de Gruyter. Retrieved April 8, 2008, from http://www.stanford.edu/dept/HPST/TimLenoir/ Publications/Lenoir-Lowood_TheatersOfWar.pdf

Lukes, G. (1997). DARPA STOW synthetic environments. *Proceedings of the 19th Interservice/Industry Training, Simulation and Education Conference* (pp. 450–460). Arlington, VA: National Training Systems Association.

Marine Corps Systems Command, PM Training Systems (2006, November). *Program Manager for Training Systems, products & services information handbook* (pp. 36–38). Orlando, FL: Author. Available from http://www.marcorsyscom.usmc.mil/ trasys/ trasysweb.nsf/All/21A46B83F0CFFBB085256FC4005BBA61

National Defense Authorization Act of 2001, Pub. L. No. 106-398, 177 (2000).

Riddell, R. (1997). Doom goes to war. *Wired, 5*(4), pp. 1–5.

Technical Director, Training and Education Command. (2001, February). *Report to Congress: Shipboard simulators for Marine Corps operations.* Quantico, VA: Author.

Zeswitz, S. (2001). Shipboard simulation system for Naval combined arms training. *Proceedings of the Interservice/Industry Training, Simulation and Education Conference* [CD-ROM]. Arlington, VA: National Training Systems Association.

Chapter 5

INFANTRY AND MARKSMANSHIP TRAINING SYSTEMS

Roy Stripling, Pete Muller, Richard Schaffer, and Joseph Cohn

Today's military must train greater numbers of individuals more quickly than in the past, and these learners must master a list of knowledge, skills, and abilities (KSAs) that are continually adapting in response to evolving threats. As part of an overarching training strategy, virtual environments (VEs) may offer one of the most potent tools. A VE is any set of technologies that allows a user to interact with a computer-simulated environment. Because they are primarily software driven, VEs can be quickly updated. Additionally, the footprint for the hardware supporting these applications is generally small enough to be deployed with troops. Moreover, many new KSAs involve maneuvers that are difficult or too risky to train in live environments or that require significant instructor intervention to impart. VEs allow a level of safety and instructor oversight that cannot be duplicated using live training. The challenge lies in understanding the content and interaction requirements that infantry VEs must satisfy in order to deliver on their promise of effective training.

The basic KSAs of infantry warfighting include marksmanship, room entry, rapid decision making, team communication, team coordination, and situational awareness. While many of these KSAs are similar to those required for other domains, developing useful infantry training simulations is more challenging. This is because infantry tasks involve direct interaction between the warfighter and the real world. By contrast, vehicle simulations place artificial instruments between the user and the simulated environment. It is relatively easy to build vehicle VEs that are similar to those of the actual system. It is much more difficult to do this for dismounted infantry.

INFANTRY VIRTUAL TRAINING SYSTEMS

Marksmanship Systems

Marksmanship training was one of the earliest applications of infantry-oriented VEs. In their simplest forms, these training systems focus on acquiring and removing targets. Other skills, such as room clearing, communicating, and maintaining situation awareness, are not covered. Consequently, the associated range of interaction requirements need not be addressed. Even with this seemingly basic skill, however, many interaction challenges need to be addressed, such as field of view, atmospheric variables, and weapon dynamics.

The earliest successful system was the Indoor Simulated Marksmanship Trainer (ISMT) developed by FATS, Inc. The early FATS systems used demilitarized weapons that were instrumented with a coded laser that was initiated by the trigger pull. The coded laser enabled the system to distinguish shots fired from different weapons at the same screen. The weapon was also instrumented with sensors that detected other important marksmanship attributes, such as the position of the safety switch. The weapons were given firing recoil powered by compressed carbon dioxide (CO_2) gas from an external supply.

Early VE systems were video based, and the scenarios presented shoot/no shoot situations. The videos were hand-coded frame by frame to indicate where in each video frame the humans were located. The video was projected on a screen and a laser detector determined if and where the people in the video were hit. Eventually, the FATS system added a full computer-generated VE in additional to the video. The latest versions untether the weapons by providing CO_2 gas for recoil through a replaceable magazine. Bluetooth communications is used to communicate from the weapon to the computer system.

The U.S. Army procured a similar marksmanship trainer known as the Engagement Skills Trainer 2000, produced by Cubic Corporation. Functionally, it is very similar to the current generation of ISMT, and several other systems on the market are also functionally similar. The largest differences between them are their specific VEs and their weapons implementation.

Immersive Systems—Training More Complex KSAs

Marksmanship is a critical skill that all infantry must have, but dismounted combat requires many more complex skills as well. These include the ability to maneuver, maintain situation awareness, team coordination, communication, and decision making in dynamic and uncertain environments. The army pioneered VEs for complex skills training with the Soldier Visualization Station (SVS) by Reality By Design (now Advanced Interactive Systems). This system combines a rear-projection VE with weapon and head position tracking via InterSense inertial-acoustic trackers. This technology does not have the marksmanship level accuracy of laser hit detection systems, but does allow for a much more interactive experience. Unlike the marksmanship trainers in which the trainee stays in one place in the virtual world, this class of system allows the individual and team to move through the VE.

A careful review of a sample of the different VE systems available provides a glimpse of the many different interaction technologies that are necessary to ensure that these more complex effects are enabled within the training environment. These can be analyzed in terms of a range of characteristics including operating system, human sensory modalities stimulated, navigation and/or interaction methods, footprint, and so forth (see Table 5.1).

As Table 5.1 suggests there are many different approaches to delivering effective interactions in dismounted infantry VE systems. The main challenge with each of these is demonstrating that a selected approach contributes directly to enhanced performance once the trainees are faced with actual combat or combat-like situations.

RESEARCH ON TRAINING EFFECTIVENESS

The U.S. military has a long history of pursuing and supporting technology advancement. Over the past few decades, this focus has both encouraged and benefited from the rapid advances in computer technologies and led to, among other things, multiple generations of simulator and VE training systems. However, this rapid advance has come at a cost. The pace of new system development and deployment often exceeds the rate at which these systems can be evaluated for their training effectiveness. This raises the risk that training systems developed for and purchased by the military may go unused or underused by the training community. For these reasons, the general approach taken in researching these systems is not to focus on specific applications or on specific pieces of equipment, but rather to identify the relative limitations that different VE approaches have on training (see Table 5.2). For example, rather than determine if a specific head-mounted display (HMD) supports a specific training objective, researchers may focus on the impact of limited field of view in a training objective so that any HMD can be evaluated based on this criterion.

One example of this approach is the experiments that were conducted under the Office of Naval Research program Virtual Technologies and Environments (VIRTE). VEs for dismounted infantry may make use of handheld controllers such as joysticks and gamepads; they may involve optically tracking body movements as the user walks across a monitored space or as he or she walks in place, or they may involve allowing the user to walk naturally across a moving platform (examples of these include the omnidirectional treadmill (Darken, Cockayne, & Carmein, 1997) and the VirtuSphere [VirtuSphere, Inc.]). Rather than test all of these interfaces and all of the commercial systems that make use of them, researchers sample a cross-section of these interfaces and evaluate them while users undertake the same set of tasks (see Figure 5.1).

In some cases, the results of these experiments have been mundane (once mastered, users seem to be able to achieve the same level of precision and accuracy with any of these interfaces), or fairly narrow in scope (proprioceptive/kinesthetic feedback for locomotion is relatively unimportant EXCEPT where visibility is poor AND the user's movement in this portion of the VE will include rotations).

Although the number of unique locomotion interfaces means that the number of experiments needed is still relatively large, the hope is that this approach will reduce the need to evaluate each and every system. Once general principles are

Table 5.1. Partial Summary of VE Infantry and Marksmanship Trainers

Category	Product	Visual	Locomotion	Tracking	360/180 View?	Rifle Prop?	6OF Aiming	Footprint
PC only	VBS-1	*Operation Flashpoint*	Mouse & keyboard		Yes	No	No	Desktop
PC only	VBS-2		Mouse & keyboard		Yes	No	No	Desktop
PC only	RealWorld		Mouse & keyboard		Yes	No	No	Desktop
PC only	*America's Army*	*Unreal Tournament*	Mouse & keyboard		Yes	No	No	Desktop
Projection	Flatworld	Rear, Gamebryo	Real	Inertial	Yes			Projection space
Projection	SVS	Rear, Proprietary	Joystick on weapon	Inertial	No	Yes		Projection space
Projection	VCCT	Front, Proprietary	Joystick on weapon	Inertial	No	Yes		Projection space
Projection	VICE	Front, Proprietary	Joystick on weapon	Inertial	No	Yes	Laser	Projection space
Projection	VIRTE Screen Shooter	Front, Gamebryo	Joystick on weapon	None	No	Yes	Laser	Projection space

Immersive VE	VCCT	Iglasses	Knee joystick	Inertial	Yes	Yes	No	
Immersive VE	IGS	V8 HMD	Joystick on weapon	Inertial	Yes	Yes	No	6′ × 6′
Immersive VE	Expedition DI	eMagin HMD	Joystick on weapon	Inertial	Yes	Yes	No	Limited
Immersive VE	VIRTE Pointman	HMD, screen or monitor	Gamepad and foot pedals	Inertial	Yes	No	Yes	Desktop
Immersive VE	VIRTE Pod	NVisor	Joystick on weapon	Inertial or optical	Yes	Yes	Yes	10′ × 10′
Immersive VE	VIRTE Gaitor	NVisor	Walk-in-place	Optical	Yes	Yes	Yes	12′ × 12′
Immersive VE	Virtu-Sphere	Iglasses	Walk inside sphere	Acoustic and inertial	Yes	Yes	No	10′ × 10′

Table 5.2.   Partial Summary of Recent VE Effectiveness Evaluations

Aspect Investigated	Determining Effect of …	Conclusions	Reference	See Chapter
Locomotion interface	Different locomotion interfaces for equivalency of precision and accuracy and of spatial knowledge encoding/ recall.	Once sufficiently trained, precision and accuracy are equivalent across tested systems. However, pattern of movement in the VE may differ notably from in the real world. Also, body-tracked systems improve user position awareness when performing rotations in visually impaired environments. Body-tracked VEs also support better spatial recall.	Cohn, Whitton, Razzaque, Becker, & Brooks, 2004; Farrell et al., 2003; Grant & Magee, 1998, Stripling et al., 2006; Whitton et al., 2005	Temple-man, Sibert, Page, and Denbrook, Volume 2, Section 1, Chapter 7
Field of view (FOV)	FOV on performance	Mixed: no measurable effect or benefits from wider FOV	Arthur, 2000; Browse & Gray, 2006; Johnson & Stewart, 1999	Bolas and McDowall, Volume 2, Section 1, Chapter 2
Passive/ active haptics	Haptic feedback on performance or sense of presence	Haptics can enhance sense of presence in VE and improve performance of infantry skills	Insko, Meehan, Whitton, & Brooks, 2001; Lindeman, Sibert, Mendez-Mendez, Patil, & Phifer, 2005	Başdoğan and Loftin, Volume 2, Section 1, Chapter 5
Auditory cues	Auditory cues on task performance and/or sense of presence	Audio cues can enhance sense of presence and performance in memory and localization tasks	Larsson, Vastfjall, & Kleiner, 2002; Sanders & Scorgie, 2002; Shilling, 2002	Sadek, Volume 2, Section 1, Chapter 4
Multimodal cues	Multimodal cueing on performance and/or sense of presence	Multimodal cues improve sense of presence and response times	Hecht, Reiner, & Halevy, 2006; Milham, Hale, Stanney, & Cohn, 2005a; Milham, Hale, Stanney, & Cohn, 2005b	Başdoğan and Loftin, Volume 2, Section 1, Chapter 5

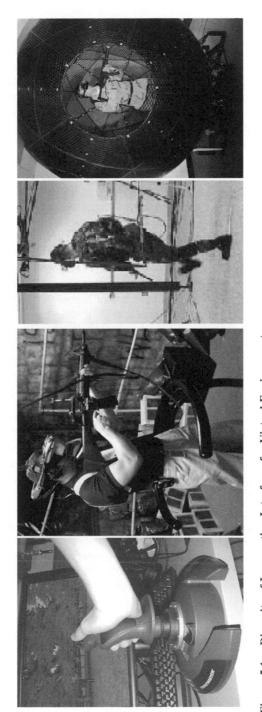

Figure 5.1. Diversity of Locomotion Interfaces for Virtual Environments

extracted, purchasers of new systems will be able to make reasonable assessments of the likelihood that a system will meet their training needs. This experimental approach should also be useful for evaluations of visual and auditory interfaces as well. Increases in screen resolution, field of view, refresh rates, and reduction in lag are likely to continue in the ensuing years; however, by following this human centric approach, the assessment of training systems should provide developers, buyers, and users with a better sense of which new developments will make a difference and which ones will not.

CONCLUSIONS

Development of VEs for infantry and marksmanship have been driven by several factors, including the need to reduce cost, increase safety, increase training throughput, and provide a greater diversity of experiences during training. In the foreseeable future, shifting mission requirements will likely drive new developments into more cognitive and complex skill domains, such as basic foreign language skills and the awareness and understanding of cultural practices, local customs, and local laws. The successful VE infantry trainer of the future will likely include avatars driven by advanced behavior models that incorporate this type of information, as well as interfaces that permit the user/trainee to interact with the avatar equally at this level. Many of these needs may still be met using lower cost desktop or laptop interfaces. However, high end immersive VEs that provide a more complete and interactive representation of the full environment will also be in demand. These systems will allow advanced students to train in an environment where they must contend with both the physical and mental challenges of infantry duties.

REFERENCES

Arthur, K. (2000). *Effects of field of view on performance with head-mounted displays.* Unpublished doctoral dissertation, University of North Carolina, Chapel Hill.

Browse, R. A., & Gray, D. W. S. (2006). Display conditions that influence wayfinding in virtual environments. *Conference on Human Vision and Electronic Imaging XI. Human Vision and Electronic Imaging XI.* (Article No. 605713; pp. 5713). Bellingham, WA: International Society for Optical Engineering.

Cohn, J., Whitton, M., Razzaque, S., Becker, W., & Brooks, F. (2004). Information presentation and control method impact performance on a complex virtual locomotion task. *Proceedings of the 48th Annual Meeting of the Human Factors and Ergonomics Society.* Santa Monica, CA: Human Factors and Ergonomics Society.

Darken, R. P., Cockayne, W. R., & Carmein, D. (1997). The omni-directional treadmill: A locomotion device for virtual worlds. *Proceedings of the 10th Annual ACM Symposium on User Interface Software and Technology* (pp. 213–221). New York: Association for Computing Machinery.

Farrell, M. J., Arnold, P., Pettifer, S., Adams, J., Graham, T., & MacManamon, M. (2003). Transfer of route learning from virtual to real environments. *Journal of Experimental Psychology-Applied, 9*(4), 219–227.

Grant, S. C., & Magee, L. E. (1998). Contributions of proprioception to navigation in virtual environments. *Human Factors, 40*(3), 489–497.

Hecht, D., Reiner, M., & Halevy, G. (2006). Multimodal virtual environments: Response times, attention, and presence. *Presence: Teleoperators and Virtual Environments, 15* (5), 515–523.

Insko, B. (2001). *Passive haptics significantly enhances virtual environments.* Unpublished doctoral dissertation, University of North Carolina, Chapel Hill.

Insko, B., Meehan, M., Whitton, M., & Brooks Jr., F. P. (2001). Passive haptics significantly enhances virtual environments. *Proceedings of the 4th Annual Presence Workshop.*

Johnson, D. M., & Stewart, J. E. (1999). Use of virtual environments for the acquisition of spatial knowledge: Comparison among different visual displays. *Military Psychology 11*(2), 129–148.

Larsson, P., Vastfjall, D., & Kleiner, M. (2002, June). Better presence and performance in virtual environments by improved binaural sound rendering. *22nd International Congress of the Audio-Engineering-Society: Virtual, Synthetic, and Entertainment Audio* (pp. 31–38).

Lindeman, R. W., Sibert, J. L., Mendez-Mendez, E., Patil, S., & Phifer, D. (2005). Effectiveness of directional vibrotactile cuing on a building-clearing task. *Proceedings of the SIGCHI Conference on Human Factors in Computing Systems* (pp. 271–280). New York: ACM.

Milham, L., Hale, K., Stanney, K., & Cohn, J. (2005a, July). *Selection of metaphoric and physical fidelity multimodal cues to enhance virtual environment (VE) performance.* Paper presented at the VR International Conference, Las Vegas, NV.

Milham, L., Hale, K., Stanney, K., & Cohn, J. (2005b, July). *Using multimodal cues to support the development of situation awareness in a virtual environment.* Paper presented at the 1st International Conference on Virtual Reality, Las Vegas, NV.

Sanders, R. D., & Scorgie, M. A. (2002, March). *The effect of sound delivery methods on a user's sense of presence in a virtual environment.* Unpublished master's thesis, Naval Postgraduate School, MOVES Institute, Monterey, CA.

Shilling, R. D. (2002, June). *Entertainment industry sound design techniques to improve presence and training performance in VE.* Paper presented at the European Simulation Interoperability Workshop, London.

Stripling, R., Templeman, J. N., Sibert, L. E., Coyne, J. T., Page, R. G., La Budde, Z., & Afergan, D. (2006, May). *Identifying interface limitations for virtual environment training systems.* Paper presented at the 55th meeting of the DoD Human Factors Engineering Technical Advisory Group (DoDHFETAG), Las Vegas, NV.

Whitton, M., Cohn, J., Feasel, J., Zimmons, P., Razzaque, S., Poulton, S., McLeod, B., & Brooks, F. (2005). Comparing VE locomotion interfaces. *Proceedings of IEEE Virtual Reality* (pp. 123–130). Los Alamitos, CA: IEEE Computer Society.

Chapter 6

FIELDED NAVY VIRTUAL ENVIRONMENT TRAINING SYSTEMS

Daniel Patton, Long Nguyen, William Walker, and Richard Arnold

The U.S. Navy has invested significantly in virtual environment (VE) simulation technology to address contemporary training requirements. VE simulation has replaced many standing training devices (TDs) and tactical training equipment (TTE), which historically were expensive to design, develop, and sustain throughout their lifecycles. TDs and TTE typically required significant building modifications and encumbered a large "footprint." Many of these devices required high power consumption, as well as other ancillary equipment to operate and maintain.

In many instances, these training suites also required extensive manpower to monitor student performance during training events. Often four to six additional personnel were needed for the complementary watch stations when training a single student. Also, since only one student could be trained at a time, a significant bottleneck occurred, causing many courses to be longer than necessary.

Because it solves those problems associated with the "large footprint" of earlier TD and TTE efforts, VE simulation has been used to reduce training course durations and annual costs for sustainment of these training capabilities. The implementation of VE training capabilities is preceded by a front-end analysis to ensure the application is addressing appropriate, contemporary training requirements. Post-implementation is followed by a training effectiveness evaluation (TEE) to validate the benefits of virtual training and identify any performance "deltas" that may remain.

The Office of Naval Research (ONR) and Naval Air Warfare Center Training Systems Division, through the Virtual Environment Training Technologies and Virtual Technologies and Environments (VIRTE) programs, made some of the first efforts to research, test, and develop VE training technologies and field them into operational military training settings. Critical development issues in the design and fielding of these three currently fielded systems cultivated under these programs are described in this chapter: Virtual Environment Submarine

(VESUB), Conning Officer Virtual Environment (COVE), and Virtual Environment Landing Craft, Air Cushion (VELCAC).

VIRTUAL ENVIRONMENT SUBMARINE

Submarine piloting and navigation skills are taught at the Submarine Training Facility in Norfolk, Virginia, and the Naval Submarine School in Groton, Connecticut. Traditionally, the classroom based training is augmented with TTE, TDs, or simulation based exercises. While this provides basic ship-handling skills, many of the finer points of submarine ship handling can be internalized only through hands-on experience. However, for dangerous maneuvers, such as navigating a surfaced submarine through a harbor or channel, opportunities for on-the-job training are limited and are generally reserved for the more experienced officers of the deck (OOD). For such tasks, a high fidelity simulation based trainer provides a safe and effective training solution (Nguyen, Cohn, Mead, Helmick, & Patrey, 2001; Hays, Seamon, & Bradley, 1997).

One potential simulation based training solution is the VE, which can provide for a high sense of presence that is comparable to the critical on-the-job experience that is unavailable to the novice junior officer (JO). The effectiveness of such a training system depends heavily on the level of fidelity for the training cues (for example, auditory and visual) and the intuitiveness of the trainee interface, which in turn, depends upon the degree to which VE technology areas have matured. Fortunately, VE technology has matured rapidly in the areas of three-dimensional (3-D) visualization and audio cues, the most critical sensory modalities in the submarine handling domain.

Similarly, speech recognition and position-tracking VE interface technologies exist; they have been successfully used in other application domains, including training. In order to determine if VE was a viable training technology for submarine handling, in the late 1990s, ONR began a three-stage process: (1) identifying the training requirements for submarine piloting, (2) developing a prototype VESUB system, and (3) evaluating the effectiveness of this VE system for training.

The first stage of the VESUB program was identifying training requirements. VESUB sought to enhance novices' skills in rare, difficult, and dangerous submarine maneuvers, such as harbor and channel navigation. A task analysis and determination of the required training cues were conducted. Submarine commanding officers and other senior officers were interviewed to gain a clear understanding of how the operational task was accomplished and how they decided when a JO was qualified to be designated as an OOD. The consensus answer for qualification was when a JO developed the "seaman's eye": the total situation awareness of the ship-handling environment and the ability to safely maneuver the vessel in all conditions (Hays et al., 1997).

Additional submarine subject matter expert (SME) interviews and focused group discussions further defined this seaman's eye concept as having 8 perceptual and 12 cognitive task components (Tables 6.1 and 6.2). These perceptual

Table 6.1.   Eight Perceptual Components of the "Seaman's Eye" (Hays et al., 1997)

Perceptual Components
1. Locating navigation aids
2. Judging distance
3. Identifying turn start/stop
4. Avoiding obstacles
5. Sense of ship's responsiveness
6. Recognizing environmental conditions
7. Recognizing equipment failures
8. Detecting and filtering communications

and cognitive tasks were mapped to 73 overall training objectives. The training objectives provided quantifiable measures of performance to facilitate training evaluation and optimization of the training system design (Hays et al., 1997).

The 8 critical perceptual elements and associated training objectives focused on supporting the training task, which was primarily visual. Extensive measures were taken to establish adequate visual fidelity in the VESUB system. Adequate visual resolution, instantaneous field of view, scene refresh rate, unlimited gaze angle, cultural features, detailed ship models, marine environment cues, weather and time of day cues, and ship visual motion cues were essential factors. VE technologies in the areas of high resolution head-mounted display (HMD), eye or head tracking, high end image generation, visual models, hydrodynamic

Table 6.2.   Twelve Cognitive Components of the "Seaman's Eye" (Hays et al., 1997)

Cognitive Components
1. Understanding visual cues and their representations on navigation charts
2. Understanding relative size, height, range relations, and angle on the bow
3. Understanding advance and transfer
4. Understanding the effects of tides, currents, wind, and seas
5. Understanding rules of the road
6. Understanding relative direction and speed
7. Understanding methods to differentiate and prioritize traffic contacts
8. Understanding ship's operation
9. Understanding methods to deal with uncooperative traffic
10. Understanding operation of ship's systems
11. Understanding how to take corrective actions
12. Understanding communication procedures

modeling, and environmental models were areas that warranted investment for VESUB.

While not as dominant as visual, audio cues were also essential in all 8 perceptual tasks identified in the seaman's eye concept. Since piloting the vessel is conducted exclusively through voice commands from the OOD or trainee to the crew, a speech interface was also essential. Accordingly, high fidelity audio cues, such as spatial audio for fog signals, environmental sounds, and engine sounds added to the realism of the virtual environment.

The human computer interface (HCI) involves primarily visual and audio cues and eye-position tracking and voice-recognition interactions. The user interacts primarily through speech commands, and the VE responds with verbal feedback. Critical developmental issues include determining the level of necessary immersion, defining speech recognition requirements, designing the hydrodynamic models, identifying visual/auditory cues, and determining the computer platform sufficient to run the entire simulation.

For the VESUB simulation, the 12 cognitive elements and associated training objectives drove the design of the training scenario and development of instructional tools. The trainee's level of mastery and understanding of the task had to be measured and the scenario had to be controlled in real time. Accordingly, an instructor operator station (IOS) with numerous features, including scenario generation, scenario control, performance measurement, and performance tracking, was warranted for VESUB.

With the training task analysis completed and a good understanding of the perceptual and cognitive tasks required, VESUB's second stage, the prototype development stage, began. The analyses indicated that a high resolution (high res) HMD technology was needed in order to accommodate the visual cue fidelity required for the submarine handling task. nVision Inc. developed a fully immersive, $1,280 \times 1,024$ pixels, 40 horizontal (H) $\times$ 30 vertical (V) deg field of view (FOV), cathode ray tube (CRT), monoscopic view HMD, which was integrated into the VESUB prototype. While this was the highest resolution available for HMDs, it did not provide enough detail for 20/20 vision. However, this HMD resolution was adequate to meet the training objectives. Since the visual scene is considered a far-field environment, stereoscopy is not required and a monoscopic 100 percent overlap HMD was used, which renders only one visual image that is the same for both eyes.

While the HMD provides for 40H $\times$ 30V degree instantaneous FOV (Figures 6.1 and 6.2), the training and system requirement was full range FOV, that is, 360° horizontal, vertical, and rotation (pitch, yaw, and roll). The high level design also required the viewpoint to be anywhere within the submarine bridge. Therefore, a 6 degrees of freedom (6 DOF) position and angle tracking system was necessary to determine trainee eye position and gaze angle. A Polhemus, magnetic based, 6 DOF, head tracking system was integrated into the prototype to meet this requirement.

Since the primary method for the OOD (trainee) to pilot the submarine is through speech commands to the crew below deck, a highly accurate voice-

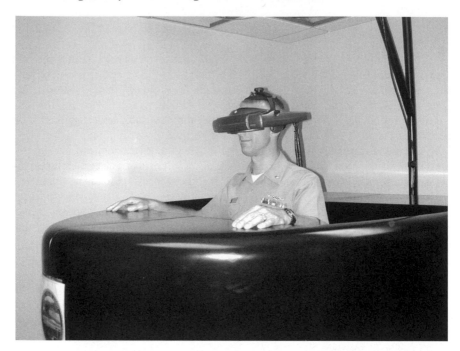

Figure 6.1.   Naval Officer in VESUB Trainer. Courtesy of B. Walker.

recognition system was required. To increase accuracy and reliability, the proto-type system design was optimized for a task specific "lexicon" of only 250 words. The voice synthesis feedback was similarly constrained.

A Silicon Graphics, Inc. (SGI) Onyx provided the image generation capability to handle the high graphics demand. Two SGI Indy desktops were included in the instructor operator station to author, monitor, and control the scenario and track the trainee's performance real time.

The virtual geospecific area included all training relevant visual and audio cues, including the submarine features, ship traffic, ocean waves and wakes, cultural objects, buoy sounds, weather effects, and so forth. The underlying software used hydrodynamics models to approximate ship dynamics accounting for such variables as wind and current.

With a working prototype, VESUB's third stage, the evaluation stage, began. VESUB underwent two system evaluations, formative evaluation and TEE. Formative evaluations using submarine SMEs provided for iterative improvements to the prototype. TEEs determined the effectiveness of VESUB and offered recommendations on how the technology could be effectively integrated into navy training (Seamon, Bradley, & Hays, 1990).

Formative evaluations of the prototype were performed, with daily guidance from submarine SMEs. Inputs from the active fleet and SMEs at the submarine school were also collected periodically as the prototype matured through the

Figure 6.2.   View from the VESUB Bridge. Courtesy of B. Walker.

iterative design phases. These provided continuous improvements in functionality and trainee interface (Hays et al., 1997).

A TEE found several areas in which VESUB improved performance, such as an increase of 57 percent in contact management skills, a decrease of 44 percent in reaction time during a man-overboard event, an increase of 40 percent in using commands during a yellow-sounding event, an increase of 39 percent in checking range markers, an increase of 39 percent in visually checking the rudder, an increase of 29 percent in using correct commands during a man-overboard event, and an increase of 13 percent in issuing correct turning commands (Hays, Vincenzi, Seamon, & Bradley, 1998).

VESUB is currently fielded in the following six locations: Groton, Connecticut; Norfolk, Virginia; Kings Bay, Georgia; San Diego, California; Bangor, Washington, and Pearl Harbor, Hawaii. The VESUB system today still uses an SGI computer for the visual image generation and a personal computer (PC) IOS. Newer hardware includes a lighter NVIS, Inc. hi-res liquid crystal on silicon HMD and an Intersense 6 DOF position tracker. IBM's ViaVoice is used for the speech recognition and speech synthesis. Also, the system has evolved to include a radar, voyage management system (VMS), and global positioning system (GPS) displays. These displays are switched to the student HMD view when looking at the display in the virtual scene and pressing a button. Planned improvements continue and include changing the image generator from an SGI computer to a high end PC. Finally, VESUB requires geospecific harbors for students to practice navigating the peculiarities of particular waterways. Currently 32 harbors from around the world are modeled for the VESUB system. These models are compatible and directly shared with COVE, the next VE training system discussed.

CONNING OFFICER VIRTUAL ENVIRONMENT

The COVE training system at Surface Warfare Officers School (SWOS), Newport, Rhode Island, is a derivative of the VESUB trainer (Figure 6.3). The goals of the COVE research project were to demonstrate a high fidelity ship simulation on a PC platform, with speech recognition and an intelligent tutor. The training goal in the COVE system was for the OOD student to develop the cognitive skills with the virtual reality system supplying the requisite visual cues to safely command the ship's movement. These factors drove the need for a high fidelity visual environment.

COVE arose from several of the lessons learned from the VESUB prototype. Specifically, the challenges of cost, instructor workload, training transfer, and deployability were addressed. The goals of COVE included reduced instructor operation requirements, a performance-driven system for training "seaman's eye," and development of both schoolhouse and deployed applications.

Figure 6.3.   Officer Practicing Maneuvers in COVE Trainer. Courtesy of MC1 David Frech, *Surface Warfare* Magazine.

A central COVE concept was the interaction of artificial intelligence techniques with a VE designed for PCs. Initial applications included the DDG-51 class destroyer and the AOE-6 class supply ship for simulating underway replenishment (UNREP). Visual rendering was executed in PowerScene. The system was operable in either high fidelity immersive HMD mode or by use of a CRT and mouse. The primary interface for communication from the user to the system was through BBN Hark's speech recognizer, and communication from the system to the user incorporated audio playback of prescripted responses to realistically simulate the helmsman or other bridge members.

While the first iteration of COVE focused on an UNREP task, other tasks, such as harbor transit and pierwork, were being incorporated. This expansion allowed the navy to employ a novel method for determining the relevant cues necessary to support effective scenario development, data-driven knowledge engineering (DDKE) (Cowden, Burns, & Patrey, 2000). VESUB relied solely on interviews or verbal reports to establish these elements. However, this method is typically time consuming, costly, and error prone. DDKE utilizes fuzzy logic tools to represent those elements of the task that directly affect performance (Cowden et al., 2000). Once identified, these task elements can serve as a guide for designing a high fidelity VE training scenario (Cohn, Helmick, Meyers, & Burns, 2000).

The embedded dynamic intelligent tutoring system (OMAR, with a Java interface) was designed with a newly commissioned officer in mind, but provides scenarios that even prospective commanding officers may find challenging. The tutor was derived from extensive interviews with surface warfare officers, shiphandling instructors, and several task analyses, both goals, operators, methods, and selection rules and Soar based (Norris, 1998; Tenney, 1999). The development was iterative in that the simulation was constructed first, with continual SME feedback. The tutor provided instant feedback and direction during the simulation, concentrating on critical decision points that the student must note and actions that must be taken to perform the evolution successfully. Student performance was recorded throughout the session and compared with a set of validated performance metrics. Students were provided grades that compared their performance with that of a prototypical expert ship handler.

At the time of transition to a production COVE system, a market survey found the VShip had improved and met most of the COVE design requirements. Therefore, VShip was chosen as a cost-effective solution for COVE with the thought of continued development to meet all the research goals.

Twelve systems were installed at the SWOS in Newport, Rhode Island. These systems are referred to as COVEs 1 and 2. They have a single visual channel powered by a high end PC to a HMD. Also included are PC-driven radar and VMS displays. COVEs 1 and 2 are used to train JOs in the basics of ship handling.

Six more stations are installed as the COVE 3 systems. These have a three-channel, flat-panel display in addition to and as an alternative to the HMD. The COVE 3s are used to train senior officers who are prospective commanding or executive officers in the finer points of ship handling for their particular class of ship.

Leveraging off the installed base of the COVE systems, a full mission bridge (FMB) was installed using the COVE software and 12 visual channels. It has a circular screen and displays a full 360°. The FMB is used to train antiterrorism force protection to middle- and senior-grade officers. It can be linked to the other COVE systems for combined multiship tactics training.

The latest training system installed is a bridge navigation trainer for the littoral combat ship (LCS). It again uses the underlying COVE software and has a five-channel, flat-panel visual display. A HMD channel is provided for simulating driving the LCS from the ship's side. It is being upgraded to be reconfigurable for both the LCS 1 and LCS 2 class of ships.

VIRTUAL ENVIRONMENT LANDING CRAFT, AIR CUSHION

The landing craft, air cushion (LCAC) hovercraft transports marine amphibious forces materiel and personnel from ship to shore. Its core manning consists of a three-man crew: craftmaster, engineer, and navigator. The LCAC's 17-week training course focuses primarily on simulation based training in the LCAC full-mission trainer (FMT), supplemented by live missions. After completion of preliminary training, students report to their operational commands where they continue in advanced qualification training.

In 2002 the LCAC fleet was undergoing a service life extension program (SLEP), which included significant cockpit redesign. The SLEP effort was scheduled to upgrade three to four crafts per year, with a fully SLEP fleet planned by 2015. The FMT would not be upgraded to reflect the SLEP configuration until half the LCAC fleet had undergone SLEP, which was projected to occur in 2009 (Muller, Cohn, & Nicholson, 2003). Thus, a requirement was established for interim training for LCAC crews designated to operate SLEP LCAC prior to 2009. VELCAC was developed primarily to address this requirement.

The development of the VIRTE VELCAC prototype trainer was accomplished via an iterative process of user input to refine training requirements, design, development, and user feedback informing each research and development spiral (Schaffer, Cullen, Cohn, & Stanney, 2003). It was determined during the process of requirements refinement that the SLEP changes would have the greatest impact on the engineer position, followed by the navigator position. The subsequent iterative design and feedback approach focused on the specific cockpit and interface features and functions that represented the most significant changes at the engineer and craftmaster stations.

Schaffer et al. (2003) have described the high level design considerations for the PC based VELCAC science and technology prototype trainer. The development environment was Microsoft Visual Studio .NET. Configuration management was via Concurrent Versions System. The system graphics engine NetImmerse by Numerical Design Limited was selected due to its well-structured application programming interface (API) and its ability to track the latest graphics hardware and the lower level API capabilities via updates. For the synthetic natural environment, the VELCAC used technologies developed by the Defense Advanced Research Projects Agency (DARPA) synthetic theater of war (STOW) and the Defense Modeling and Simulation Office environmental federation. The terrain database was the U.S. Army Topographic Engineering Center's Camp Lejeune terrain database. VELCAC also included ephemeris and illumination models. The ephemeris model was reused from the DARPA STOW program. The output of the ephemeris model was used to feed the illumination model, which was based on the U.S. Army Research Laboratory's ILUMA (illumination under realistic weather conditions) model. Surface wave and ocean dynamics were based on models developed by the Naval Surface Warfare Center, Carderock Division, and craft dynamics were based on the LCAC full mission trainer craft dynamics model. To meet the requirement of distributed training, VELCAC was developed in accordance with the Department of Defense high level architecture, version 1.3. Two run-time infrastructures (RTIs) were used to implement HLA: RTI-1.3NG and RTI-s. VELCAC used the MCO2 federation object model (FOM) for FOM attribute data sharing (Schaffer et al., 2003).

The VIRTE program used an iterative integration and transition approach. Since the VELCAC transition involved use of the trainer for interim SLEP "differences" training, the system was installed locally at the training site, Coastal Systems Station, Panama City, Florida. The product of each development spiral was pushed to the local site, with usability, functionality, and configuration feedback solicited from instructors and incorporated into subsequent development spirals. The transitioned prototype thus was the product of extensive user testing based design input.

Task analysis was used to elicit the data required to enable task and mission performance standards to be incorporated into system design. Cognitive task analyses were performed to elicit training objectives, to support elements for the objectives, and to provide guidance on training scenario design. For the SLEP interim training, the key high level elements elicited by the task analyses were an interactive three-dimensional environment and live instrumentation for all three crew stations (Muller et al., 2003).

A HCI evaluation was conducted to identify the sensory modalities that needed to be represented in the VELCAC and to provide a review of available technologies capable of representing the identified sensory modalities. Ultimately the evaluation identified two critical sensory modalities for the VELCAC system, vision and haptics (Muller et al., 2003). A system usability analysis was conducted to assess three critical usability factors: effectiveness, intuitiveness, and subjective perception. The results were in the form of ranked redesign recommendations elicited from users and SMEs, which informed design decisions during the iterative design and integration spirals.

The VELCAC S&T prototype (Figure 6.4) addressed the LCAC community's most urgent training requirements for the SLEP LCAC. Specifically, the engineer station, and to a lesser extent the navigator station, received greater attention than the craftmaster station. Post-transition, the Naval Sea Systems Command has

Figure 6.4.   VELCAC Training System. Courtesy of R. Wrenn, Unitech.

funded development of full functionality of all three starboard cabin crew stations. VELCAC has continued to be incorporated into the SLEP differences course after the course was relocated from Coastal Systems Station, Panama City, Florida, to the operational units. The SLEP VELCAC software architecture, originally adapted from the legacy LCAC full mission trainer, is now serving as the baseline software architecture for the SLEP full mission trainer, scheduled for completion in fiscal year 2009 (FY2009). With VELCAC now on site at the operational LCAC units, other uses are now under consideration, such as currency training and mission rehearsal.

Virtual environment training devices, such as VESUB, COVE, and VELCAC, have demonstrated the utility and flexibility of VE to address emergent naval operational training needs. VE component technologies have continued to rapidly improve in terms of cost and performance. As operational costs rise, the use of VE based training solutions to provide effective, affordable training is likely to increase significantly. The legacy of pioneering VE trainers such as VESUB, COVE, and VELCAC will be to provide a foundation of lessons learned and technologies to underpin the next generation of VE training systems.

REFERENCES

Cohn, J. V., Helmick, J., Meyers, C., & Burns, J. (2000). Training-transfer guidelines for virtual environments (VE). *Proceedings of the 22nd Annual Interservice/Industry*

Training Systems Conference (pp. 1000–1010). Arlington, VA: National Training Systems Association.

Cowden, A., Burns J., & Patrey, J. (2000). Data driven knowledge engineering. *Proceedings of the 22nd Annual Interservice/Industry Training Systems Conference* (pp. 11–12). Arlington, VA: National Training Systems Association.

Hays, R. T., Seamon, A. G., & Bradley, S. K. (1997). *User-oriented design analysis of the VESUB technology demonstration system* (Tech. Rep. No. 97-013). Orlando, FL: Naval Air Warfare Center Training Systems Division.

Hays, R. T., Vincenzi, D. A., Seamon, A. G., & Bradley, S. K. (1998). *Training effectiveness evaluation of the VESUB technology demonstration system* (Tech. Rep. No. 98-003). Orlando, FL: Naval Air Warfare Center Training Systems Division.

Muller, P., Cohn, J., & Nicholson, D. (2003, November). *Developing and evaluating advanced technologies for military simulation.* Paper presented at the 2003 Interservice/Industry Training, Simulation and Education Conference, Orlando, FL.

Nguyen, L., Cohn J., Mead A., Helmick J., & Patrey, J. (2001, August). Real-time virtual environment applications for military maritime training. *Proceedings of the HCI International Conference* (pp. 864–868). Mahwah, NJ: Lawrence Erlbaum.

Norris, S. D. (1998). *Task analysis of underway replenishment for virtual environment ship-handling simulator scenario development.* Unpublished master's thesis, Naval Postgraduate School, Monterey, California.

Schaffer, R., Cullen, S., Cohn, J., & Stanney, K. M. (2003, November). *A personal LCAC simulator supporting a hierarchy of training requirements.* Paper presented at the 2003 Interservice/Industry Training, Simulation and Education Conference, Orlando, FL.

Seamon, A. G., Bradley, S. K., & Hays, R. T. (1999). VESUB Technology Demonstration: Project Summary (Tech. Rep. No. 1999-02). Orlando, FL: Naval Air Warfare Center Training Systems Division.

Tenney, K. R. (1999). *A virtual Commanding Officer, intelligent tutor for the underway replenishment ship-handling virtual environment simulator.* Unpublished master's thesis, Naval Postgraduate School, Monterey, California.

Chapter 7

VIRTUAL TECHNOLOGIES FOR TRAINING: INTERACTIVE MULTISENSOR ANALYSIS TRAINING

Sandra Wetzel-Smith and Wallace Wulfeck II

The Interactive Multisensor Analysis Training program, called *IMAT,* is a series of developmental efforts that have explored the use of virtual technologies for training and performance aiding since the early 1990s.[1] This work has been conducted jointly by the Space and Naval Warfare Systems Center in San Diego, California, and by the Naval Surface Warfare Center, Carderock Division in West Bethesda, Maryland.[2] The term *IMAT* refers to these development efforts and also refers to some products of these development efforts that are now used in day-to-day training and operations in the U.S. Navy.

IMAT development has been done in the context of antisubmarine warfare (ASW). ASW is a branch of naval warfare in which many teams of people on submarines, ships, and aircraft employ sensors to detect, locate, classify, and, if necessary, interdict opposing submarines, while avoiding counterdetection or counterattack. The tasks involved in ASW include the following:

- Choosing, configuring, placing, and operating sensors so as to detect an opposing submarine;
- Locating or "localizing" the opposing submarine and determining speed, course, and depth—a process called "target motion analysis";
- Classifying or identifying the submarine once it is detected;
- Maintaining contact with the submarine and, if necessary, attacking it.

Supervisory tasks include the following:

- Planning operations so as to assure detection, while minimizing both search time and the number of ships, aircraft, and consumable sensors required;
- Coordinating ASW operations with other naval tasks, such as air and missile defense;
- Monitoring and adjusting tactics during ASW operations;
- Analyzing or "reconstructing" completed ASW events; and
- Planning exercises to develop and maintain skill among ASW practitioners.

These tasks are incredibly complex (Wulfeck & Wetzel-Smith, 2008) because they involve dynamic abstraction, multiple interacting sources of nonlinear variation, and both ambiguity and uncertainty. Further, these tasks may involve many people and platforms: a recent exercise in the Pacific involved about 30,000 individuals (3,000 of them directly participating in ASW operations), nearly 30 ships, dozens of aircraft, and cost many millions of dollars. Clearly, extensive training and experience is required to prepare people for the massive complexity of such exercises and, more importantly, for the possibility of real warfare operations.

VIRTUAL ENVIRONMENTS IN THE IMAT PROJECT

There is no generally accepted definition of the expression *virtual environment*. Characterizations range from "systems that use computers" to complete virtual reality systems that provide an immersive computer representation of a space in which users can move, change their point of view, interact with objects, and interact or collaborate with other users or simulated actors. The IMAT project has developed several different learning and performance support systems, including four distinct efforts with different goals and with different underlying technology developments:

- Instructor- or student-controlled visualization tools for classroom learning,
- Deployable systems for operational training,
- Collaborative systems for collective training in multiship ASW operations, and
- Command level training and performance support systems for senior level staff.

Shore School Based IMAT Training (1994–1998)

Initial work developed visualizations (for example, for acoustic properties of rotating machinery) to explain physical phenomena that are the basis for passive acoustic detection and classification. This work began in the aviation warfare (AW) apprentice school and was subsequently extended to apprentice and advanced courses in all the ASW communities (air, surface, subsurface, and surveillance). These technologies have since transitioned into over 20 different training programs in the surface, subsurface, and air communities. Many are still in use, and some of the laboratories have been reimplemented to provide Web based individual interactive training.

Figure 7.1 shows one view of a computer-modeled laboratory for understanding sources of acoustic energy emitted by submarines. The propulsion and other mechanical systems are animated, and their characteristics (such as speed or gearing arrangement) can be varied to illustrate their effect on the generation of acoustic signals. The sounds can be played aloud and shown on the sound spectrogram in the bottom part of the display. The acoustic laboratory includes modeled engines, gears, pumps, motors, generators, compressors, turbines, blowers, clutches, and other devices. They are coupled to a high fidelity acoustic

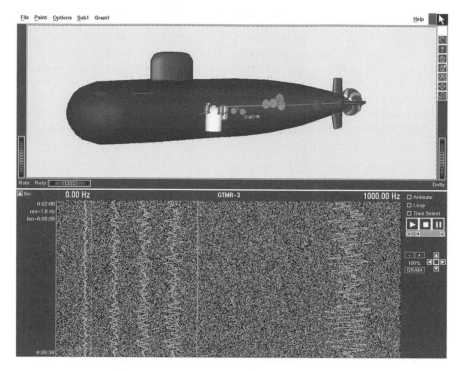

Figure 7.1.   Acoustic Laboratory

simulator that can be "operated" to show how changes in operating mode and speed are related to changes in the visual and auditory displays.

In addition to this acoustic laboratory, many other laboratories were developed that similarly illustrate such concepts as sound propagation in the ocean and the properties of acoustic sensors and sensor arrays. These laboratories were implemented on high end graphics workstations, which at the time provided the needed computation and display power. It is important to note that all these developments relied on the use of navy-validated models (for example, for propagation loss) and approved databases of oceanographic and atmospheric parameters. These included, for example, bathymetric "maps" of the ocean floor (because sound reflects from the bottom), and bathythermographic information (variation in temperature with depth, because temperature affects sound speed, which in turn affects ducting or focusing of sound due to refraction).

A good example is the propagation loss laboratory shown in Figure 7.2, which allows a visual exploration of sound propagation paths due to reflection and refraction. In this display, the leftmost panel is a color code for the amount of loss in decibels (dB) (here coded in gray scale due to printing limitations). Next, a sound speed profile (SSP) is displayed on the left edge of the main display. (Sound speed is inversely related to refractivity, which is a function of pressure,

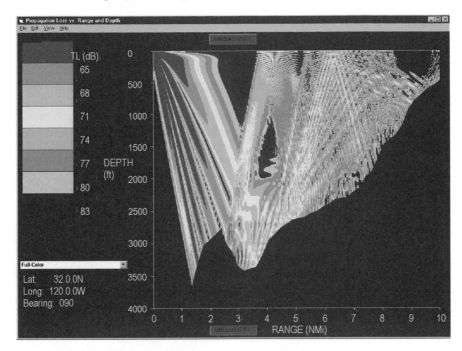

Figure 7.2.   Propagation Loss Laboratory

temperature, and salinity variation with depth.) The bottom type, SSP, and bottom contour data can be manually entered or extracted from high resolution databases. The top right panel shows an example full-field plot of energy loss. All the factors that affect transmission loss, such as spreading, absorption, or reflection by the bottom, and scattering at the bottom and surface, are modeled and contribute to the interactive display. In this example, it is easy to see several different ways that acoustic energy may propagate from a source at the upper left, including direct spreading, bottom bounce, and refraction. The striations in the display are a typical pattern resulting from in-phase or out-of-phase multipath interference.

Aside from building the visualizations themselves and implementing the physics based computational models that drive them, the main developmental questions involved whether such advanced visualization technologies could improve individual learning in school, for example, by providing context and underlying physical explanation. Evaluations were conducted in six different training courses ranging in length from 3 to 65 days, with the number of subjects ranging from 47 to 117. In all these courses (and in all other IMAT work), care was taken to apply modern principles of instructional science in the instructional design. These included (a) use of cognitive objectives, (b) scenario and context based explanations and examples, (c) development of appropriate mental models, (d) use of the laboratories for "what-if" explorations, and (e) tests involving problem solving as opposed to mere multiple-choice recognition (see, for example, Ellis, Knirk, Taylor, & McDonald, 1992; Reigeluth, 1999).

Results indicate that these visualization technologies, together with the new curricula based on improvements in instructional science, yielded test-score improvements of one to over three standard deviations compared to conventional training, while reducing time to train and training development costs (Czech, Walker, Tarker, & Ellis, 1998; Ellis, Devlin, & Allen, 1999; Ellis & Parchman, 1994; Ellis, Tarker, Devlin, & Wetzel-Smith, 1997; Wetzel-Smith, Ellis, Reynolds, & Wulfeck, 1995). Evaluations of training effectiveness in shore schools indicated that IMAT technologies are among the most successful classroom training technologies ever introduced in the navy. In 1997, the Naval Studies Board of the National Academy of Sciences noted (Committee on Technology for Future Naval Forces, 1997) the following:

- IMAT students outperform students in conventional classroom instruction, and in many cases score higher than qualified fleet personnel with 3 to 10 years experience. Improved performance has been observed in apprentice and advanced training in aviation and submarine ASW courses. Evaluations consistently show gains of two to three standard deviations on comprehension, reasoning, and problem solving tasks. Overall, the IMAT approach is much more effective than conventional lecture instruction, or technologies such as interactive video or computer based training.

- Instructors report that IMAT increases their ability to teach difficult topics, respond to student questions, and reinforce critical principles.

- IMAT students score higher on attitude scales measuring attention, relevance, confidence, and satisfaction than students in standard Navy classrooms or students in specially designed individualized computer based training.

- IMAT development costs for initial courses are equivalent to or less than conventional courses, and less expensive than other new-technology courses. Subsequent development of related training is up to 90% less expensive.

Deployable Sonar Operations Training (1997–2003)

The initial goal of the next phase of IMAT was to develop new technologies for platform level team training on tasks involving real world performance (rather than school knowledge). The goal expanded in two different directions as it became clear that IMAT visualization and modeling technologies had tactical value for performance aiding and that they could also be used for simulation based training and performance support. We were therefore particularly interested in both deployable technologies that could be used at sea and in development of simulation based training coupled with visualization techniques.

As a result of the IMAT schoolwork, flag officers and senior officials from the submarine community challenged the team to apply IMAT training methods at sea and to determine whether improvements in at-sea performance could be obtained. This led first to initial development of a personal computer (PC) version of the prior workstation based IMAT visualization and modeling programs and then to 10 at-sea developmental tryouts of the evolving software system and associated training. These at-sea tryouts yielded rapid development and revision of the software, and new features were built in that directly supported real world tasks. With at-sea instruction, use of these systems resulted in demonstrable improvement in at-sea performance (Chatham & Braddock, 2001). For example, in some at-sea sub-on-sub exercises, the propagation modeling capabilities described above were used to make better predictions about the detectability of acoustic signals from opposing submarines than had previously been available.

Following this developmental period, the system was independently tested by the submarine force. PC-IMAT then became an ASW mission support and training system approved by Submarine Development Squadron Twelve (2000) as a navy-standard tactical decision aid for submarines and was used aboard all submarines and most surface combatants. The sonar tactical decision aid (STDA), based on the same technologies but integrated with the acoustic rapid commercial off-the-shelf insertion (ARCI) combat system, has since replaced PC-IMAT on submarines, while development has continued on PC-IMAT for other ASW platforms. It is currently approved for use on integrated shipboard network systems (ISNS), the OCONUS (outside contiguous United States) Navy Enterprise Network (ONEnet), and the Navy Marine Corps Internet (NMCI).

Meanwhile, development also continued on upscaled versions of IMAT sensor performance prediction programs (Beatty, 2000) and on development of real time acoustic simulation and fast propagation modeling for purposes of simulation for operator team training. This led first to the development of the sonar employment trainer (SET; Wetzel-Smith & Wulfeck, 2000), which later transitioned into acquisition by the Naval Sea Systems Command (Wulfeck, Wetzel-Smith, Beatty, & Loeffler, 2000). The SET combines operator console simulations, a real time propagation simulation engine, and visualizations like those described earlier. It provides an "immersive" virtual environment, even though it does not actually submerge. The primary purpose of the SET is to provide instructor-controlled scenario based training with what-if capabilities for submarine sonar teams (four operators plus sonar supervisor), coupled with a highly visual explanation and debriefing capability. This training supports development of reasoning concerning sonar systems employment and tactics by exposing trainees to experiences that might have been encountered only opportunistically during mission deployments. The SET includes a large number of acoustic contacts to provide multiple contact experience, including such effects as contact merging/masking, and tracker sharing. In order to avoid mere memorization of target characteristics, targets support operating mode changes, appropriate steady state and speed-related components, and changes in signature appropriate for speed, course (aspect), and depth changes. Target simulations also support transients and other nontraditional acoustic events related to significant changes in target mode, speed, aspect, and depth.

The SET presents to the trainee the acoustic effects of complex ocean environments. The ocean environment includes variable resolution bathymetry, sound velocity, salinity, and ambient noise databases. Scenarios control surface wind speed; sea state; ambient noise, such as rain or shipping noise; and appropriate local biological, physical, and man-made noise effects. The ocean modeling system properly models effects from shallow water bathymetry, such as steep slope effects, ridges, and trenches; short-range effects; and computes an adequate number of paths for complex shallow water environments. This level of performance from the ocean-environment model is necessary to show how and why environmental conditions affect sensor performance and to provide what-if evaluation of alternative tactics to deal with environmental variability. The SET also provides scenario control capability for use by instructors or exercise controllers, including scenario startup, pause/resume, and backup/resume. These features are necessary so that explanations for physical effects on sensor performance can be given and so that alternative courses of action can be explored (see Figure 7.3).

The SET is now in place at Naval Submarine School. More importantly, the technologies demonstrated in the SET led directly to a new version of the submarine multimission team trainer (SMMTT). The SMMTT is a full mission simulator for the submarine combat center. Versions of the SMMTT are being developed for the Los Angeles, Seawolf, and Virginia classes of attack submarines and for SSBN (nuclear-powered ballistic missile submarine) and SSGN (nuclear-powered cruise missile submarine) Trident variants. SMMTT simulators are installed at submarine training facilities at Norfolk, Groton, San Diego, Pearl Harbor, Bangor, and Kings Bay (Lotring & Johnson, 2007).

Figure 7.3.   Sonar Employment Trainer

Finally, the sensor performance prediction and visualization systems developed in the IMAT work led to development of the STDA incorporated in the ARCI process for combat-systems development and now installed on many submarines (replacing the stand-alone PC-IMAT). Versions of the STDA are also being transitioned to surface ships and the surveillance community.

Surface Platform, Strike-Group (Battle-Group), and Network Level Training (FY2002–2007)

The third major phase of IMAT work has focused on ASW on surface ships operating in battle groups, and on using network connectivity to achieve multi-platform collaborative ASW. The objective was to provide training and performance support systems for network centric ASW—systems that did not then exist. Specific products to the fleet include shore based and at-sea training, exercise support, and feedback systems for multiplatform undersea warfare planning and tactical execution that support network-enabled collaboration. Supporting products included task and mission analyses for ASW operations and tactics, training curricula for surface active ASW, new-technology network based training systems, tactical planning and prediction systems for use as planning tools, tactical decision aids, and assessment/reconstruction tools at the platform and strike-group levels.

In general, the development approach was to provide early versions of performance support systems and training to fleet users and then progressively to refine them through heavy fleet interaction. An example of this approach was our initial development of the first multiplatform version of PC-IMAT that was developed and tested with Destroyer Squadrons Fifteen and Seven. The following is an early report from a DESRON (destroyer squadron) commodore on the use of the initial multiplatform version of PC-IMAT: during development PC-IMAT was used as a planning tool for Composite Training Unit Exercises (COMPTUEXs). Commanders found that PC-IMAT superbly supported tactical planning and that it provided a basis for a common understanding among the DESRON team. The associated mobile training team taught and reinforced tactical sensor employment strategies and complex relationships among threat characteristics, the properties of sensors, and environmental variables. Second, the at-sea use of the PC-IMAT helped to verify and validate the accuracy of the physics models underlying the system. In addition, a new capability allowed PC-IMAT to ingest current environmental data in situ, so as to further increase the fidelity of prediction. PC-IMAT allowed individual ships' sensor capabilities to be integrated to form a comprehensive and understandable Strike-Group ASW picture. Users noted this is key for optimum asset placement for both offensive and defensive postures. Third, users were enthusiastic about the ease of system operation. PC-IMAT developers tailored the user interface to support the search-planning and tactical-execution tasks. The system is designed to allow users to access automatically network resources such as threat intelligence, low level sensor system characteristics, and environmental data.

Based on experiences like this, the IMAT program initiated the IMAT mobile training teams, which transitioned to the Naval Mine and ASW Command in 2004. Detachments of the training teams are now located in Norfolk and Jacksonville on the East Coast and in San Diego, California, the Pacific Northwest, Pearl Harbor, Hawaii, and Yokosuka, Japan. The IMAT mobile training teams and in many cases IMAT scientists and researchers over the past five years have provided training and fielded decision-aiding systems for every carrier strike group and most expeditionary strike groups deploying to the western Pacific and Indian oceans.

Subsequent development and refinement continued to 2008. Products produced include the following:

- **Active-sonar optimization:** Our colleagues at the University of Texas Applied Research Lab have developed a new active-sonar processing system and associated training, called the Advanced Acoustic Analysis Adjunct for IMAT (A4I). The system provides enhanced sonar processors and digitized displays for surface ships to improve their ability to detect and track submarines. On the basis of at-sea trials, the program executive officer for integrated warfare systems has adopted A4I as an adjunct processor for active acoustic data and installed it on 23 surface combatants (Ma, 2005). The A4I combines active signal processing and display software along with IMAT.Explore A4I lessons. The A4I processing is based upon advanced echo tracker classifier processing software, which is the core of AN/SQQ-89(A)V15 active hull functional segment signal processing. The A4I software contains real time and faster than real time signal processing and display of fleet hull-mounted active recordings. Over 24 hours of recorded AN/SQS-53C data packaged into 17 scenarios are included.

The A4I team has been collecting evaluative data on the effectiveness of A4I training. The training is given to shipboard ASW personnel already fully qualified and trained on ASW tasks who have had all the training the navy offers for their ratings/grades. Earlier studies on the effectiveness of IMAT approaches have consistently shown effects sizes between experimental and control or pre-test–post-test groups ranging from .84 to 2.0 standard deviation units (Wulfeck, Wetzel-Smith, & Dickieson, 2004). This compares very favorably with typical effects sizes on principle and conceptual tasks in problem based learning of about .78. In the current A4I study (Wulfeck & Wetzel-Smith, 2008), we have pre-test–post-test data on 50 subjects. The effect size is 2.62, and the difference between tests is highly significant [paired $t(49) = 15.37, p < .001$].

- **Incorporation of IMAT products with composable FORCEnet (CFn).** PC-IMAT visualization technologies provide the ASW portion of the CFn system currently installed on CTF 74, CTF 72, CTF 72.2, USS *Blue Ridge* (C7F), USS *Kitty Hawk,* USS *Ronald Reagan,* and at NOPF (Naval Ocean Processing Facility) Whidbey Island, Washington. Adm. Gary Roughead (then Commander, United States Pacific Fleet, now the Chief of Naval Operations) spearheaded the initial test of CFn. CFn resulted in a major improvement in the ASW warfighting capability. In a subsequent interview, Admiral Roughead described IMAT products as follows:

"We are very good at antisubmarine warfare, but we can be better," Roughead said Sept. 16 in a telephone interview with *Navy Times.* Roughead wants to expand the basic antisubmarine warfare skills of the fleet. "There's always much more to learn about the ocean and the environment in which we operate," he said. The ASW exercises will put some new technologies to the test.

(Fuentes, 2005)

One such technology, composeable FORCEnet, is a new networking tool designed to generate and integrate information and intelligence and move both quickly. Roughead wants to know how to incorporate technologies to move information, especially with submarines moving at speed undersea, so commanders and operational planners can quickly make decisions. "This is a great tool being used at multiple levels in the antisubmarine warfare game," (Fuentes, 2005, p. 3) Roughead said.

In August 2006, the Chief of Naval Operations, Adm. Mike Mullen (now Chairman, Joint Chiefs of Staff), commended SPAWAR (Space and Naval Warfare) Systems Center, San Diego for meritorious service for work on composeable FORCEnet for antisubmarine warfare. The award cited "unparalleled improvements in the Fleet Commander's understanding of the tactical situation and ability to protect carrier strike groups from submarine attack." The award further cited the Center's "tenacious dedication to the mission and unsurpassed technical acumen in developing and implementing a new network centric means for performing Anti-Submarine Warfare Command, Control, Communications, Intelligence, Surveillance, and Reconnaissance." Elements of the Composeable FORCEnet concept were developed and installed at key intelligence and command and control nodes in the Pacific Theater, enabling substantially improved operational management of antisubmarine warfare forces and tactical antisubmarine warfighting.

- **Development of a Web based online learning system.** In order to support the learning and retention of critical knowledge related to effective ASW performance after initial shore training, the IMAT project has developed a new system, called **IMAT.Explore,** for training development and distribution based on the IMAT family of products (for example, it uses PC-IMAT to generate visualizations automatically during instruction run time). This system is currently used to teach the integrated ASW course for strike group training, as part of the STDA A-RCI program, and is fielded to most of the air warfare classrooms through dedicated installations and as part of NAVAIR's air-crew online.

Current navy requirements for development and transition of online performance aiding or learning systems are complex and severe. They include requirements for Sharable Courseware Object Reference Model Conformance, the Functional Area Manager, and Department of the Navy Application Database Management System processes, a formal Authority to Operate from the fleet or a systems command, and then formal certification testing for security and network compliance for three different networks: NMCI, ONE-net, and ISNS, each of which has its own requirements. Without these certifications, no computer based training or performance-aiding system of any kind can be connected to a ship or navy network, even for testing. IMAT.Explore v.2.0 completed ONE-net and NMCI "Ready to Deploy" certification in July 2007, with ISNS pending (IMAT Development Team, 2007).

During 2006 and 2007, IMAT.Explore courseware was developed for over 20 online training courses, on such topics as submarine characteristics, directional noise measurement, ASW mission planning, and on oceanographic environmental characteristics for interesting areas of the world's oceans. These courses have been provided for transition customers and have been briefed extensively to theater warfare commanders, the Naval Meteorology and Oceanography Command, and the Naval Mine and ASW Command. In addition, many pre-deployment briefings have been given to individual ships and helicopter squadron weapons and tactics units, and this development contributed heavily to new senior ASW seminars at the Naval Mine and ASW Command. All courses are being made available over the Secret Internet Protocol Router Network during 2007 and 2008.

For our next effort, the IMAT.Explore platform will serve as the foundation for an integrated training and performance support system. Although it began as a training tool, it will become an interactive repository and intelligent support system for all planning factors that go into theater level planning and mission execution.

Integrated IMAT Training and Performance Support for Theater Level ASW Operations (2008–2012)

The IMAT work above has focused on the knowledge requirements for effective performance at various levels of command (for example, individual sensor operator, sensor team, platform command team, squadron, and strike group). The culmination of this approach is to extend the effort to the highest levels of command that deal with ASW—the theater/force level.

The new effort directly supports navy requirements. The chief of naval operations's recent ASW task force, Team Bravo, recommended the development of a high fidelity physics based training and mission support environment to prepare commanders and senior staff at the theater and force levels for ASW operations using modern C4ISR (command, control, communications, computers, intelligence, surveillance, and reconnaissance) systems. Such an environment is required to provide practice on the full range of tasks among all levels of command, at realistic levels of complexity, against competent and alerted opponents, in highly complex multithreat ASW scenarios. No realistic planning, practice, and reconstruction environment currently exists.

The overall objective of this effort is the development of mission-rehearsal-quality training systems and mission support applications for theater level commanders and their staffs. The approach will be to identify knowledge requirements for effective performance at the force/theater command levels, to characterize high level decision making, and to identify critical command tasks that can be addressed by training and decision support tools. The effort will then develop simulation based training systems to support expert performance throughout the warfighting continuum of training, mission planning, execution, and reconstruction. It will also develop physics based simulation and visualization tools to create theater commander–level views of, and intelligent decision aids to support the management of, the detect-to-engage sequence for dynamic, complex, ASW against multiple opponents.

For this effort, Anti-Submarine Warfare in the Pacific Theater was directed as the immediate application and training domain by several Chiefs of Naval Operations and the last several Commanders of the Pacific Fleet. Theater and Force-Level Anti-Submarine Warfare is the most critical warfighting priority for the Pacific Fleet.

LESSONS LEARNED FROM IMAT EXPLORATIONS IN VIRTUAL ENVIRONMENTS FOR TRAINING

User Involvement

The history of IMAT involvement with the fleet is long and extensive. The strategy of conducting development, test, and revision of products directly in concert with operational users of IMAT products leads directly to task-centered system design and to highly relevant training. In addition it greatly facilitates the transition of developmental products into fleet use.

Fidelity Requirements of Underlying Physical Models

Early in training development, or to save money, it is tempting to adopt cartoon or notional or fake visualizations or display environments to depict relationships or even to "simulate" system operation. This is an extremely bad idea, for several reasons. First, it leads to oversimplification in the training process. This reductive bias (Feltovich, Hoffman, Woods, & Roesler, 2004; Feltovich, Spiro, & Coulson, 1991) leads to poor learning. Second, it often omits key underlying parameters because their importance to outcomes may not be realized. High fidelity physical models, however, force the inclusion of all variables known to affect the phenomenon under study and allow their exploration during system refinement. Third, the lack of physical rigor makes further transition more difficult because most of the system ultimately needs to be reimplemented. A related issue involves the adoption of probabilistic models of system outcomes, rather than fully modeled predictions. This hides variation and uncertainty in statistical error, rather than making them objects of study where, for training complex tasks, they belong.

SUMMARY

The IMAT program has developed virtual environments for training and performance support systems designed to make difficult scientific and technical concepts comprehensible to the operational users of advanced sensor systems. Products of the effort integrate computer models of physical phenomena with scientific visualization technologies to demonstrate interactive relationships for training, to simulate sensor/processor/display systems for team training, and to provide tactical decision-aiding systems for use in the fleet. The IMAT vision is to integrate training, operational preparation, tactical execution, and post-mission analysis into a seamless support system for developing and maintaining mission-related critical skills. In many ways, IMAT is a prototype for future human-performance support systems that transcend the traditional dichotomy between formal school training and actual live performance, to span career-long skill development and expert performance from apprentice to master levels, across missions, platforms, and communities.

ACKNOWLEDGEMENT/DISCLAIMER

The Interactive Multisensor Analysis Training (IMAT) program is a joint effort of the Space and Naval Warfare Systems Center, San Diego, and the Naval Surface Warfare Center, Carderock Division, supported by several contractor companies. Over 50 people have worked on the IMAT effort, and the authors in particular thank Bill Beatty and Rich Loeffler (NSWC-CD), Eleanor Holmes (Rite Solutions, Inc., Middletown, RI), Kent Allen (Anteon Corp.), and Joe Clements (Applied Research Laboratories, University of Texas at Austin) for their contributions to the program. The Capable Manpower Future Naval Capability program at the Office of Naval Research, Code 34, supports portions of the work described in this chapter. The views and opinions expressed herein are those of the authors and should not be construed as official or as reflecting those of the Department of the Navy.

NOTES

1. The principal investigators' initial work on ASW-related issues began in the 1980s, during which time Ms. Wetzel-Smith conducted a number of studies concerning knowledge and skill retention in ASW operations, and Mr. Beatty conceived and managed a classified sensor/signal processing program for maritime patrol ASW that led to initial visualization techniques for sensor performance.

2. In the 1990s predecessor organizations were the Navy Personnel Research and Development Center, San Diego, California, and the Naval Surface Warfare Laboratory, White Oak, Maryland.

REFERENCES

Beatty, W. F. (2000). The design of a high-speed transmission loss server and its tactical implications (U). *U.S. Navy Journal of Underwater Acoustics, Winter.* [SECRET]

Chatham, R., & Braddock, J. (2001). *Training superiority and training surprise* (Report of the Defense Science Board Task Force). Washington DC: Defense Science Board. Available online: http://www.dtic.mil/ndia/2001testing/chatham.pdf

Committee on Technology for Future Naval Forces. (1997). *Technology for the United States Navy and Marine Corps, 2000–2035 becoming a 21st-Century force: Volume 4. Human resources.* Washington, DC: National Academy Press. Available online: http://books.nap.edu/openbook.php?record_id=5865&page=48

Czech, C., Walker, D., Tarker, B., & Ellis, J. A. (1998). *The Interactive Multisensor Analysis Training (IMAT) system: An evaluation of the airborne acoustic mission course* (Rep. No. NPRDC TR 98-2). San Diego, CA: Navy Personnel Research and Development Center (ADA338076).

Ellis, J. A., Devlin, S., & Allen, K. (1999). *The Interactive Multisensor Analysis Training (IMAT) system: An evaluation in Sonar Technician Submarine (STS) "A" School* (Tech Rep. No. 99-3). San Diego, CA: Navy Personnel Research and Development Center.

Ellis, J. A., Knirk, F. G., Taylor, B. E., & McDonald, B. A. (1992). The course evaluation system. *Instructional Science, 21*(4), 313–334.

Ellis, J. A., & Parchman, S. (1994). *The Interactive Multisensor Analysis Training (IMAT) system: A formative evaluation in the Aviation Antisubmarine Warfare Operator (AW) Class "A" School* (Tech. Note NPRDC TN 94-20). San Diego, CA: Navy Personnel Research and Development Center.

Ellis, J. A., Tarker, B., Devlin, S. E., & Wetzel-Smith, S. K. (1997). *The Interactive Multisensor Analysis Training (IMAT) system: An evaluation of acoustic analysis training in the Aviation Antisubmarine Warfare Operator (AW) Class "A" school* (Rep. No. NPRDC-TR-97-3). San Diego, CA: Navy Personnel Research and Development Center. (DTIC AD No. ADA328827).

Feltovich, P. J., Hoffman, R. R., Woods, D. R., & Roesler, A. (2004). Keeping it too simple: How the reductive tendency affects cognitive engineering. *IEEE Intelligent Systems,* 90–94. Available online: http://www.ihmc.us/research/projects/EssaysOnHCC/ReductiveExplanation.pdf

Feltovich, P. J., Spiro, R. J., & Coulson, R. L. (1991). Learning, teaching, and testing for complex conceptual understanding. In N. Frederiksen, R. J. Mislevey, & I. I. Bejar (Eds.), *Test theory for a new generation of tests* (pp 181–217). Hillsdale, NJ: Lawrence Erlbaum.

Fuentes, G. (2005, October 3). Pacific Fleet Commander keeps eyes below water: Roughead stresses anti-sub technology. *Navy Times, 3.* Available online: http://www.navytimes.com/legacy/new/0-NAVYPAPER-1140361.php

IMAT Development Team. (2007). *IMAT.Explore NMCI documentation.* West Bethesda, MD: Naval Surface Warfare Center Carderock Division.

Lotring, A.O., & Johnson, E. A. (2007). Improving Fleet ASW training for submariners. *Proceedings of the US Naval Institute, 133*(6), 34–39.

Ma, J. (2005, February 7). Enhanced processors, displays to help surface ships' ASW capability: LaFleur Finds Money For A4I. *Inside the Navy* [Online].

Reigeluth, C. M. (Ed.). (1999). *Instructional-design theories and models: A new paradigm of instructional theory.* Mahwah, NJ: Lawrence Erlbaum.

Submarine Development Squadron Twelve. (2000). *IMAT tactical employment manual (U)* (TM-FZ1460-1-00). Groton, CT: Author. [Confidential]

Wetzel-Smith, S. K., & Czech, C. (1996, August). *The Interactive Multisensor Analysis Training system: Using scientific visualization to teach complex cognitive skills* (Rep. No. NPRDC TR 96-9). San Diego, CA: Navy Personnel Research and Development Center. (ADA313318)

Wetzel-Smith, S. K., Ellis, J. A., Reynolds, A. M., & Wulfeck W. H. (1995). *The interactive multisensor analysis training (IMAT) system: An evaluation in operator and tactician training* (Tech. Rep. No. NPRDC TR 96-3). San Diego, CA: Navy Personnel Research and Development Center.

Wetzel-Smith, S. K., & Wulfeck, W.H. (2000). Tactical sensor employment training (U). *U.S. Navy Journal of Underwater Acoustics, Winter.* [SECRET]

Wulfeck, W. H., & Wetzel-Smith, S. K. (2008). Use of visualization tasks to improve high-stakes problem solving. In E. L. Baker, J. Dickieson, W. H. Wulfeck, & H. F. O'Neal (Eds.), *Assessment of problem solving using simulations* (pp. 223–238). New York: Lawrence Erlbaum.

Wulfeck, W. H., Wetzel-Smith, S. K., Beatty, W. F., & Loeffler, R. (2000). *Military characteristics for the Sonar Employment Trainer (SET)* (MC No. N87-MC-S-20-00-03).

Wulfeck, W. H., Wetzel-Smith, S. K., & Dickieson, J. L. (2004). Interactive Multisensor Analysis Training. In *Advanced Technologies for Military Training* (RTO Meeting Proceedings No. MP-HFM-101, pp. 4.1–4.14). Neuilly-sur-Seine Cedex, France: Research and Technology Organisation.

Chapter 8

A VIRTUAL ENVIRONMENT APPLICATION: DISTRIBUTED MISSION OPERATIONS

Dee Andrews and Herbert Bell

U.S. AIR FORCE TRAINING AND OPERATIONAL CHALLENGES[1]

Operation Enduring Freedom and Operation Iraqi Freedom have shown clearly that military forces must remain flexible as they conduct new types of warfare. In Iraqi Freedom, after the first phase of the war, it became clear that the advantage allied troops held in maneuver warfare was greatly affected by the type of insurgent tactics used by the enemy. Fresh approaches to training assure that coalition forces can optimally adapt to new battle conditions. Virtual environments provide some of the new approaches to training required. This chapter describes a form of virtual environment training, distributed mission operations (DMO), that has provided the United States Air Force (USAF) with an effective method for training required new skills and competencies. While our focus will be on the use of DMO principles in USAF training, it is important to point out that distributed virtual environments for training are being used by all U.S. services and, indeed, by all coalition forces to increase readiness.

Current U.S. Air Force warfighter training and operational needs are driven by a number of different factors. There are increased operations and constant deployments as the United States fights wars in a number of countries and conducts both wartime and peacetime missions in many more (Andrews, 2001). These increases in operations not only put strain on personnel and equipment, but they decrease training opportunities because personnel are engaged in real world missions. They also take warfighters away from many training resources at their home bases. In addition, the increased operations tempos put more hours on aging equipment (some airframes are being flown by the grandchildren of the original aircrews), and there is a desire to limit training time on these equipment

---

[1]The opinions expressed in this chapter are those of the authors and do not necessarily represent the official views of the Department of the Air Force or the Department of Defense.

sets. Also, there is growing pressure on training ranges due to population growth, environmental concerns, and competition for airspace. These pressures make it more difficult to expand training ranges that exist and even to maintain the training range areas that currently exist. Increasing fuel costs have caused decision makers to seek less expensive ways to train than in the actual equipment at least part of the time. At the same time these constraints are being felt, the need for better and more frequent training has speeded up as complex, perishable skills have increased and the need for refresher training accelerates. Finally, USAF senior management would like to use current modeling and simulation technology to break down organizational "stovepipes" that prevent different U.S. Air Force organizations from training and operating with organizations in other Departments of Defense and coalition allies.

DISTRIBUTED MISSION OPERATIONS

The Air Force Research Laboratory developed a construct, and attendant methods and technologies, called distributed mission training (DMT), that has helped the USAF to overcome many of the problems discussed above (Grant, Greschke, Raspotnik & Mayo, 2002). After DMT showed its capability to solve those training problems, senior USAF management determined that the DMT construct could also be used to improve actual operations and the term distributed mission operations was coined. DMO connects live, virtual, and constructive environments to form a synthetic battle space for training and for operations. DMO helps break down stovepipes between military units so they can have better communication and understanding of how best to work together.

DMO TECHNOLOGIES

A DMO links virtual and constructive technologies with live equipment (for example, actual aircraft) via interconnection technologies. Virtual technologies include human-in-the-loop, immersive capabilities, such as flight simulators. Constructive technologies include computer-generated entities and wargames.

The goal is to allow USAF warfighters to train as they intend to fight. This imposes performance requirements on participating simulators. For example, if the time required to send information from one simulator to another is too long, simulator performance may appear unrealistic and may negatively impact training effectiveness. Therefore, a performance goal is to keep the transmission delays between simulators to 100 ms or less.

DMO technologies include communication technologies for brief/debrief of the missions. These include typical video and telephonic devices, as well as electronic whiteboards that allow instructors and trainees to transmit photos, PowerPoint slides, and maps. A key feature of the electronic whiteboards is the capability for instructors or trainees to immediately communicate with all other participants on the network. Experience has shown that because DMO participants may well have never worked together before, any means by which they can rapidly develop a shared mental model that builds the trust necessary for effective teamwork (Crane, 1999). It is also very helpful to capture data as the exercise unfolds so the entire exercise can be played back for the trainees after the exercise is finished. Freeze features and the capability to replay the exercise at slower or faster speeds also are very helpful (Bennett., Schreiber, & Andrews, 2002).

DMO technology improves the capability to measure training performance in two ways. The first involves capturing objective data by embedding measurement technologies in the computers that run the DMO exercises. These measurement capabilities allow digital data to be captured from fast-moving training exercises as they happen. Second, measurement technologies that allow instructors and observers to record subjective data in real time can be invaluable in helping to highlight key performance failures during the debriefing and later for analysis (Schreiber, Carolan, MacMillian, & Sidor, 2002; Rowe, Schvaneveldt, & Bennett, 2007).

It is important to again note that all of the technologies and methods that make DMO viable for training can also be applied to operational purposes. The USAF believes that eventually these technologies will be used to carry out operational missions. So, in many cases the human-in-the-loop equipment, although installed in a building, could be used not only to train warfighters, but also to let them perform their missions on the same suite of equipment. While embedded training (providing training exercises on operational consoles) has been used for some time, DMOs would now allow that concept to include exercises even for equipment not tethered to a fixed facility. For example, as unmanned aerial systems (UAS) have been introduced into the inventory, they are now flown over the operational theater half way around the world by operators in fixed sites who can use their control consoles to both train and operate.

DMO METHODS AND ISSUES

When warfighters first start to use a DMO capability, they have a tendency to revert to the same training methods that they are accustomed to using on a live training range. While DMO can make use of those training methods, trainees soon learn that DMO can support additional methods that can provide better learning and learning retention. A few examples include the following:

- DMO can allow for exercises to be frozen in midflight so that points can be made by the instructor and errors corrected before they become reinforced;
- DMO provide real time kill removal capability, which means that synthetic and human-in-the-loop entities can be taken out of the scenario as soon as they become casualties. This has an important learning benefit because it means that trainees will not spend time worrying about entities that are no longer germane to the training exercise. Currently, when an aircrew is informed by a range controller that it has been hit, the aircrew flies the aircraft to a "regenerate" zone, from which it is then allowed to come back into the exercise. However, while it is transiting to the regenerate area, it may be mistaken for an active player by aircrews that do not know it has already

become a casualty. In that case, it may be attacked again, which takes the trainees who do the attacking away from a part of the exercise that is still active and relevant.

- DMO allow exercises to be flown so that an aircraft that is hit does not suffer real time kill removal. That is, the aircrew in that aircraft is allowed to keep flying in the exercise, but it is signaled that the aircraft has been hit (usually by flashing the out-the-window displays red) and the aircrew continues to fly. This feature is used when an instructor believes it would harm the integrity of a multiteam exercise to take out one of the aircraft early in an exercise. This condition is often referred to as "shields up."

- Because of the digital nature of DMO exercises, the same conditions for exercises can be re-created over and over again. In training range exercises it is very difficult to re-create exact conditions, and therefore it is difficult to measure progress from one exercise to the next.

A key issue that affects DMO training is the need to have a training strategy that systematically trains new concepts and measures results, rather than merely a practice strategy. This problem plagues all simulator based training, but is particularly pronounced in DMO because DMO may have a live component (Andrews, 1988). When the "practice strategy" is used, the general feeling is that all the instructors have to do is set up a realistic scenario with high fidelity entities and let the trainees fight in the way they normally would in an actual operation (Allesi, 1989). There is no doubt that such practice exercises do produce some learning in a discovery mode; however, the learning is generally haphazard, unsystematic, and unpredictable. DMO instructors who instead follow the instructional system development approach in developing the training exercises find that considerably more learning takes place when training is designed and conducted in that mode (Rothwel and Kazanas, 2003). Prerequisite skills should be defined before the training starts, and then clear training objectives must be stated based upon training requirements. Using this front-end analysis, the scenarios then are planned with appropriate measurement of process and product stated. Trainees should be given time to familiarize themselves with the simulators and constructive models before the exercise begins again. Then, and only then, should the actual training exercise start. Instructors must decide beforehand about the following issues: whether or not freeze will be used ("freeze" refers to the strategy of stopping the scenario at certain points for instructional purposes), if and how new entities will be introduced once the exercise is under way, and whether real time kill removal will be used. Significant DMO experience has shown that these systematic steps are crucial to the instructional effectiveness of the training exercise.

DMO EVALUATION

Evaluating the effectiveness of DMO is a complex undertaking. Metrics are necessary for assessing the trainees' process as well as mission effectiveness on the actual battlefield (Bell, Bennett, Denning & Landrum, 2003). Such DMO evaluations follow many of the same procedures and use many of the same process and product measures as are used when training is delivered in non-DMO modes. These include process and product metrics such as number and types of communications between teammates, degree of coordination, accuracy of situational assessments, correctness of command and control decisions, and impact of the mission's effects on the simulated battlefield (Schvaneveldt, Tucker, Castillo, & Bennett, 2001).

In addition to the evaluation of trainee and operator actions in the DMO environment, technologists can also measure the effectiveness of the technology in providing a realistic synthetic battle space. Examples include the following:

- **Interdevice transport delay**—How long does it take for an output of one DMO device to be distributed to other nodes on the DMO network?

- **Adequacy of communication quality over the DMO network**—This is typically measured subjectively by instructors who are listening to the DMO exercise. The criteria for measuring quality of communications have to do with the type of communication, the actual message, and the timeliness and completeness of the message.

- **Network security**—Is the DMO network protected from external intrusions? Is the network protected from internal intrusions; that is, can all the sites inside the DMO network be sure that other sites inside the network do not intrude into parts of their computer complex for which they do not have authorization?

- **Mission planning**—Can the warfighters who are planning the missions access and send relevant information in a time frame consistent with the mission requirements? What is the quality of the missions that are planned?

IMPACT OF DMO VIRTUAL ENVIRONMENTS

Training

The DMO construct has had a considerable positive impact on USAF training. In addition, over the years that impact has spread to the other U.S. military services and to coalition allies. DMOs are used in many of the air force's major commands, including Space Command, Air Mobility Command, and Special Operations Command. To provide an example of how DMO works in the U.S. Air Force, we now examine briefly DMO use in the Air Combat Command.

The Air Combat Command has installed "mission training centers" (MTCs) at many of its fighter bases. These MTCs consist of two- or four-flight simulators and attendant instructional support systems. The simulators have wraparound domes with out-the-window 360° visual scenes. The cockpits have high physical and high functional fidelity. The trainees can fly as a two- or four-ship formation just as they do in the real world. The MTC can be linked to other simulation centers that might simulate command and control platforms, U.S. Army or Navy units, or coalition partners.

Training exercises can include air-to-air and air-to-surface missions. The DMO simulators can be used to provide a range of training opportunities for the warfighters: individual procedural training, two-ship or four-ship element level team training, as well as team of teams training with other DMO sites including coalition partners. Instructors are provided an instructor/operator station that allows them to view the formation from a "God's eye view," as well as the entire mission evolution. In addition, they can see what the trainees are seeing out of their front windscreens. They hear all communication between the pilots and other entities on the network. When the exercise is completed, the trainees can debrief, as was described above, by replaying the data from the exercise. The trainees or instructor can stop and start the exercise debrief as needed.

Operations

The full DMO concept has not yet been realized in operations. Perhaps the best current application is in mission rehearsal. In mission rehearsal, various DMO entities can be linked to create a synthetic mission environment that closely mimics the environment the warfighters will be encountering when they perform their mission. Terrain, cultural features, humans, weather, threat effects, and many other elements can be highly modeled to present a very close approximation to what the warfighters will see when conducting the mission. As mentioned above, a good example of this principle is the operation of a UAS over a battlefield while its operator is many thousands of miles away at a control station. The difference between the UAS control station as an operational control device versus a training simulator is difficult to differentiate physically; only in purpose do we see differences, those differences of purpose being actual operational control versus training.

FUTURE OF DMOS

The DMO concept has been adopted (often with different names) across the DoD and in many allied countries. We expect increased use of DMO technologies and methods as budgets become tighter, the military seeks to relieve stress on personnel, mission deployment training needs increase, and alliances increase in size and complexity. These four factors are explained in more detail below.

DMOs can provide significant cost savings for both training and operations. Current fuel costs and wear and tear on actual equipment can make flying even a one-ship training mission very expensive. When the costs are combined to train entire multiplayer exercises, the costs can easily be in the millions of dollars for a large exercise. DMO allow the trainees to train in relatively inexpensive-to-operate simulators and constructive models as opposed to actual equipment. Orlansky and Chatelier (1983) provide an excellent framework for determining the cost-effectiveness of single simulators for training. It is believed that that model can be used to determine the cost-effectiveness for DMOs. While the DMO concept allows for live equipment entities (for example, aircraft) to be part of DMO exercises, it is expected that their role will become more limited in the DMO future as simulators and constructive models improve. Having said that, it is important to note that there will always be a place for live equipment in those exercises. DMO is expected to allow simulators and constructive models to be used more and more in actual operations as supports and/or replacements for some actual equipment. That will potentially save lives.

Since the end of the Cold War, coalition forces have deployed at a much greater frequency because many of the forward-deployed bases used in the Cold War were closed. That means that the warfighters are generally away from home base more often than before. Not only does this frequency of deployment affect the warfighters' personal and family lives, but this makes it much more difficult to meet their training goals. Therefore, the training they do get at home must be as effective as possible with high skill retention. DMO technologies can assist in increasing the effectiveness of training time they do get. In addition, DMO assets are becoming more deployable and will be going with the warfighters to their deployed bases more often.

Warfighters will rely more and more on DMO to help them prepare for and carry out missions. Mission rehearsals at home and in the area of operations will rely increasingly on DMO technologies. These include rapidly updatable databases that will give DMO scenarios remarkable fidelity for the missions that will be conducted. These database updates can include real time weather changes, as well as new threats. Virtual and constructive DMO technologies are used more and more to actually support the mission, including having warfighters conduct their mission at very long distances through the use of weapon systems such as the UAS.

Finally, the U.S. military forces will seldom conduct operations, especially large operations, by themselves anymore, but instead will fight with coalition partners. Obviously, physically bringing together large units from many different countries to train together is limited due to distance and cost. This DMO coalition concept for distributed mission training across countries, continents, and oceans has already been tested, and this approach will become much more widespread in the future. In like manner, entire operations of coalition partners will see the positive impact of DMO as virtual and constructive entities work with live operational equipment in the theater to support the mission (McIntyre and Smith, 2002). For all of these reasons, DMO will be a major factor in future training and operations around the world.

REFERENCES

Allesi, S. M. (1989). Fidelity in the design of instructional simulations. *Journal of Computer-Based Instruction, 15,* 40–47.

Andrews, D. H. (1988). Relationships among simulators, training devices, and learning: A behavioral view. *Educational Technology, 28,* 48–53.

Andrews, D. H. (2001). Distributed mission training. In W. Karwowski (Ed.), *International Encyclopedia of ergonomics and human factors* (Vol. II, pp. 1214–1217). New York: Taylor and Francis.

Bell, J. A., Bennett, W., Jr., Denning, T. E., & Landrum, L. (2003). Tactics development and training program validation in distributed mission training a case study and evaluation with the USAF weapons school. *Proceedings of the 2003 Interservice/Industry*

Training, Simulation and Education Conference [CD-ROM]. Arlington, VA: National Training Systems Association.

Bennett, W., Jr., Schreiber, B. T., & Andrews, D. H. (2002). Developing competency-based methods for near-real-time air combat problem solving assessment. *Computers in Human Behavior, 18*(6), 773–782.

Crane, P. M. (1999, April). *Designing training scenarios for distributed mission training.* Paper presented at the 10th International Symposium on Aviation Psychology, Columbus, OH.

Grant, S., Greschke, D., Raspotnik, B., & Mayo, E. (2002). A complex synthetic environment for real-time, distributed aircrew training research. *Proceedings of the 2002 Interservice/Industry Training, Simulation and Education Conference* [CD-ROM]. Arlington, VA: National Training Systems Association.

McIntyre, H. M., & Smith, E. (2002, April). *Collective training for operational effectiveness.* Paper presented at the NATO RTO Studies, Symposium on Air Mission Training Through Distributed Simulation (MTDS)—Achieving and Maintaining Readiness, Brussels, Belgium.

Orlansky, J., & Chatelier, P. R. (1983). The effectiveness and cost of simulators for training. *International Conference on Simulators* (Publication No. 226, pp. 297–305). London: London Institution of Electrical Engineers.

Rothwell, W., & Kazanas, H. (2003). *Mastering the instructional design process: A systematic approach* (3rd ed.). San Francisco: Pfeiffer.

Rowe, L. J., Schvaneveldt, R. W., & Bennett, W., Jr. (2007). Measuring pilot knowledge in training: The pathfinder network scaling technique. *Proceedings of the 2007 Interservice/Industry Training, Simulation, and Education Conference* [CD-ROM]. Arlington, VA: National Training Systems Association.

Schreiber, B. T., Carolan, T. F., MacMillan, J., & Sidor, G. J. (2002, March). *Evaluating the effectiveness of distributed mission training using "traditional" and innovative metrics of success.* Paper presented at the NATO SAS-038 Working Group Meeting, Brussels, Belgium.

Schvaneveldt, R., Tucker, R., Castillo, A. R., & Bennett, W., Jr. (2001). Knowledge acquisition in distributed mission training. *Proceedings of the 2001 Interservice/Industry Training, Simulation and Education Conference.* Arlington, VA: National Training Systems Association.

Chapter 9

VIRTUAL ENVIRONMENTS IN ARMY COMBAT SYSTEMS

Henry Marshall, Gary Green, and Carl Hobson

This chapter provides an overview of past and current efforts to embed virtual environments (VEs) into operational army combat vehicles for training and augmented reality. It addresses embedded VEs for vehicles, such as the M1 Abrams tank, the Bradley fighting vehicle, and the Stryker family of vehicles. Discussion includes the requirement for embedded training (ET) and the technology challenges that have been and are being addressed to use VE in military systems.

TASK ANALYSIS

Performance Standards

The Army Training and Doctrine Command (TRADOC) defines ET as training capabilities built into or added onto operational systems, subsystems, or equipment to enhance and maintain the skill proficiency of personnel (Department of the Army, 2003). Vehicles equipped with ET may be employed as stand-alone trainers to sustain crew or individual skills or, when networked together with other vehicles or simulators, for combined arms training and mission rehearsal. The operational requirements documents for the army's future combat system (FCS), Abrams, Bradley, and Stryker all require ET. Providing the technologies to support these requirements has been a major army research focus for the past 10 years.

Technology Requirements

Virtual simulation is one of the primary ET technologies. This term includes numerous supporting technologies, such as computational systems, image generation, rapid terrain database development, computer-generated forces (CGF), local and wide area networks, position and orientation tracking, and miniaturization. ET can be integrated into vehicles at different levels. To define these levels TRADOC provided the following criteria (Department of the Army, 2003).

Fully Embedded

"Fully embedded" means built into the operational systems with no training unique hardware. This is the ultimate goal of ET. Currently FCS is pursuing this level of integration as it is being designed. ET is a key performance parameter for FCS and is a mandatory capability before fielding. The FCS program mandates horizontal integration between the vehicle and training system developers to capitalize on innovative methods to integrate ET with operational needs. This horizontal integration is an ongoing effort and will continue through the life of the FCS vehicles.

Appended

"Appended" means added to an existing operational system. Most current force ET systems fall into this category. This level of integration typically adds one or more line replaceable unit (LRU) computer systems to the vehicle to perform ET functions and provide virtual imagery to the crew members' displays. The current Stryker ET system is an example of an appended system and is the only ET system fielded in current force vehicles. This ET system provides remote weapon station gunnery training. The system uses an appended ET module as the simulation host and interfaces to the video display terminal, a device used by the training manager as the instructor operator station (IOS) to select a gunnery training exercise and monitor trainee progress.

Sustainment/opportunity gunnery training has also been demonstrated in the M1A2 Abrams tank and in the Bradley fighting vehicle. Its focus is keeping soldiers proficient on gunnery skills that have been proven to be perishable over time. Currently the army has an ongoing effort to develop and field a common embedded training system for the Abrams M1A2 system enhancement package tank and the Bradley M2A3 fighting vehicle in 2009. The common solution will initially provide sustainment opportunity gunnery training capability, followed by a mission rehearsal and live training by 2012.

Umbilical

"Umbilical" means connected to the operational system as needed. An example is the Abrams full crew interactive simulator trainer (AFIST) (Department of the Army, 2000). AFIST appends monitors outside the Abrams' vision blocks for trainee visualization and attaches sensors to the vehicle controls to capture trainee operation of vehicle systems during simulation. Data are routed to computers outside of the vehicle to drive the simulation software, and training sessions are controlled using an IOS.

SYSTEM HIGH LEVEL DESIGN

Fully embedded, appended, and umbilical approaches each present implementation issues. The fully embedded approach requires a significant redesign of the vehicle electronics (vetronics) architecture. The appended approach requires the addition of one or more LRUs to an already space-constrained vehicle, increases the thermal dissipation requirements, and requires some vetronics modification. The umbilical approach requires significant add-on hardware, connections to vehicle control devices and LRUs, increases the logistical footprint, and is time consuming to set up and tear down.

The fundamental issue in providing ET to any current force system is the routing of information from vehicle controls and subsystems to an ET system for simulation of the vehicle's operation in the virtual environment, which is typically called ownship simulation among ET developers, and generation of data to replicate the vehicle's sensor imagery. Typically, analog-intensive vehicles are the most problematic for ET and require appended devices to capture their state information. Examples include steering wheels and braking systems that are linked directly to hydraulic systems, switches and dials that are purely analog, and direct view optics, such as vision blocks and degraded mode weapon sights, which are the direct view optics to back up the digital optical systems in case of failures. A multifunctional vision block concept that can switch between direct view and simulated view modes is under development to address this latter issue (Montoya et al., 2007). Multifunctional vision blocks also provide an operational enhancement with a mixed view mode that overlays the outside view of the real world with synthetic imagery, such as a heads-up display for system state information and for augmented reality.

Safety is a major concern since ET systems must interoperate with vehicle software that controls potentially dangerous components. Examples of common safety concerns include weapon computers, laser range finders, automated ammunition systems, and movement of a turret or ramp. Typically these systems must be disabled in preparation for training. At the same time vehicle cautions and warning systems should not be inhibited since the trainee could become confused between what is simulated and actual, when a potential dangerous condition could be occurring.

There is a notional high level design for ET in combat vehicles (Department of the Army, 2003). This architecture shows how the various vehicle components are related to ET and how an ET system could interface to external systems. In operation, a crew member receives simulated information from the vehicle displays and sensors and reacts to this stimulus by making menu selections, pushing buttons, flipping switches, firing a weapon, and so forth. The ET application senses the crew actions and injects a stimulus that causes the displays to transition, which then requires the crew member(s) to react to the new situation. In a combined arms training exercise when actions by the crew change the world state by moving the vehicle, destroying another vehicle, and so forth, the world state is updated in all the other players in the exercise.

EMBEDDED TRAINING RESEARCH

There have been three distinct army ET technology programs as follow:

The first, called inter-vehicle embedded simulation and training (INVEST; Institute for Simulation and Training, 2002), focused on appended ET for the

Abrams and Bradley (A/B) during the period 1998–2002. It examined issues of architectures, communications, live-virtual correlation, and mixed reality for live and virtual ET. The two areas of primary emphasis of the INVEST Program were the development of an A/B kit architecture for the appended ET hardware/software and the exploration of methods for reducing the bandwidth requirements of the CGF, which are the computer-controlled simulated forces that typically play the opposing forces and any friendly forces not controlled by a manned simulation. A/B kit architecture employs the concept of a B kit that contains common hardware/software applications that can be used across multiple vehicle platform ET applications, such as CGF or after action review. The A kit interfaces to items unique to the vehicle configuration of a given platform, such as handles and displays. The CGF research focused on the exploration of synchronization and meta level representation of crew interactions and vehicle dead reckoning. The program concluded with a networked demonstration of Abrams and Bradley vehicles and an Abrams testbed with various degrees of ET technology performing a combined arms training exercise.

The follow-on to INVEST was named embedded combined arms team training and mission rehearsal (ECATT/MR; Research Development and Engineering Command, 2007). ECATT spanned the years 2002–2006 and shifted the research focus from current force vehicles to FCS. At the time FCS was considered the replacement for both the Abrams and Bradley vehicles in the 2014 time frame. This program investigated issues for fully embedding ET applications into the vehicle. It researched ET architectures based on the army's standard semi-automated forces program, the use of intelligent structured training as a replacement for on-site instructors, and interfaces between mounted and dismounted ET and augmented reality. This program culminated with a demonstration of dismounted ET for mission planning, mission rehearsal, and after action review at the Army Aerial Expeditionary Force Experiment in November 2006 (Marshall, Garrity, Roberts, & Green, 2007).

The most recent program, scalable ET and mission rehearsal (SET-MR), began in December 2006 and again focused on current force vehicles. SET-MR is researching common ET components applicable to multiple vehicles. This program is also seeking to develop highly accurate, miniaturized sensors to determine weapon location and orientation for pairing of shooters and targets without line of sight. In addition to the army's Abrams, Bradley, and Stryker vehicles, the U.S. Marine Corps expeditionary fighting vehicle is also participating in this ET research. SET-MR developed software requirements for ET on each of the vehicles of interest (Oasis Advanced Engineering, 2007). It also defined a variety of use cases for ET on these vehicles.

USER CONSIDERATIONS

The INVEST program primarily examined issues of embedded gunnery training. Since an objective of the program was to minimize the addition of new hardware to the vehicle, INVEST used existing vehicle displays to provide information to the crew. For the Abrams and the Bradley, this limited involvement to the vehicle commander and gunner stations as they have the only displays capable of displaying virtual imagery. In actual operations, the driver also plays a major role in target detection, but at the time there was no means to support the driver's visualization of the virtual environment. ECATT and SET-MR expanded the research to include the entire crew for mission rehearsal.

PROTOTYPES

Each of the research programs built a variety of prototypes. Recent prototypes included an FCS infantry carrier vehicle (ICV) for demonstration and experimentation, an FCS command and control vehicle for robotics management experimentation, and several dismounted soldier systems ranging from man-wearable, fully immersive systems to handheld computer versions. SET-MR is also prototyping simulation software for the future force warrior program.

DEMONSTRATIONS

The ET research programs have conducted a number of demonstrations over the years. Most recently, in April 2007 the SET-MR research program participated with the program manager for Stryker in an ET interoperability experiment (Optimetrics, Inc., 2007). The goal was to show the potential of collective ET and mission rehearsal for the Stryker vehicle and obtain early feedback from the user community on the path forward for ET objective requirements. In order to move from the current gunnery ET simulation to a collective capability involving the full crew and supporting dismounted soldiers, the experiment required that the vehicle's driver, the commanders C2 (command and control) capabilities, and the infantry soldiers that ride in the vehicle be included in the experiment.

The experiment included two ICVs, one command and control vehicle (C2V) and one mobile gun system (MGS). The systems were networked using high level architecture (HLA) and a gateway between HLA and dismounted soldier simulation systems using the distributed interactive simulation (DIS) protocol. To support the collective training scenario the existing Stryker ICV ET software was modified to accept ownship control from a joystick in the driver's station, and a video feed was routed to the driver visualization enhancement device to provide the driver's view. The remote weapons station, which permits the commander to fire the top-mounted weapon from inside the vehicle, is supported by the current ET system and was used without modification. A similar configuration was designed for the main gun on the MGS. To replicate the dismounted forces several models were explored. One of the more successful used a system built around the *Half-Life* game engine that was interoperable with the DIS protocol (Institute for Simulation and Training, 2002; Research Development and Engineering Command, 2007). The exercise scenario was based on a breaching tactic where the MGS fired rounds to create a hole for the infantry to enter the building. The ICVs then moved into position where the infantry soldiers dismounted and moved through the breach to conduct a raid. When the opposing forces were cleared, the soldiers remounted the ICV and departed the area.

The demonstration generated a number of lessons learned that are being incorporated into the research (Optimetrics, Inc., 2007). Among them was the inability of current computer-generated forces systems to realistically play opposing forces in urban environments and the cost and difficulty of maintaining the existing HLA implementation. In general, the feedback from soldiers was positive. They stated that the collective ET system could be a great tool to use during home station training and rehearsals, but had concerns for its effectiveness in a theater of operations because small units typically have little time for rehearsals between the receipt of orders and the start of a mission. Other areas of concern include setup time and ease of use.

SUMMARY

Program managers for each of the vehicles of interest are active participants in the ongoing SET-MR research. Plans call for technologies developed by SET-MR to transition directly to the vehicle program managers. SET-MR sensor development has already transitioned to the program manager for tactical engagement simulation. Dismounted soldier prototypes have also transitioned to various organizations.

A final demonstration of SET-MR technologies is planned for 2009. This will be a field demonstration of small unit mission rehearsal with a mix of current force vehicles working with dismounted soldiers. It is likely that additional ET research will be approved after SET-MR ends in 2009.

REFERENCES

Department of the Army. (2000, May). *Tank gunnery training devices and usage strategies* (Field Manual No. 17-12-7). Washington, DC: Author, Headquarters.

Department of the Army. (2003, June). *Objective Force Embedded Training (OFET) users' functional description* (TRADOC Pamphlet 350-37). Ft. Monroe, VA: Author, Headquarters.

Institute for Simulation and Training. (2002, April). *Distributed embedded simulation and training research, Summary of Findings—2001–2002* (Final Rep. No. IST-CR-02-02). Orlando, FL: Author.

Marshall, H., Garrity, P., Roberts, T., & Green, G. (2007, November). *Initial real-world testing of dismounted Soldier embedded training technologies.* Paper presented at the Interservice/Industry Training, Simulation and Education Conference, Orlando, FL.

Montoya, J., Lamvik, M., White, J., Frank, G., McKissick, I., & Cornel, G. (2007, November). *Switchable vision blocks: The missing link for ET.* Paper presented at the Interservice/Industry Training, Simulation and Education Conference, Orlando, FL.

Oasis Advanced Engineering, Inc. (2007, October). *Software requirements specification for Embedded Mission Rehearsal (eMR)—Version 1.1.* Auburn Hills, MI: Author.

Optimetrics, Inc. (2007, April). Stryker ET interoperability experiment (Final Rep.). Ann Arbor, MI: Author.

Research Development and Engineering Command, SFC Paul Ray Smith Simulation & Training Technology Center. (2007, September). *Embedded simulation testbed research and FCS technology integration—Consolidated Final Report 2005–2006* (Final Rep.). Orlando, FL: Author.

Stryker ET systems layout—Remote weapons station [Briefing]. (n.d.). Sterling Township, MI: General Dynamics Land Systems.

Chapter 10

DAGGERS: A DISMOUNTED SOLDIER EMBEDDED TRAINING AND MISSION REHEARSAL SYSTEM

Pat Garrity and Juan Vaquerizo

Since the late 1990s, the visual simulation community has been in the midst of a revolution in terms of price, performance, size, scalability, and features available from personal computer (PC) based visual image generation systems. In 2003 this revolution entered a new accelerated phase of system evolution where size, portability, power consumption, and overall system performance were being stretched to the current limits of technology to meet the requirements for fully embedded, battery-operated, man-worn, embedded training systems required by such army programs as land warrior (LW) and future force warrior (FFW). These programs have embedded training key performance parameters written in their respective capability development documents that specify embedded training applications to conduct virtual training exercises and mission rehearsals in the live, virtual, and constructive domains anywhere, anytime.

The army lacked a comprehensive training and simulation system for the dismounted soldier to meet these requirements. Such a man-portable system would provide an immersive synthetic environment for the soldier to practice and improve individual and team collective skills with simulation on demand. Packaging, weight, miniaturization, and power requirements have in the past been the primary technology obstacles to building such a system. Rapidly advancing technologies for mobile computing, three-dimensional/two-dimensional (3-D/2-D) graphics, interactive simulation environments/models, and immersive virtual reality now make it possible to redress this shortfall. The U.S. Army Research, Development and Engineering Command (RDECOM) Simulation and Training Technology Center (STTC) had been involved in researching dismounted training both in virtual and augmented reality applications to address the challenge of a man-portable simulation system for training the next generation of soldiers. Under the embedded training for dismounted soldiers (ETDS) science and technology objective (STO), RDECOM spearheaded the research with a small team of private industry partners to develop the capabilities required to train dismounted infantrymen (DI) using wearable computers similar to the operational equipment contemplated for the current LW system and the FFW system in the 2010 to 2012 time frame. As part of this development, STTC needed to research capabilities and test prototypes in a testbed. The DAGGERS (distributed advanced graphics generator and embedded rehearsal system) project supported STOs in this area through the ETDS STO and subsequent embedded combined arms team training and mission rehearsal army technology objective. During the course of early research, the ETDS team investigated several hardware technologies, including commercial off-the-shelf computing platforms, wireless communications, motion trackers, head-mounted displays, battery technologies, and human factors for an advanced warfighter-machine interface prototype.

Although the goal of this research was to provide a realistic and highly immersive environment for the soldier to train anywhere and at anytime, it was never considered as a replacement for the live training that the dismounted soldiers perform at military operations in urban terrain sites or training ranges.

Current situations in both Afghanistan and Iraq have identified a persistent need for a virtual training simulation to build and sustain the training readiness of dismounted infantry units. These units are required to conduct close combat operations in complex/restrictive terrain against asymmetric forces. Virtual training can prepare and build confident and adaptive dismounted infantry leaders and soldiers to dominate battlefield situations in varying conditions (combined arms, joint interagency and multinational, and contemporary operational environment with urban and complex terrain).

The DAGGERS system offers a unique system to research, evaluate, and develop new training doctrine for the twenty-first century warfighters. Historically, there has never been a widely accepted virtual/immersive dismounted soldier training system due to computer form factor, weight, power, display resolution, wireless connectivity limitations, and other human factors restrictions. While these limitations remain a challenge and merit further development, the DAGGERS system successfully illustrates sufficient performance capabilities to offer the warfighter a completely untethered dismounted soldier training and mission rehearsal system.

EARLY STO OBJECTIVES

The ETDS STO, managed by the U.S. Army Simulation, Training and Instrumentation Command, was a three-year (fiscal year [FY] 2002–FY2004) research program to develop and demonstrate revolutionary embedded training and simulation capabilities for the dismounted soldier.

DAGGERS was one of the main projects under the ETDS STO, and its main objective was the development of a proof of concept embedded training system, completely untethered, soldier worn, battery powered, and requiring no external facilities or infrastructure to operate. Critical advances in graphics processor technology and low power, high performance central processing units provided the foundation for development of the man-portable visual computing system called Thermite, the heart of the DAGGERS system.

CONCEPT OF OPERATION

The DAGGERS system is intended to provide dismounted soldiers with an embedded virtual training and mission rehearsal capability. Using geospecific or geotypical synthetic environment databases, the system provides the ability to move, shoot, and communicate in a combined-arms virtual battlefield. The system is utilized in a distributed (networked) configuration. When the distributed interactive simulation network interface is activated, the system can interface with several other distributed simulation systems, such as other virtual simulators and computer-generated forces (that is, DI-SAF [Dismounted Infantry Semi-Automated Force], OneSAF [One Semi-Automated Forces], and so forth).

DAGGERS ARCHITECTURE

The basic DAGGERS system is composed of two hardware systems: the soldier station and an 802.11g (802.11b compatible) wireless broadband router. The router can support multiple simultaneous soldier stations.

The soldier station software receives orientation updates from the helmet-mounted tracker wired via serial input. The weapon-mounted wireless tracker sends orientation updates via the wireless serial port. These tracker updates are received from an InterSense library using a published application programming interface. The soldier station software receives digital input information from the weapon-mounted controller via the wireless serial interface.

User Interfaces

The primary viewing device for the soldier station is the eMagin Z800 head-mounted display (HMD). All operating systems and image generation software displays will be rendered on the HMD.

Users can control their views in the virtual environment by turning their heads and/or their bodies, kneeling, standing, or going prone. By synchronizing virtual motions to corresponding physical motions and maintaining consistent real time video update, motion sickness is almost nonexistent in the DAGGERS' design. The helmet-mounted motion tracker and the weapon motion tracker are used to determine the user views in the virtual environment with a tracking accuracy of 1° root mean square or less. Users can also control their weapon aiming in the virtual environment by aiming their actual weapons in the real environment. The weapon-mounted tracker (in combination with the helmet-mounted tracker) controls weapon orientation in the virtual environment.

Once the system is running, users can maneuver through the virtual environment using the weapon-mounted controller. This device has a joystick that can be used to move the soldier forward and back by pushing the joystick in the up or down direction (relative to the up-down axis of the M4A1 weapon). By rotating their bodies (or their heads) in the real world, users can turn left or right; the same concept applies for looking up or down. In addition, to move (slide) left and right in the virtual world, the users must press the joystick left or right (again relative to the left-right axis of the M4A1 weapon).

There are a few menu selections that are available to the users to allow for sight-type selection, calibration of the helmet, and weapon motion trackers, among others. The weapon controller has a total capability for six independent buttons: one is tied directly to the weapon trigger mechanism; a second one is tied directly to the weapon magazine release; a third one acts as a menu mode toggle (as indicated above); and the other three buttons have functionality that provides for quick sight change (top button), a "reincarnate" function that is useful during tests (second from top), and changing from standing to kneeling and back (second from bottom) for configurations where the leg tracker is not used/desired. The function of the buttons might change in the future based on user feedback and might be fully programmable sometime in the future to allow users to have a system configured to their preferences. Audio cues are presented to the users via the headset that forms part of the helmet subassembly.

GAME BASED ENVIRONMENTS AND APPLICATIONS

As part of the research for the DAGGERS project, several different game based training environments were evaluated on the DAGGERS prototype system. One of the more interesting games that was integrated and tested on the DAGGERS platform was the *America's Army* game (Unreal Tournament–U.S. Army).

America's Army (AA)

Developed by the army as a recruiting tool, *AA* can be downloaded free of charge via the Internet. The game was built using the *Unreal Tournament* game engine and has become one of the most played first-person-shooter games over the last few years. Since the game's concept was based on introducing players to the basic training skills required for becoming a U.S. Army soldier (using a first-person perspective), it is a natural platform for the development of a training system.

Part of the research that the ETDS STO undertook was to integrate the DAGGERS prototype system with one of the commercial versions of *AA*. The goal was to test the flexibility of the DAGGERS design by integrating its controls with the game controls and then evaluating the usability of the resulting merger of technologies. Initially the interface was limited to a simple keyboard/mouse emulation that supported the commercial game interface without modifications. This initial effort exposed basic issues related to the design inconsistencies between the DAGGERS' and the *AA*'s user input paradigm. DAGGERS' design implements all user-motion tracking as a highly accurate, absolute measure of the orientation of the trainee's head, body, and weapon. *AA*'s design simplifies the user controls and the level of control granularity available to the player. *AA* uses a computer mouse's input to control the avatar's direction and gaze angle. In order to simplify the interface for the average PC user, the avatar's weapon and head (view) move in unison using the mouse's motion. The game also supports a joystick that essentially acts as a more precise mouse. While these simplifications improved the game playability for PC users with standard interfaces, they minimized and in some cases defeated the usability and functionality of the DAGGERS system design. The main issues created by these simplifications centered on the integration of the mouse-driven user interface and the reduction of the functionality of the DAGGERS system by "locking" together the motion of the avatar's head and weapon.

AA's User Weapon-Control Integration

Like most modern first-person-shooter games, *AA* portrays the avatar's view of self and whatever weapon it holds as one unit that moves together throughout the environment. This means that the weapon is always in the ready to fire position and is always pointing in the same direction as the avatar's view (implied head). This approach works well for these types of games since they require fast, coordinated motion to point and shoot and because most PC users have a mouse (or joystick) that they can use to control the avatar and its weapon. That approach increases the target audience since anyone with a PC can play the game.

In contrast, the DAGGERS system was designed to increase the level of virtual immersion that a user experiences in part by allowing the avatar's view direction to be separate and independent from the weapon's direction. By using two independent motion trackers (one for the head and one for the weapon) and incorporating a display (HMD) that physically follows the user's head motion, the DAGGERS system creates a very realistic feeling of "being there" that allows the user to experience a deep level of immersion.

FUTURE APPLICATIONS OF THE TECHNOLOGY

Although the DAGGERS system was designed as a self-contained dismounted soldier training device using today's technologies, the result was a system with a vast field of potential applications (commercial and government) and large potential for expansion and growth. As identified in the gaming technology experiments, by mixing the right components of system hardware, software, and human-machine interface designs, the DAGGERS concept has the potential to provide very immersive and engaging virtual environments. In addition to the obvious use of this system to train dismounted soldiers, this system could easily be adapted to provide an immersive environment for almost every station of a mounted crew trainer, with the added advantage of having the ability to dismount and mount from the vehicles at will, all while remaining immersed in a full surround 3-D scenario. Use of this system, or an upgrade of it in this way, could mean that the Department of Defense services, such as the army, the marines, the navy, and Special Operations Command could have complete crew training systems (9 to 12) fielded anywhere in the world, using a single small transport crate and requiring essentially no support infrastructure (see Figure 10.1). Other

Figure 10.1. Small Unit Training Using DAGGERs

potential applications of this technology are for training police, emergency response, and SWAT (special weapons and tactics) teams. By replacing the weapon subsystem with a firefighting tool(s) subsystem, teams of firefighters could train side by side to learn how to work as a team to contain large urban or wild fires.

Beyond regular training applications, the DAGGERS system design is ideally suited for in-theater mission rehearsal applications. Due to its ability to be used anywhere at anytime and its full 3-D immersion capabilities, this system represents the ultimate way for a team to rehearse details of an upcoming mission in a way that allows all participants to experience the environment as if they were there. Participants can analyze all moves, identify all relevant structures around them, and play "what if?" games while rehearsing the mission rather than guessing once they are there. The system design allows individuals outside the mission team to participate from anywhere in the world while relating relevant experiences about certain locations or structures and even guiding the mission team virtually through the environment.

THE FUTURE OF DAGGERS' SUCCESSORS

In looking toward the future and analyzing some of the deficiencies in the technologies or components used by the DAGGERS system, some of the future concepts look very promising. One of the areas where the current system is lacking is in the area of HMD technology. While the HMD technology used in the system is the best you can buy (today) within a reasonable budget, it is still very deficient in image resolution (800 × 600), field of view (40° diagonal FOV), and optical distortion effects. Future DAGGERS' successors should provide at least 1,280 × 1,024 image resolution per eye (perhaps even 1,600 × 1,200), all while providing a FOV greater than 200° horizontal and 100° vertical covering the entire field of human vision with distortion-free, very fast (>60 hertz) and deterministic video portraying highly detailed visual representations of virtual environments. Although the tracking technology used by the current system design works relatively well, newer motion tracking technologies and human machine interfaces will have to be used to provide a system that can support a wide variety of communication metaphors, such as hand, body, facial gestures, and eye movement. Future successor designs will have to do all of this while maintaining or lowering the cost of the system, making user operation of the system natural and intuitive, and maintaining most if not all of the basic design principles of the original DAGGERS system.

Chapter 11

MEDICAL SIMULATION TRAINING SYSTEMS

M. Beth Pettitt, Michelle Mayo, and Jack Norfleet

BENEFITS AND NEED OF SIMULATION IN MEDICAL TRAINING

Medical training has changed significantly throughout the years. Traditional classroom lectures are still useful for teaching the fundamentals of medicine, but hands-on training is critical for developing lifesaving skills. Through the years, hands-on training has progressed from live animals to cadavers, to high tech patient simulators and haptic interfaces, to virtual environments throughout the military and civilian medical communities. The old model "see one, do one, teach one" is effective in controlled environments, such as universities and teaching hospitals, but gaining the necessary numbers of patient contacts in military training is nearly impossible given the number of students needing training. Simulation has emerged as an important player in training medical personnel to master those skills needed to preserve human life (Mayo, 2007).

HISTORY OF THE USE OF SIMULATION IN MEDICAL TRAINING

Although medical professionals have used simulation devices for hundreds of years, the military really began institutionalizing computerized simulation for medical training only approximately 10 years ago. Even at that time, simulation was primarily used for verification and validation of the simulation technology's training effectiveness. The stated model for trying to employ medical simulation and training in existing military medical curriculums was the aviation simulation community. The operation of actual flights is hazardous and potentially fatal without proper hands-on training. Didactic training is still important, but it does not teach the tactile awareness and the psychomotor skills necessary to fly an aircraft. The military realized this and addressed this problem as early as World War I when pneumatic motion platforms powered mock-up cockpits for hands-on training. Pilots were able to learn controls and movement of the aircraft without ever having to become airborne. Medical training faced some of the same challenges as aviation training, but it took much longer for simulation to be accepted by the military. Hands-on training with patients was nearly impossible to achieve given the large number of students versus the small number of available patients. Until a decade ago military medical students (especially the prehospital caregivers, such as the combat medic) were forced to practice their skills on animals, on cadavers, and occasionally on other students and current patients in military medical facilities.

Over the past 10 years, the army's research and development community has striven to improve medical training through the use of simulation and has changed the way the army trains combat medics and combat lifesavers. In 1997, the Simulation, Training and Instrumentation Command's technology base, now part of the Research, Development and Engineering Command (RDECOM) Simulation and Training Technology Center (STTC), developed the first distributed medical simulation training system called the combat trauma patient simulation (CTPS). The CTPS system was the first attempt by the army to introduce distributed interoperable simulation products into the military medical training community that up to that time had relied heavily on didactic and live tissue training. The CTPS program yielded significant research and user data for its time that guided an additional 10 years of research and development of simulation technologies for military medical training (Rasche & Pettitt, 2002).

The purpose of the CTPS program was to provide a simulated capability to realistically assess the impact of battlefield injuries on the military medical care structure. Commercial off-the-shelf and government off-the-shelf components provided a network infrastructure to simulate medical handling and treatment for combat injuries through every echelon of care on the battlefield. Capabilities include simulating, replicating, and assessing combat injuries by type and category, monitoring the movement of casualties on the battlefield, capturing the time of patient diagnosis and treatment, and comparing interventions and outcomes at each military treatment level and passing injury data to and from warfighter simulation systems (Carovano & Pettitt, 2002).

STAND-ALONE VERSUS NETWORKED MEDICAL TRAINING SYSTEMS

CTPS is a networked medical training system designed with an open system architecture to allow interoperability with any simulation system, including other patient simulators, distance learning systems, constructive simulations, and instrumentation systems (Figure 11.1). The system consists of six treatment nodes, a casualty transfer network, pre-programmed clinical scenarios, and an after action review capability. Computerized mannequins representing treatment nodes are the most recognizable CTPS components as they provide the primary student interface with hands-on training for users. Sophisticated physiological models drive the computerized mannequins, forcing users to assess and treat the casualty. The transfer network electronically moves casualties through each echelon of care. Clinical scenarios provide standardized training on specific medical

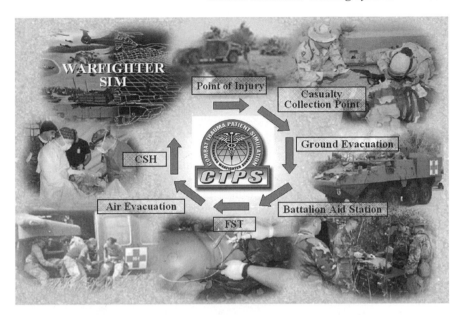

Figure 11.1.   Combat Trauma Patient Simulation System Cycle of Care

skills and can be varied for trainees of differing skill levels. The mannequins have the ability to accurately represent hundreds of injury and disease conditions, both battlefield specific as well as those primarily seen only by civilian care providers.

The CTPS program set out to develop the end-all medical simulation training system, which unfortunately, with a large and costly footprint, was not reasonable for training the huge numbers of combat medics and combat lifesavers. The CTPS program did, however, generate vast amounts of data that identified training gaps and user requirements. These data were used as guides to develop smaller, lighter, cheaper, and more portable stand-alone training system solutions.

A prime example of technology that emerged from the CTPS program is the mannequin known commercially as the Emergency Care Simulator (ECS). The ECS, requiring only compressed air and a single laptop for operation, is a stand-alone system with a much smaller footprint than the networked CTPS system or the more complex mannequins originally used in the CTPS system. As part of the ECS development, it was incorporated into the CTPS system to bolster the pre-hospital nodes that trained combat medics, combat lifesavers, and first responders. The pre-hospital levels of care encompass the vast majority of military medical caregivers.

This new technology introduced a new capability that was affordable and the right fidelity for training large numbers of students in both classroom and non-classroom settings for more realistic and effective training. These users, both military and civilian medical personnel, were finally testing the limits of current simulation training systems, which subsequently led to new requirements. These requirements include increased portability, more realistic physiological attributes, and ruggedness ideal for movement in atypical training environments. The result was the Stand-Alone Patient Simulator (SAPS) (Figure 11.2), which blends physiologically accurate injuries, sensor technologies, miniaturization/packaging technology, and wireless networking technologies with state-of-the-art patient simulation technologies. Designed primarily to meet military training needs derived from user test data, the SAPS is a rugged, full body patient simulator that is physiologically based and completely wireless, enabling soldiers to move their patients to truly train as they fight. The original CTPS system transferred patients through an electronic network, but SAPS requires the physical evacuation of patients from the battlefield. This enhanced capability allows for more accurate resource management data to be collected from the simulation. For example, instead of hitting a button and sending the virtual patient on its way, the SAPS continues to deteriorate during transport. With the possibility of the patient dying, the evacuation system is exercised and en route care is required. This capability opens a training realm that has not been possible in the past and has the potential to better prepare warfighters to save lives and resources.

Other simulation technologies that have emerged from the CTPS program include partial task trainers developed to train specific skills (Figure 11.3), game based systems for patient management and basic care, as well as surgical trainers for more advanced procedures, such as the crichothyroidotomy. As CTPS research has proven over the years, there is no end-all training system for medical simulation. Stand-alone systems, partial task trainers, and networked,

Figure 11.2.   A Stand-Alone Patient Simulator (SAPS)

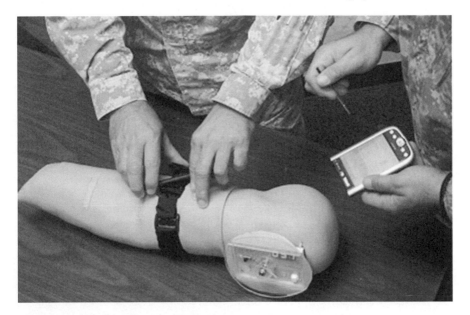

Figure 11.3. A Partial Task (Tourniquet) Trainer

interoperable systems all serve specific purposes and contribute to the overall success of training medical personnel for the difficulties of providing care on the battlefield.

SUCCESS STORIES

Transitioning medical research and development efforts to actual funded programs has been a government challenge for years. Because of the disbursed nature of medical simulation and training facilities and, consequently, medical simulation research and development, this has been a particularly difficult challenge for the medical simulation community. There are, however, some success stories that have emerged from medical simulation research at RDECOM and other organizations. None of these simulation success stories have resulted from any type of formal or typical acquisition program. As an example of this, the CTPS system was developed with congressional research funding by a commercial vendor, Medical Education Technologies, Inc. This research has been performed through the years by academia, industry, and government partners, including the University of Central Florida Institute for Simulation and Training in Orlando, Florida; the Tekamah Corporation in Fairfax, Virginia; the National Center for Simulation in Orlando, Florida; and the U.S. Army Medical Research and Materiel Command (MRMC) located at Fort Detrick, Maryland. RDECOM STTC has provided oversight and management for this government initiative over the past 10 years through numerous congressional awards. User based design and a spiral engineering methodology guided development throughout all nine phases of the CTPS program. Early phases yielded a test and evaluation program that ensured functionality and usability of the components. By phase 4, a full system test and evaluation were implemented that measured training efficacy of the system in a simulated operational environment. The last few phases of the program incorporated stand-alone trainers and refined the overall design. Currently, user tests are still ongoing at Fort Gordon, Georgia; Fort Polk, Louisiana; Camp Pendleton, California; and the Defense Medical Readiness Training Institute, Fort Sam Houston, Texas.

After several successful installations and user evaluations, the CTPS system became a catalyst for driving more medical simulation efforts within the army. The army funded three science and technology objectives to build upon the CTPS architecture and develop advanced medical training technologies for combat medics and combat lifesavers. The most recent technological development lies in the SAPS. The SAPS is changing the way the military medical community conducts training, while finally gaining appropriate support from the army leadership. This is a direct result of the revolution in medical training that CTPS started 10 years ago.

Throughout the years of CTPS user tests, research, and development, the army has realized the positive impact medical simulation devices have on training. In 2005, Lt. Gen. Kevin C. Kiley (Kiley, 2005), under orders from the Vice Chief of Staff of the Army (VCSA), signed an operational needs statement (ONS) for medical simulation training centers (MSTC) to be established for combat lifesavers (CLSs) and tactical combat casualty care (TC3) training. This ONS is the first time the army has stated that medical simulation is critical for the success of the army's mission. It also established the very first medical simulation acquisition program that is currently being managed by program executive office simulation training and instrumentation. Although the military has been training with medical simulation devices for many years, the MSTC program provides lifecycle support for the fielded training devices. Sites are also provided instructors, a building with all classroom equipment, and the necessary training devices to train combat medic advance skills training, TC3, and CLSs.

Some researchers speculate that the CTPS program jump-started the MSTC acquisition program. Without the CTPS system and its derivative technologies being used at installations for user tests and training, soldiers would not have known the capability existed. As a result, units and installations would not have procured and sustained systems on their own, and the VCSA would not have realized the power of simulation in medical training or seen the need to establish the MSTC. Furthermore, the training device that was selected and fielded as the army standard patient simulator for the MSTC was the ECS.

CHALLENGES TO THE USE OF SIMULATION IN MEDICAL TRAINING

As with all simulation and training systems, verification and validation (V&V) must be performed. However, there must be well-established test objectives and criteria, and the V&V must be based on current training practices and objectives. It is obvious that there is no 100 percent solution for any simulation and training system. To continue to validate a system with no well-defined performance criteria is futile, and all of the results become null and void as soon as the system changes. Some in the military research community have used government V&V as the end goal and ignored the need to use technology to improve training and ultimately save lives. Many of the technologies in question have been proven effective training tools in the civilian sector by improving test scores and by providing training opportunities that would not otherwise exist. Many times, these systems are not new at all; they are simply new to the government. V&V should not be used as a roadblock, but as a way to measure the strengths and weaknesses of a system.

REQUIRED BREAKTHROUGHS FOR THE GROWING USE OF SIMULATION IN MEDICAL TRAINING

One of the next big challenges of medical simulation and training devices will be to improve the realism of the simulation injury or disease state. The goal is to create training devices that are so realistic that they are indistinguishable from real skin, flesh, blood, and bone, including smells and textures. To be successful as training devices, the new technologies must also be durable and reusable to be able to meet the training needs. To meet these goals, the army has established an advanced technology objective jointly funded by RDECOM STTC and MRMC to investigate the next generation of simulated injuries and injury repair. As with so many other military medical simulation developments, these results will serve multiple medical professionals from frontline medics to surgeons.

WHAT THE CIVILIAN AND MILITARY COMMUNITIES CAN LEARN FROM EACH OTHER

Technologies developed for military medical simulation are directly applicable for civilian use. Additionally, civilian developments in medical simulation often meet military needs with little or no additional research. For example, the mannequins that were the main components of the original CTPS system came from industry. The military adapted those devices for its use, adding such capabilities as trauma and chemical and biological scenarios that have become indispensable in the post 9-11 world. There are also a number of game based platforms that have easily flowed between military and homeland defense uses. Medical simulation is a global requirement as all caregivers, regardless of uniform, must master their skills in order to save lives. Saving lives is the ultimate goal.

REFERENCES

Carovano, R. G., & Pettitt, M. B. (2002, December). *The combat trauma patient simulation system: An overview of a multi-echelon, mass casualty simulation research and development program.* Paper presented at the Interservice/Industry Training, Simulation, and Education Conference, Orlando, FL.

Kiley, K. C. (2005, April). *Operational needs statement for medical simulation training Centers for Combat Lifesavers (CLS) and Tactical Combat Casualty Care (TC3) Training* [Memorandum to Deputy Chief of Staff, G-3].

Mayo, M. (2007). *The history of the Army's research and development for medical simulation training.* Paper presented at the Fall 2007 Simulation Interoperability Workshop, Orlando, FL.

Rasche, J., & Pettitt, M. B. (2002, December). *Independent evaluation of the Combat Trauma Patient Simulation (CTPS) system.* Paper presented at the Interservice/Industry Training, Simulation, and Education Conference, Orlando, FL.

Chapter 12

AVIATION TRAINING USING PHYSIOLOGICAL AND COGNITIVE INSTRUMENTATION

Tom Schnell and Todd Macuda

Effective flight training requires a pedagogical approach that provides training scenarios and quantitative feedback. The goal of flight training is to provide the student with a lasting knowledge base of relevant information, a set of cognitive and motor skills, and the tools to exercise judgment in situation assessment. Learning a skill, such as flying, is a cognitively involved process. Today, instructors use objective measures of task performance and additional estimated, subjective data to assess the cognitive workload and situation awareness of students. These data are very useful in training assessment, but trainees can succeed at performing a task purely by accident (referred to as "miserable success"). Additionally the student can be in a less than optimal state for learning when the instructor/operator applies brute force training tasks and methods with little regard to the learning curve, which can result in the training being too easy or more often too difficult thereby inducing negative learning. By using neurocognitive and physiologically based measures to corroborate objective and subjective measures, instructors will be much more able to diagnose training effectiveness.

At the Operator Performance Laboratory (OPL), we quantify cognition using a system that we call the quality of training effectiveness assessment (QTEA) tool. QTEA is based on the cognitive avionics tool set (CATS) that has been developed over the last four to five years at the OPL. QTEA is a systems concept that allows the instructor pilot to assess a flight student in real time using sensors that can quantify cognitive and physiological workload. Using QTEA, the instructor can quantify the student's workload level in real time so that the scenarios can be adjusted to an optimal intensity. The cognitive and physiological measures also serve as a measure of a student's learning curve, making it possible for the trainer to detect plateaus in learning. Using QTEA, the trainer will be able to assess the need for further training in a student. The basic idea of QTEA is to give the instructor a real time picture of the performance of a student based on physiological and cognitive data, flight technical data, and mission-specific data. With the help of the data collected by the instrumentation, the instructor is able to provide the student with a detailed after action review. This provides the student with unambiguous knowledge of results (KR), a key ingredient in closing the learning loop.

The deployment of the QTEA instrumentation goes hand-in-hand with instrumented flight training assets, such as those that are available at the OPL and the National Research Council Flight Research Laboratory (NRC-FRL). The OPL has focused its instrumented training paradigms on fixed-wing simulators and aircraft, and the NRC-FRL has specialized on rotary-wing aircraft. This chapter illustrates a training system that can be developed from those core capabilities.

QTEA leverages years of work in basic research in augmented cognition (Schmorrow, 2005) and applies it to the exciting field of flight training. Wilson et al. (1999, 2003, 2005) generated a sound foundation of workload measurement using neurophysiological measures in the aviation context. Schnell, Keller, and Macuda (2007a, 2007b) described their workload assessment research in actual flight environments. The introduction of human neurocognitive data to quantify the effectiveness of training represents a breakthrough application in the operational field of flight training.

INSTRUMENTED TRAINING SYSTEMS REQUIREMENTS

There are several methodologies that have evolved in flight training over the years, and effective instructor pilots are familiar with the broad principles of learning (FAA-H-8083-9, 1999). A critical component of these principles is motivating students to learn. It is also well known that exercise and repetition are a required element of flight training, but care must be taken not to "burn out" the student with mindless rote memory and recall exercises. Through cause and effect, students learn skills more quickly, especially if the effect was associated with a positive feeling of accomplishment. In the training context, it is not productive to expose students to scenarios for which they are not ready and in which they are guaranteed to fail. The feeling of defeat and failure that comes from such exposure is sure to hamper the learning effect. The principle of primacy is important as well, and it is the instructor's job to ensure that a skill is taught correctly the first time. Habit-forming mistakes are hard to eliminate once they become engrained. An instrumented quantitative approach such as our QTEA system concept provides the instructor with the required tools to detect bad habits before they become engrained. Training scenarios need to have an intensity that is appropriate to the level of expertise in the student. Instrumented training systems such as QTEA should provide the instructor with real time cognitive loading data of the student, thus allowing for real time adaptation of scenario intensity. Figure 12.1 shows an example of how QTEA can be integrated into training station software, such as the Common Distributed Mission Training Station (CDMTS). Figure 12.1 shows CDMTS as the platform that is used to generate and manage the training scenarios and analyze the results for after action review. Through a plug-in software interface inside of CDMTS, QTEA provides such components as a timeline with traces of workload, level of fatigue, and

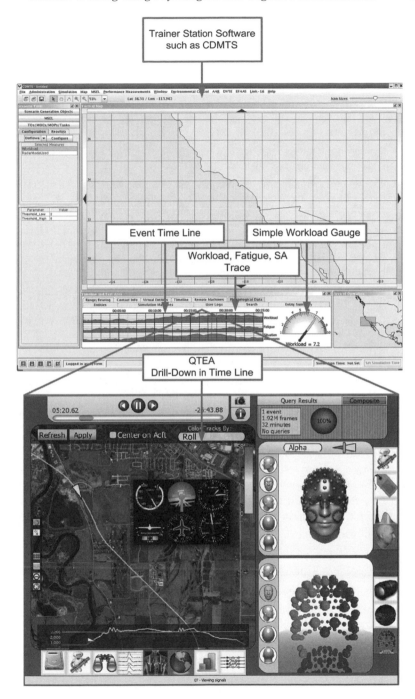

Figure 12.1. Quality of Training Effectiveness Assessment System Incorporated in the Common Distributed Mission Training Station

situation awareness, as well as a workload meter. In this concept, the user also has the capability to drill into the timeline at a specific time in the scenario to launch the QTEA application as shown in Figure 12.1. This drill-down capability allows the user to find detailed information about pilot and aircraft states at selected locations on the timeline. In conversations with potential users, we identified the following core requirements for the user interface:

1. **User Interface**
   a. Synchronized timeline of scenario events—This allows events to be shown as flags on a timeline for easy visualization.
   b. Quality of training metric based on student neurophysiological and aircraft state data—This metric is described in Schnell, Keller, Cornwall, & Walwanis-Nelson (2008).
   c. Ability to drill down into the sensor and aircraft state data to determine root causes of training problems—This is illustrated in Figure 12.1. The user should be able to click any location along the timeline to explore the underlying aircraft and human performance data.
   d. Graphical workload meter (real time)—This meter should represent an overall score of workload, such as the National Aeronautics and Space Administration (NASA) Task Load Index (Hart & Staveland, 1988). The overall score may consist of such subscores as task difficulty, time pressure, flight technical performance, sensory effort, physical effort, frustration, and so forth. The subscores may be revealed in the drill-down mode.
   e. Summary of performance (pass/fail)—This score is based on flight and mission technical performance (Schnell et al., 2008).
   f. Real time adaptation of scenario intensity based on operator state and aircraft state characterization—This system capability is intended to keep the student optimally stimulated. Scenario intensity and difficulty may be increased by adding tasks such as clearance changes (rerouting, holding, and approaches), weather changes, or non-normal situations. Scenario difficulty may be simplified by removing additional tasks and by providing steering vectors.
   g. Quantitative after action review using hard sensor data from the instrumented flight training system—This is one of the main purposes of QTEA.
   h. Ability to interact with established training tools, such as CDMTS.
2. **Operator State Classification**
   a. Reliable sensor and classification system.
   b. Simple to train/calibrate with little or no need for sensor preparation.
   c. Real time workload and situation awareness gauge.
   d. Indicator of available cognitive resources.
3. **Architecture**
   a. PC based and low cost.
   b. Ability to connect to multitude of sensors of different sampling rates.
   c. Sensor manufacturer independent—This means that the training system should be flexible enough to accommodate sensors from different providers.

   d. Rugged and robust system for use by flight instructors without expertise in neural measurement.

   e. Networked, high level architecture (HLA) federation, and Ethernet. This will provide the training system with the ability to participate in live virtual training exercises involving several participants across a networked federation of flight simulators and flight assets.

   f. Automatic synchronization of data with robust protocol.

   g. Distributed mission operations capable for team training and assessment of team cognition.

   h. Easy to integrate in a virtual environment.

   i. Integrate and fuse neural, physiological, aircraft, and mission state data.

   j. Rugged and tested.

4. **Usability**

   a. Sensors must be easy to set up.

   b. No burden on trainer; easy to learn and use.

   c. Instrumented system must save training time without loss in quality of training.

   d. Better KR.

In our work toward an instrumented training platform, we addressed these user requirements by translating them to engineering requirements as embodied in our QTEA concept. For example, to achieve a high system level of acceptability early on, we feel that a phased approach of sensor deployment may be sensible (Schnell et al., 2008). Sensors that are ready for deployment with high payoff from a flight training point of view include eye tracking, electrocardiogram (ECG), and respiration amplitude and frequency. Past research by Schnell, Kwon, Merchant, Etherington, & Vogl (2004) indicated the value of eye tracking systems in determining the scanning patterns of instrument pilots. Also, as a certified instrument flight instructor, Schnell found that some instrument flight students who have problems flying accurate instrument approaches used inappropriate instrument scanning techniques. Detailed after action review using eye scanning records will provide such instrument flight students with evidence that their scans can be improved. On this basis, we are confident that these sensors can benefit the training community today with considerable cost savings. Next in line would be neural sensors, such as electroencephalogram (EEG), that could be developed to field hardened robustness in a matter of a few years. The key is to use an underlying core architecture such as QTEA that can grow with maturing sensor technology.

SYSTEM HIGH LEVEL DESIGN

The QTEA architecture concept (see Figure 12.2) was developed by OPL to interface with the airborne and simulation assets at the OPL and the NRC-FRL. Additional connectivity to established training systems, such as CDMTS, was enabled through a plug-in architecture, and connectivity to distributed training

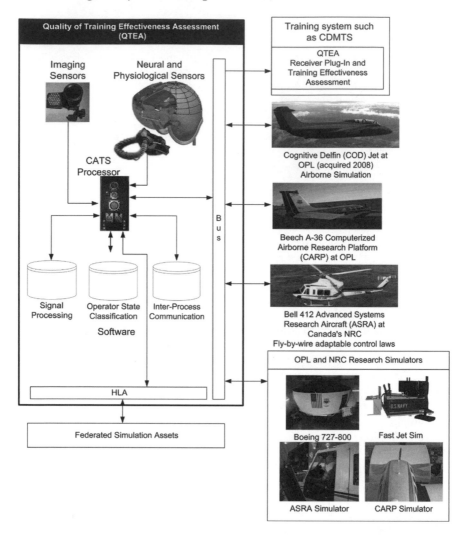

Figure 12.2.   Quality of Training Effectiveness Assessment System Architecture

systems, such as the deployable virtual training environment, was enabled through an HLA gateway.

QTEA uses several neural and physiological sensors, including dense array EEG, ECG, galvanic skin response, pulse oximetry, respiration (amplitude and frequency), and noncontact measures, such as facial feature point location, facial temperature differences, and eye tracking. Not all sensors need to be used to perform operator state characterization. During a transition from the research to the training world, it may be best to deploy sensors that are ready for field use, such as eye tracking and ECG first and gradually migrate the other sensors as they mature to fieldable products.

DEVELOPMENT OF QTEA SYSTEM PROTOTYPES

QTEA is an instrumented aviation training systems concept developed by the OPL. The instrumented aircrafts that OPL and NRC-FRL use in their cognitive avionics research are the Computerized Airborne Research Platform (CARP) at OPL and the Bell 412 Advanced Systems Research Aircraft (ASRA) at NRC-FRL. The CARP is a Beech A-36 Bonanza equipped with the full QTEA suite of sensors, a tactile cueing seat, and liquid crystal display overlay displays that can be controlled to systematically manipulate pilot situation awareness, spatial orientation, and workload. The Cognitive Delfin (COD) is an Aero Vodochody L-29 Delfin training aircraft that OPL acquired in February 2008. Once fully equipped with QTEA, this aircraft will support research in advanced cognitive avionics and concepts of airborne simulation for training of aviators. Airborne simulation refers to a concept where part of the scenario is simulated and part of it is real. That is, the aircraft is flown in real airspace with real dynamics, but in simulated battle space. Through a high speed datalink, the COD will be able to federate with ground based training assets.

The ASRA facility at NRC-FRL is a single-string (simplex), full authority, fly-by-wire (FBW) system (Gubbels, Carignan, & Ellis, 2000). The ASRA facility has a unique capability to change the control laws of this airborne platform to allow the simulation of other helicopter types and to vary the workload experienced by the evaluation pilot. This provides test pilot school participants with a rotorcraft platform that can emulate a wide range of handling qualities. The native handling quality of normal commercial helicopters is so refined that test pilot school participants cannot experience the range of poor handling qualities the Cooper-Harper rating scale can offer. By manipulating the control laws of the FBW system, the NRC-FRL can give students a sense what a helicopter with poor handling qualities would feel like. Combined with our operator state classification technology, it is becoming feasible to systematically manipulate pilot workload by manipulating handling quality and simultaneously monitoring operator state for the purpose of providing KR in pilot training programs.

OPERATIONAL CONSIDERATIONS OF INSTRUMENTED AVIATION TRAINING SYSTEMS

QTEA uses a battery of neurocognitive and physiological sensors on the student, fuses these data with aircraft and mission data, and applies sophisticated signal processing and classification techniques to gain a fuller picture of training effectiveness. The aviation training community could benefit from such quantitative tools that measure the effectiveness of training on the basis of human performance data. Using this real time data, training scenarios can be adapted to optimal intensity so as to maximize the effectiveness of learning. This will likely save money in terms of shorter training times, less time spent on actual airborne training assets, higher operational success levels, and improved performance based specifications for flight simulators.

ITERATIVE INTEGRATION AND TRANSITION

Transitioning a tool such as QTEA may be best accomplished through an initial deployment of the mature components, such as eye tracking and ECG based operator state characterization measures of training effectiveness. Other sensors including EEG can be transitioned in iterative steps when they reach operational readiness that can satisfy the user requirements. In the naval aviation context we identified the CDMTS framework as the most effective way for an initial transition of QTEA to the fleet. CDMTS is an instructor software package that handles all aspects of training, ranging from scenario development to after action review. Through a plug-in mechanism, it is possible to integrate the neurocognitive measures from QTEA into the training timeline for dynamic adaptation of scenario intensity and for quantitative after action review. An early demonstration of the QTEA system was held at the 2007 Interservice/Industry Training, Simulation, and Education Conference in Orlando, Florida. The system was integrated in a fixed-base jet procedure training flight simulator and connected to the CDMTS station.

TEST AND EVALUATION

Tests of the underlying neurocognitive operator state characterization system have been conducted over the past few years at OPL and NRC-FRL. Research on the ASRA and CARP was discussed by Schnell, Keller, and Macuda (2007a, 2007b) and Schnell, Macuda, and Poolman (2006). The detailed architecture of the CARP and the ASRA are discussed by Schnell, Macuda, and Keller (in press).

Going forward, we intend to test QTEA in environments that will iteratively approach operational status. Of particular importance will be the development of detailed performance measures that the instructor pilot can use to assess the student's performance. These measures are discussed in detail by Schnell et al. (2008). There are two types of performance measures in QTEA, mission specific and physiological ones. The mission specific measurers will change with the use case of QTEA. For naval aviation in a close air support task, they will include airmanship measures (flight technical), administrative, and tactical. The physiological measures describe the body's reaction to the task and include cognitive measures quantifying stress, cognitive resources, and attention (alertness). The physiological measures subsystem of QTEA will function automatically and in real time, providing the trainer with a metric that indicates cognitive loading. The mission-specific performance measures will be integrated and partially automated with trainer override. QTEA will provide the trainer with a leading indication of impending task saturation, thus providing a measure of the student's cognitive bandwidth.

CONCLUSIONS

Instrumented aviation training systems, such as QTEA, provide instructors with quantitative data of the student's performance. These data can be used by automated scenario generation systems to adjust scenario intensity in real time to maximize learning by keeping stimulation at its optimal level. The quantitative data generated by QTEA also provide for superior after action review, offering the instructor and the student the ability to review deviations in mission or flight technical domains, as well as the occurrence of cognitive (workload) bottlenecks, poor control manipulation, or ineffective eye scanning technique. Through review and discussion of such quantitative data, the instructor and the student can develop training strategies that achieve the training goal in a shorter time than would be possible without such advanced tools.

REFERENCES

FAA-H-8083-9. (1999). *Aviation instructor's handbook.* U.S. Department of Transportation, Federal Aviation Administration (FAA), Flight Standards Service.

Gubbels, A. W., Carignan, S., & Ellis, K. (2000). *Bell 412 ASRA safety system assessment* (Tech. Rep. No. LTR-FR-162). Ottawa, Canada: Flight Research Laboratory.

Hart S. G., & Staveland, L. E. (1988). Development of NASA-TLX (Task Load Index), results of theoretical and empirical research. In P.A. Hancock & N. Meshkati (Eds.), *Human mental workload* (pp. 239–250). Amsterdam: North Holland Press.

Schmorrow D. (Ed.). (2005). *Foundations of Augmented Cognition.* Mahwah, NJ: Lawrence Erlbaum.

Schnell, T., Keller, M., Cornwall, R., & Walwanis-Nelson, M. (2008). *Tools for virtual environment fidelity design guidance: Quality of Training Effectiveness Assessment (QTEA) tool* (Rep. No. N00014-07-M-0345-0001AC, Contract No. N00014-07-M-034). Arlington, VA: Office of Naval Research.

Schnell T., Keller, M., & Macuda, T. (2007a, October). *Application of the Cognitive Avionics Tool Set (CATS) in airborne operator state classification.* Paper presented at the Augmented Cognition International Conference, Baltimore, MD.

Schnell, T., Keller, M., & Macuda, T. (2007b, April). *Pilot state classification and mitigation in a fixed and rotary wing platform.* Paper presented at the Aerospace Medical Association (ASMA) annual conference, New Orleans, LA.

Schnell T., Kwon J., Merchant S., Etherington, T., & Vogl, T. (2004). Improved flight technical performance in flight decks equipped with synthetic vision information system displays. *International Journal of Aviation Psychology, 14*(1).

Schnell, T., Macuda, T., & Keller, M. (in press). Operator state classification with the cognitive avionics tool set. In D. Schmorrow & K. Stanney (Eds.), *Augmented Cognition: A practitioner's guide.*

Schnell, T., Macuda, T., & Poolman, P. (2006, October). *Toward the cognitive cockpit: Flight test platforms and methods for monitoring pilot mental state.* Paper presented at the Augmented Cognition International Conference, San Francisco, CA.

Wilson, G. F., & Russell, C. A. (1999). Operator functional state classification using neural networks with combined physiological and performance features. *Proceedings of the Human Factors and Ergonomics Society 43rd Annual Meeting* (pp. 1099–1102). Santa Monica, CA: Human Factors and Ergonomics Society.

Wilson, G. F., & Russell, C. A. (2003). Real-time assessment of mental workload using psychophysiological measures and artificial neural networks. *Human Factors, 45,* 635–643.

Wilson, G. F. (2005). *Operator functional state assessment in aviation environments using psycophysiological measures. Cognitive Systems: Human Cognitive Models in System Design Workshop,* Santa Fe, NM.

Chapter 13

VIRTUAL ENVIRONMENT LESSONS LEARNED

Jeffrey Moss and Michael White

This chapter provides an overview of lessons learned from integration and transition of research prototypes into joint interoperable training systems. Specific examples we cite include joint interoperability special events from the Interservice/Industry Training, Simulation, and Education Conference (I/ITSEC), as well as the virtual at sea trainer, virtual technologies and environments, virtual fire support trainer, virtual simulation training toolkit, and the deployable virtual training environment programs. The term "joint" is used to describe actions, events, or operations within or associated with the military that involve more than one service. Highlights of our activities included the following:

1. Innovative reuse of existing technologies,
2. Development of new capabilities that allow multiple simulation systems to interact with one another along with live battle command systems, creating a training environment that allows military members to train as they fight.
3. Replacement of proprietary software solutions with open source applications,
4. Use of rapid prototyping, allowing users to experiment with and provide feedback on systems that are still under development, and
5. Integration of science (that is, psychologists, industrial hygienists, and educators conducting usability studies, training effectiveness evaluations, and so on) and technology (discussed in items 1–4).

TASK ANALYSIS

Performance Standards

The process used for performing task analysis was generally the same for all programs. In each instance, a thorough understanding of the military knowledge domain was indispensable in defining the capabilities required of the systems. The knowledge acquisition process began with investigation of existing military doctrine, as well as current tactics, techniques, and procedures. The initial analysis culminated in the preparation of a systems requirements document, the basis for rapid prototyping, tracing requirements to training needs, and providing the foundation for iterative testing and integration.

Lesson Learned

Tether performance standards to military doctrine.

Technology Requirements

Technology requirements were derived from a survey of current and emerging technologies. Desired capabilities were based on performance requirements and tempered with system optimization and, of course, budgetary considerations. Hardware and software system configurations were selected based on intended use. Low cost hardware alternatives were employed when using facilitated simulation goals (Stripling et al., 2006). For prototype systems targeted toward transitioning to the Marine Corps' deployable virtual training environment, we chose to adopt a notebook computer configuration. Early virtual technologies and environments experimental systems (demonstration I and demonstration II) employed rack-mounted computers as well as desktop and notebook computer configurations. At least one of the virtual technologies and environments demonstration II systems, the Warfighter Human Immersive Simulation Laboratory's Infantry Immersive Trainer (see Stripling, Muller, Schaffer, and Cohn, Volume 3, Section 1, Chapter 5), adopted a combination of single-person pods, high end head-mounted displays, and infrared tracking cubes, all mounted within a medium room–sized tubular framework.

Innovative reuse of existing technologies was a common theme throughout all the efforts. For instance, the multihost automated remote command and instrumentation application, originally developed under the synthetic theater of war advanced concept technology demonstration, served as the core element of the battle master exercise control component in virtual technologies and environments demonstration I and again as the core element of the common distributed mission training station in virtual technologies and environments, virtual fire support trainer, virtual simulation training toolkit, and deployable virtual training environment, where it is used to launch and control multiple computer applications across computers interconnected by a local area network or a wide area network.

Another example of technology innovation common to these projects was development of the joint live virtual constructive data translator. Initially funded as a research and development effort for Joint Forces Command, joint live virtual constructive data translator allows multiple disparate simulation systems using such communication protocols as distributed interactive simulation, high level architecture, and training and testing enabling architecture to interact with one another. Beyond facilitating communications across multiple simulation protocols, the team adapted joint live virtual constructive data translator, enabling stimulation of live battle command systems by virtual and constructive simulations to create a comprehensive training environment, allowing military personnel to train as they fight. For instance, the development team created a joint live virtual constructive data translator module that translates digital messages between the advanced field artillery tactical data system and the computer-generated forces application know as Joint Semi-Automated Forces, resulting in appropriate, realistic interactions between the live battle command system and constructive artillery elements.

One of the challenges we sought to overcome was that of replacing proprietary software solutions with open source applications, saving the government money on licensing fees while enhancing the training experience. For example, the original virtual fire support trainer prototype employed a proprietary image generation system that imposed a $10,000 license fee per computer. Considering the implications of this expense for the Marines Corps deployable virtual training environment program's initial fielding plan of nearly 1,000 computers, the integrated product team of developers and customer stakeholders decided to adopt DEL-TA3D,[1] a technologically mature, open source gaming application developed under the sponsorship of the Naval Postgraduate School[2] as the visualization engine.

Lesson Learned

To the maximum extent possible, reuse work that has already been done for the Department of Defense because where appropriate, software reuse saves the customer money and time.

SYSTEM HIGH LEVEL DESIGN

The system architecture employed in the virtual technologies and environments' multipurpose operational team training immersive virtual environment was initially adopted in the virtual at sea trainer, improved during the virtual fire support trainer effort, and further refined in the virtual technologies and environments and virtual simulation training toolkit development efforts to be included in the deployable virtual training environment.

The linchpin of the architecture is the joint live virtual constructive data translator, which enables integration of live battle command systems with simulation components, as well as integration of simulation systems operating on disparate communications protocols. Its modular construction facilitates bidirectional translation of real world battle command system messages, usually in variable message format variants, between such applications as advanced field artillery tactical data system or command and control personal computer and such simulation components as the Joint Semi-Automated Forces. The joint live virtual

[1]http://www.delta3d.org/index.php?topic=about
[2]MOVES Institute, Naval Postgraduate School, c/o Erik Johnson, WA366, 700 Dyer Road Monterey, CA 93943-5001.

constructive data translator application also enables simulation systems operating on different protocols to interoperate as parts of a complete simulation federation.

Lesson Learned

Even when embarking on leading-edge research and development to create innovative training systems, common commercial items may be appropriate.

DEVELOPMENT OF PROTOTYPES

Although we employed different development lifecycle models for the projects we address in this chapter, a common theme was rapid development of a prototype capability (see Nicholson and Lackey, Volume 3, Section 1, Chapter 1). Worthy of note, virtual technologies and environments share with the virtual fire support trainer the unique experience of having end-user interaction with prototypes while the software and hardware was under development, providing invaluable feedback as the development progressed. User input on the prototypes enabled us to rapidly respond to customer needs.

Prototype Development

After development and delivery of the initial virtual fire support trainer prototype to the 10th Marine Regiment at Camp Lejeune, North Carolina, the marines immediately pressed it into service to assist in training forward observers. Although virtual fire support trainer implemented only a minimal set of advanced field artillery tactical data system messages, the Artillery Training School instructors integrated the device into their program of instruction, relying on it as a training aid for 2 days of the 10 day course designed to teach forward observers how to employ the advanced field artillery tactical data system to prosecute targets with field artillery.

During the virtual technologies and environments effort, the team developed a prototype amphibious assault vehicle turret trainer. Combining a refurbished amphibious assault vehicle turret that had been scrapped, an infantry skills marksmanship trainer–enhanced demilitarized .50 caliber machine gun and Mk19 grenade launcher, a projection system featuring the DELTA3D visualization engine and a deployable virtual training environment prototype effort, the multipurpose operational team training immersive virtual environment system provided the marines of Company D, 3d Assault Amphibian Battalion, 1st Marine Division with a useful part-task amphibious assault vehicle (AAV) crew trainer.

After arrival in Orlando, the salvaged AAV turret and stand were repaired, reconditioned, and fitted with the simulation hardware, software, and weapons. The amphibious assault vehicle turret trainer team conducted systems integration and testing and delivered the prototype system to the marines, who immediately put it to use. Based on the performance of the amphibious assault vehicle turret trainer prototype, the Marine Corps has decided to purchase 16 additional training devices.

One example of the integration of science and technology that permeated the virtual technologies and environments effort was the human factors research using the prototype multipurpose operational team training immersive virtual environment platform as the subject of operational field experiments conducted by the training effectiveness evaluation virtual product team. As a nominative case study, the training effectiveness evaluation virtual product team collected data on the interactions of a Marine Corps Fire Support Team engaged in supporting arms training and compared their observations with hypotheses developed in an exhaustive search of learning science and cognition literature in an effort to suggest potential remediation in techniques or technology to improve training performance. The results of this study will be forthcoming.

To support the training effectiveness evaluation virtual product team data collection, the multipurpose operational team training immersive virtual environment system was introduced to users as early as practicable and at various stages throughout the iterative development lifecycle. This spiral development and rapid prototyping enabled the training effectiveness evaluation virtual product team to collect training data while simultaneously providing system and software developers with meaningful and timely feedback that was incorporated into each subsequent release. At nearly every data collection event, marines were provided updated software based on feedback gathered during the previous event. This type of spiral development allowed the team to provide a simulation system to the marines that incorporated their user input. Normally the marines could see the results of their input and use the resultant tools in the subsequent spiral.

Lessons Learned

User input to spiral development gave the marines a simulation tool they understood how to use upon delivery.

At the lowest level of command structure, marines know what tools they need to train.

Operational or User Considerations

The primary user of the prototype virtual fire support trainer, the virtual simulation training toolkit, and the multipurpose operational team training immersive virtual environment systems, as well as the deployable virtual training environment to which they are transitioning, is the U.S. Marine Corps. From very early in the development lifecycle, marines employed the systems in existing facilities with no special provisions required for the simulation systems. Power supply and distribution, a perennial issue with multistation, personal computer based simulations, was somewhat mitigated through the implementation of laptop based technologies, thus reducing electrical power requirements. However, the ability to host multiple workstations is constrained by physical space, the number of available power receptacles, and the appropriate load distribution across the circuits servicing the outlets. The use of power strips or other current splitting devices provides a reasonable approach, but caution must be used in the distribution of electric current as the power draw can quickly overwhelm a circuit. In most cases, any expanded use of the simulation systems will need to be supported by two or more 20 amp circuits.

Of particular interest to scientists and engineers alike is system usability. The typical user of the simulation system does not have a robust background in computer engineering, but possesses sufficient computer skills to use Microsoft Windows–like products and point-and-click navigation techniques. Since the users of the simulations we developed are military members, we realize members of the training audience may have a widely divergent set of computer skills. Therefore, in most cases we found that some minimal level of system familiarization is required prior to using the systems for training. Familiarization training with first-person simulations is usually brief since the simulators are easy to use and often replicate the tools available to the marine while in the field to a reasonable level of accuracy. Anecdotal evidence indicates that a large portion of the training audience learned the systems quickly, since many of the applications either replicate real world military systems or present icons and symbology with which users are familiar.

The human-computer user interface for the instructor/operator presented a greater challenge because the workstation controls both Windows and Linux applications and offers a more robust set of controls with which to present a synthetic battle space to the training audience. As one might imagine, many instructor/operators have never used Linux or the applications that run on this operating system. We discovered that it is usually advisable to allow sufficient time (three to four days) to train prospective instructors. As with the acquisition of other computer skills, nothing takes the place of hands-on experience to achieve an adequate level of proficiency.

Lesson Learned

Even when using common interfaces, it will take a nominal amount of time to train a new user—plan for the time.

Iterative Integration and Transition

Use of Department of Defense Standards

With the exception of the voice over Internet protocol communications emulation, which operates under distributed interactive simulation standards, all the systems we have discussed use the Department of Defense high level architecture standard. High level architecture is a general-purpose architecture for simulation reuse and interoperability. The high level architecture was developed under the leadership of the Defense Modeling and Simulation Office to support reuse and interoperability across the large numbers of different types of simulations developed and maintained by the Department of Defense. The Defense Modeling and

Simulation Office has been renamed the Modeling and Simulation Coordination Office (http://www.msco.mil/).

Lesson Learned

Use industry standard interfaces and protocols; integrating with other applications will be easier later.

Demonstrations or Transitions

The virtual fire support trainer (2005, 2006), virtual technologies and environments' amphibious assault vehicle turret trainer (2005) and multipurpose operational team training immersive virtual environment (2006), and deployable virtual training environment (2006, 2007) were all feature demonstrations at the I/ITSEC[3] over the last few years. The expeditionary fighting vehicle trainer and the amphibious assault vehicle turret trainer are currently being used by marines (Schmorrow, 2005). Key system attributes showcased during the I/ITSEC demonstrations included portability, distributed training capability, simulation system interoperability, and linking virtual and constructive simulations with live battle command systems. We encountered a few technical challenges during these demonstrations, but none that were insurmountable. For instance, demonstrating the interoperability of the virtual fire support trainer and the deployable virtual training environment with the advanced field artillery tactical data system required the use of a single channel ground and airborne radio system, necessitating request for permission to broadcast on the appropriate military radio frequencies during the show. A more mundane, but nonetheless important, consideration was the power requirement for the radios, usually provided by a vehicle when mounted and portable battery packs when employed in a dismounted configuration.

In preparation for transition to the Marine Corps' deployable virtual training environment, virtual fire support trainer, virtual simulation training toolkit, and the virtual technologies and environments multipurpose operational team training immersive virtual environment systems participated in numerous demonstrations, user acceptability tests, and data collection events. Because all were designed to transition to the deployable virtual training environment, travel to and from the events was relatively painless. The hardened travel cases and modular design reduced the logistical burden of packaging, shipping, unpacking, setting up, and placing the systems into service.

SUMMARY

Through the reuse of existing technologies, the development of new capabilities, the use of open source applications, the use of rapid prototyping, and the integration of science and technology, we have assessed various aspects of integrating and transitioning research prototypes into various joint interoperable

---

[3]http://www.iitsec.org/

training systems. The purpose of this study, the lessons learned, and the overall military goal is to create a training environment that allows military members to train as they fight.

REFERENCES

Schmorrow, D. (2005, May 25). *Marine Corps modeling & simulation review.* Available from www.techdiv.org/erik.jilson/2005%20USMC%20M&S%20Review/14%2005 MCMSMOVIRTE.ppt

Stripling, R. M., Templeman, J. N., Sibert, L. E., Coyne, J., Page, R. G., La Budde, Z., & Afergan, D. (2006). Identifying virtual technologies for USMC training, information technology and communication [Electronic version]. *2006 NRL Review.* Available from http://www.nrl.navy.mil/Review06/images/06Information(Stripling).pdf

Part III: Game Based Training

Chapter 14

SO YOU WANT TO USE A GAME: PRACTICAL CONSIDERATIONS IN IMPLEMENTING A GAME BASED TRAINER

John Hart, Timothy Wansbury, and William Pike

There is tremendous interest across a wide spectrum of training domains in the use of computer based games to enhance training and education. Dr. Robert T. Hays (2005) presents an excellent summary of the use of games for education and training in K–12 education, college, and the workplace, citing over 50 examples from these arenas. In addition, the U.S. Department of Defense has invested significantly in identifying the best practices in how to effectively design, develop, and use computer based games in both institutional as well as distributed training environments. The interest in using games in training continues to grow across academia, business, and all of the military services.

Many of the books and articles devoted to the use of games for training provide different definitions of what is a "game," with many distinguishing among such terms as game, simulation, game based simulations, simulation games, nonsimulation games, and so forth. The fact is that the term "game" means different things to different people.

For purposes of this chapter, the term "games" refers to personal computer (PC) based computer games, and the discussion in this chapter focuses on describing six key recommendations that trainers should consider when preparing for the development or use of a PC based computer game for training. These six recommendations were derived from lessons learned through a number of research efforts conducted by the U.S. Army to include a recent effort to develop a PC based computer game called *Bilateral Negotiation* (BiLAT). BiLAT is a game based tool that provides an environment for soldiers to demonstrate their knowledge and skills on how to plan for and conduct successful bilateral meetings and negotiations in different cultural settings. The recommendations discussed in this chapter are of such significant importance that they should all be addressed by any trainer considering the use of "a game" as part of a future training program.

RECOMMENDATIONS

The U.S. Army Research, Engineering and Development Command, Simulation and Training Technology Center (RDECOM STTC) has led several research efforts focused on identifying the "keys to success" in how to design, develop, and use PC based computer games to train soldiers effectively. These projects involve a variety of training objectives to include instruction on traditional, small unit, infantry leadership tasks (Pike & Hart, 2003), tactical combat casualty care (Fowler, Smith, & Litteral, 2005), asymmetric warfare (Mayo, Singer, & Kusumoto, 2006), and cultural awareness training (Hill et al., 2006).

Lessons learned from these research efforts have resulted in six recommendations that should be considered by trainers considering the use of PC based computer games for training. The first four directly influence the trainer's decision regarding whether a game should, in fact, be used for training in the first place, as well as selecting the "right" game for the training exercise. The last two recommendations affect how the trainer plans to actually use the game in a training event. The importance of these recommendations has been validated in developing the BiLAT as well as other follow-on prototype efforts. This chapter will discuss the following recommendations within the context of developing the BiLAT prototype:

1. Define how the game fits into the overall instructional framework for the training.
2. Define the specific learning objectives for the training exercise.
3. Define how the game activities support the overall learning objectives.
4. Identify how performance and feedback will be accomplished during game play and following completion of an exercise.
5. Assess the relative experience of the trainees in using games, and be prepared to provide the additional time and training required to help those with little or no game-playing experience.
6. Assess the experience level of the instructors or trainers in using games for training, and provide the required assistance in order to ensure the trainers are adequately prepared to use the games effectively.

Define How the Game Fits into the Overall Instructional Framework

Identifying the right game and deciding how to integrate it into an established curriculum can be a difficult task. As Hays argues, in order to be successful, games must be "incorporated logically into an instructional framework" (Hays, 2006, p. 252). Key questions have to be addressed, such as who is the training audience, what knowledge does this target audience already possess, how much training content must be provided before the trainees participate in a game exercise, and how much content must be included within the game itself. The trainer must decide if the game should present such training content as initial concepts or should it be used solely as a practice environment allowing trainees to demonstrate their knowledge in specific training scenarios. Identifying how a game fits into the overall instructional framework is a critical first step to using a game effectively in training.

The U.S. Army Command and General Staff College, School for Command Preparation (SCP) at Fort Leavenworth, Kansas, was the first organization to use the BiLAT. SCP conducts the army's battalion and brigade commanders' pre-command course (PCC), a three week resident training course focused on preparing senior officers for key command assignments in the army. The BiLAT is a purpose-built game created specifically by the U.S. Army to support training at the SCP. Defining how the BiLAT would fit into the overall instructional framework of a training program was an essential step that drove key design and development decisions during the early phases of the project.

Instructional designers, software designers, and SCP course instructors devoted significant effort at the beginning of the project evaluating how the BiLAT could be integrated into the overall PCC course of instruction. Key decisions were made identifying the functions and tasks to be performed in the application, as well as the tasks that were to be provided by another form of training. For instance, the designers and instructors agreed that basic instruction on the core principles of cultural awareness and negotiation would be provided by other means, such as lectures, readings, and classroom discussions, and that BiLAT would serve only as a practice environment for honing skills in the art of bilateral meetings and negotiation. As a result, no preliminary instruction is provided in the BiLAT game environment. The decisions about how to integrate BiLAT in the overall instructional design laid the foundation for the overall success of the BiLAT project.

Establish Learning Objectives

Identification of clearly defined learning objectives in any training exercise is the second important step that must be addressed to successfully create or select a game for training. Identification of the learning objectives is also the single most important factor that differentiates a "training game" from an "entertainment game." As Hays states, "instructional objectives must be determined *a priori* and depend on job requirements." (Hayes, 2006, p. 260).

Oftentimes, trainers try to evaluate different games in order to select the best one for use in a training exercise. A key consideration during that evaluation process is to determine to what extent does a game support the achievement of the trainer's overall learning objectives. Finding an existing game that satisfies specific learning objectives can be difficult. As a result, a trainer must be prepared to modify how a game is used or, worst case, be prepared to develop a whole new game in order to ensure it supports the specific learning objectives to be achieved.

The design team for BiLAT devoted significant effort identifying strategic learning objectives and tested those objectives initially through the use of a paper based version of the game. The flow of the paper based version provided valuable insight that allowed the designers to validate the key learning objectives before costly software development began. Once the learning objectives were finalized, work commenced in developing the software for the PC version of the BiLAT.

Link Game Activities to Learning Objectives

The next major issue in the evaluation and selection process is to ensure that the specific activities performed by a trainee in a game scenario actually support the overall learning objectives for the exercise. Hays states that "a simulation game is ineffective if it does not directly link game events to instructional objectives and does not ensure that the learner understands whether he or she has met those objectives" (Hays, 2006, p. 252).

Again, the designers of BiLAT used the low cost, paper based prototype as a method of vetting the specific steps that should be performed by an individual preparing for and conducting a successful meeting and negotiation. We utilized a number of subject matter experts and employed a process called the Cognitive Task Analysis (CTA) to guide the development effort. "Cognitive Task Analysis (CTA) uses a variety of observation and interview strategies to capture a description of the knowledge which experts use to perform complex tasks" (Clark, Feldon, van Merrienboer, Yates, & Early, 2006). Using such techniques as the CTA ensured that the tasks performed during each BiLAT scenario were linked to and supported the strategic learning objectives for the application.

Assess Performance and Provide Feedback

Assessing performance and providing feedback is essential in determining the overall success in using a game for training. Ideally, assessment should take place within the game itself, and feedback should be provided in the form of a tutor or coach during game play. Feedback must also be provided to the trainee in the form of an after action review (AAR) at the end of the exercise. A critical factor that differentiates between "good" and "poor" assessment and feedback is the ability to link performance to the overall learning objectives. The trainer who is able to link performance in the game experience to the overall learning objectives greatly increases the likelihood of success in using the game in training.

BiLAT incorporates an enhanced set of artificial intelligence tools, methods, and technologies to provide thorough assessment and feedback throughout each BiLAT exercise. It provides a rich coaching and tutoring capability through both the meeting and negotiation stages of each exercise and provides a thorough AAR using a virtual coach at the end of each meeting engagement. These assessment and feedback capabilities make the BiLAT an effective trainer in both instructor-facilitated and distance learning environments.

Assess the Experience Your Trainees Have Using Games before Conducting the Training Exercise

A PC based computer game can be a powerfully effective tool for training individuals; however, even well designed and well-built games can result in a waste of time if they are not used effectively. Assessing the relative experience that trainees have in using games is an important step in the overall planning process. Trainers should recognize the likelihood that a significant number of their students are not "gamers" and will likely need additional training to overcome a natural fear or reluctance to use the application if additional support is not provided.

The Army Research Institute recently reported that

> fewer than 32% of 10,000 Soldiers surveyed across all ranks in the US Army admitted to playing videogames recreationally on a weekly basis (numbers vary by rank) ... Consistently, our research shows that the assumption that most Soldiers are "gamers" is exaggerated. Continuing to act on that assumption can be troublesome unless certain precautions are taken.
>
> (Belanich, Orvis, Moore, & Horn, 2007, p. 958)

The researchers go on to suggest the importance that "instructors should assess trainees' game experience" and "provide targeted opportunities to gain prerequisite experiences prior to training." (Belanich et al., 2007, p. 963).

One of the best ways to overcome this issue of game experience is to ensure that an adequate overview/introduction is provided prior to beginning the actual training event. This overview should include a discussion of the learning objectives, a demonstration of the "knobology" (for example, instruction on how to "play the game"), and an opportunity for the students to practice with the game before actual training begins. SCP instructors learned this lesson with the BiLAT and today set aside approximately one hour of additional classroom time prior to a BiLAT exercise to ensure that each student is ready to conduct the training.

Assess the Experience Level of the Instructors Who Will Be Using the Game in Training

The use of PC based computer games in training is a relatively new phenomenon, and many instructors do not have a lot of experience using games for training. Game developers are encouraged to provide support materials that can be used by new instructors and trainers to help them learn how to use a game in training before attempts are made for the first time. The developers of BiLAT addressed this issue by creating a simple to use learning support package (LSP), providing new trainers with detailed instructions in how to set up and conduct a successful BiLAT training exercise. This LSP is distributed along with the BiLAT to users across the army.

CONCLUSION

There is growing interest in schools, business, and in the military in using PC based computer games to enhance training opportunities. As a result of its experience in conducting research into the design, development, and use of PC based computer games, the RDECOM STTC has identified six key recommendations that trainers should address prior to using a game for training. These recommendations and lessons learned have been validated in the successful development of the BiLAT, a game that is being deployed and used across the army. These lessons are being consistently revalidated as new game prototypes are being developed and tested in educational settings.

REFERENCES

Belanich, J., Mullins, L. N., & Dressel, J. D. (2004). *Symposium on PC-based simulations and gaming for military training* (Army Research Institute Product 2005-01). Retrieved on May 22, 2007, from http://www.hqda.army.mil/ari/wordfiles/RP%202005-01.doc

Belanich, J., Orvis, K. A., Moore, J. C., & Horn, D. B. (2007). Fact or fiction—soldiers are gamers: Potential effects on training. *Proceedings of the Interservice/Industry Training, Simulation & Education Conference—I/ITSEC* (pp. 958–964). Arlington, VA: National Training Systems Association.

Clark, R. E., Feldon, D., van Merrienboer, J. J. G., Yates, K., & Early, S. (2006). Cognitive task analysis. In J. M. Spector, M. D. Merrill, J. J. G. van Merrienboer, & M. P. Driscoll (Eds.), *Handbook of research on educational communications and technology* (3rd ed.). Mahwah, NJ: Lawrence Erlbaum. Retrieved April 18, 2008 from http://www.cogtech.usc.edu/publications/clark_etal_cognitive_task_analysis_chapter.pdf

Fowler, S., Smith, B., & Litteral, D. J. (2005). *A TC3 game-based simulation for combat medic training.* Paper presented at the 2005 Interservice/Industry Training Simulation & Education Conference, Orlando, FL.

Hays, R. T. (2005). *The effectiveness of instructional games: A literature review and discussion* (Tech. Rep. No. 2005-004). Orlando, FL: Naval Air Warfare Center Training Systems Division.

Hays, R. T. (2006). *The science of learning: A systems theory perspective.* Boca Raton, FL: BrownWalker Press.

Hill, R., Belanich, J., Lane, C., Core, M., Dixion, M., Forbell, E., Kim, J. & Hart, J. (2006). *Pedagogically structured game-based training: Development of the ELECT BiLAT simulation.* Paper presented at the 25th Army Science Conference, Orlando, FL.

Mayo, M., Singer, M. J., & Kusumoto, L. (2006). Massively multi-player (MMP) environments for asymmetric warfare. *Journal of Defense Modeling and Simulation, 3*(3), 155–166.

Pike, W. Y., Anschuetz, R., Jones, C., & Wansbury, T. (2005). The rapid decision trainer: Lessons learned during an R & D development and fielding process. *Proceedings of the 2005 Interservice/Industry Training Simulation & Education Conference* [CD-ROM]. Arlington, VA: National Training Systems Association.

Pike, W. Y., & Hart, D. C. (2003). Infantry officer basic course (IOBC) rapid decision trainer (RDT). *Proceedings of the 2003 Interservice/Industry Training Simulation & Education Conference.* Arlington, VA: National Training Systems Association.

Chapter 15

MASSIVELY MULTIPLAYER ONLINE GAMES FOR MILITARY TRAINING: A CASE STUDY

Rodney Long, David Rolston, and Nicole Coeyman

Massively multiplayer online games (MMOGs) are one of the fastest growing forms of entertainment, with such games as *World of Warcraft* claiming more than 10 million users (*World of Warcraft,* 2008). Large-scale online social environments, such as *Second Life* (www.secondlife.com) and *There* (www.there.com), draw millions of additional users. These online environments are examples of virtual worlds—online three-dimensional (3-D) synthetic "worlds" that have the following characteristics:

- A large number of players, typically thousands;
- Large-scale geographic areas, often up to continent-sized worlds;
- Avatars (graphical characters in the virtual world) that may be controlled by a player's actions;
- Nonplayer characters that are controlled by artificial intelligence;
- Team environments that emphasize interaction among many players;
- Geographically distributed players that can log into the server from any location around the world that has a high speed Internet connection;
- Dynamic objects that can be moved, carried, placed, operated, and so forth;
- Free-form activity where avatars can move around freely, exploring and interacting with the environment; and
- Persistent environments that are intended to exist continuously and indefinitely, evolving over time.

In 2003, the U.S. Army Simulation and Training Technology Center began a research program, Asymmetric Warfare—Virtual Training Technologies, to investigate the use of MMOG technology to train soldiers for asymmetric warfare while operating in large urban areas, which are crowded with civilians.

TASK ANALYSIS

In the 2003 military operating environment of Afghanistan and Iraq, the U.S. Army was no longer fighting a force-on-force operation as had been the case in past conflicts. Soldiers were fighting an asymmetric war in urban cities, surrounded by a civilian population. As soldiers performed such dismounted infantry operations as traffic control points, patrols, buildings searches for weapons and high value targets, they were being attacked by improvised explosive devices (IEDs) and small arms fire. The goal of the research program was to create a virtual world that could be used to train soldiers for this environment and the wide variety of tasks they were required to perform.

Terrain Development

The virtual world had to reflect the type of terrain conditions where soldiers were currently operating and where they might be operating in the future. Having the virtual world reflect the current terrain not only enhanced realism and immersion, but also improved training scenarios. For example, the amount of trash and rubble on city streets made it very difficult for soldiers to detect IEDs. For our program, we provided terrain environments typical of urban areas in Southwest Asia and the United States (for Homeland Security), as well as rural desert and jungle areas. MMOG technology allowed us to provide large terrain areas and have all of these different areas available at the same time for concurrent operations.

Avatars

Using capabilities inherent in MMOGs, the role-players could change their avatars' facial features, skin/hair color, clothing, shoes, and so forth according to the culture. Civilian role-players had the clothing typically used in theater, while insurgent role-players could wear a uniform or a shemagh/ghutra to hide their faces. They also had the weapons to implement the enemy tactics being used, such as IEDs and guns. Soldiers were also provided the tools to perform their tasks. This included uniforms for identification, basic weapons, military vehicles, and radios for communications. Avatars and vehicles assessed damage based on weapon type. For military vehicles, the damage included mobility kill, fire power kill, or catastrophic kill.

Communications

Given the interactions between soldiers and civilians in urban operations, communication mechanisms, both verbal and nonverbal, were provided. Communication over simulated military radio networks enabled communications among military personnel to support operations over large geographic areas. Voice communication was implemented to reflect the verbal interactions among military and civilians in urban environments. The voice communication was tied to the speaker through lip synching and was spatially accurate in 3-D, attenuating with distance. Commonly found in MMOGs, animated gestures, or emotes, were provided to reflect culture and nonverbal communication. Role-players could select a button on the screen to have their avatars perform the Arabic gestures for greetings, thanks, to show anger, and so forth. Soldiers could use emotes to perform nonverbal commands to civilians, who might not understand English, for example, when operating a traffic control point.

Performance Standards

The virtual simulation used the scalability of an MMOG to support a large number of soldier trainees, as well as civilian and enemy role-players. Nonplayer characters could also be used, driven by artificial intelligence versus human role-players. The goal was to support approximately 200 characters/avatars in the virtual world at the same time to reflect urban clutter.

Technology Requirements

While the virtual world provided a very flexible and scalable simulation environment, there were other technical challenges that had to be solved to support effective training. Nonplayer characters were needed to create realistic, crowded urban cities, without requiring a large number of human role-players. The artificial intelligence had to reflect such typical civilian behavior as wandering and reacting to gunfire and IEDs. Also, a record and replay capability was needed to support after action review of the training exercise. Considering one of the strengths of MMOG technology is supporting multiplayers over the Internet, the after action review tool needed to be able to support a distributed after action review, including distributed playback of the recorded exercise.

SYSTEM HIGH LEVEL DESIGN

To support large numbers of simultaneous players/avatars and large geographic areas, the game engine runs on a cluster of servers. The large geographic area is broken down into sectors, with each sector assigned to a different server in the cluster. As an avatar moves through the virtual world, the modeling of the avatar is passed to different servers as it crosses sector boundaries. These boundaries also help filter the data that flow over the network to the individual client machines, as avatars in one sector do not need to know about events and interactions in the other sectors. Of course, there are exceptions to this, such as weapons and radio communications. To ensure a smooth transition, software logic handles the movement of avatars from one sector to another.

While the trainee's computer provides control of his avatar in the virtual world, it also has to display what is happening around him. In an urban area with many civilians, the client computer and graphics card could easily become overloaded.

To manage the load, the game engine uses different levels of detail, with the closest 20 avatars displayed in full detail with a fast update rate. The remaining avatars in the scene will have a much lower level of detail and will be updated only at a three second rate. As the trainee's avatar moves through the environment, the set of 20 avatars will change, and the avatars with the lower level of detail will transition to the higher level of detail and will begin updating at the higher rate.

DEVELOPMENT OF SYSTEM PROTOTYPES

With the network latency inherent in MMOGs, the simulation was designed to support training for operations that involved interactions between the soldiers and civilian population, as well as IEDs and small groups of insurgents, as opposed to high intensity combat. The commercial game engine chosen was very strong in human interactions and communication, which included integrated 3-D, spatially accurate voice. To explore MMOG technology and how it could be leveraged to improve military training, prototypes were developed that focused on specific training scenarios that were then evaluated through experimentation with soldiers.

Prototype Development

One of the first scenarios was checkpoint operations, which involve a lot of soldier-civilian interaction. Using the strength of the simulation to manipulate objects in the environment, materials used to set up checkpoints, for example, concertina wire, cones, signs, jersey barriers, and so forth, were added to the soldiers' inventories. Soldiers used these objects to practice setting up the checkpoint while receiving feedback from the instructor. Once properly set up, the checkpoints could then be operated by the soldiers, with role-players acting as civilians driving cars. Using verbal communication, as well as hand gestures, the soldiers were able to successfully operate the checkpoint (Mayo, Singer, & Kusumoto, 2006). Functionality was gradually added to the virtual world to support other scenarios, including building searches, patrols, and so forth.

Operational or User Considerations

One of the greatest strengths of MMOGs and virtual worlds is to be able to interact in a virtual environment with people across the world. However, getting permission to run the software on a military network is a long and difficult process. Due to the number of ports used, the ports needed to operate the MMOG were often blocked by firewalls. This made running exercises on a military installation challenging. As a result, exercises were often run over a local area network. This was still an issue because permission was needed to load software on the military computers. Also, most of the military computers did not meet performance specifications, especially for the graphics cards, and could not run the MMOG software without being upgraded.

Use of Department of Defense Standards

To be interoperable with other existing military training simulations and simulation support tools, the design decision was made to implement a network interface using the Institute for Electrical and Electronics Engineers Standard for Distributed Interactive Simulation (IEEE 1278.1). This standard allows heterogeneous simulations to be interoperable by standardizing the way data are shared through the network. Through this network interface, we were able to populate the virtual world with nonplayer characters from the U.S. Army's One Semi-Automated Forces simulation. This simulation allowed large numbers of nonplayer characters to be generated and controlled with a single operator and had the artificial intelligence required to support civilian and military behaviors. We were also able to integrate with the dismounted infantry virtual after action review system developed by the Army Research Institute to support exercise recording, replay, and critique of the trainee's performance.

Demonstrations

In 2005, a demonstration of the prototype was provided to the Secretary of the Army and the Army Science Board, using Iraqi role-players at the National Training Center, a National Guard unit in Orlando, Florida, and role-players in other parts of the United States. The scenario highlighted how personnel assets at the National Training Center and soldiers with recent experience in theater could all be tied into a training exercise over a wide area network with soldiers who are preparing to deploy. The Iraqi role-players spoke Arabic in the scenario, bringing in language and culture aspects into the scenario while the soldiers with experience in theater could share their observations and lessons learned. The demonstration showcased how the distributed nature of MMOG technology could improve training, providing a realistic simulation environment to support unit training at a home station.

TEST AND EVALUATION

Technical Performance

The goal of the training simulation was to be able to model a crowded urban environment on personal computers. At a simulation and training conference in December 2006, the Simulation and Training Technology Center demonstrated 200 avatars in the virtual world over the Internet, using a combination of live role-players across the United States and nonplayer characters generated by the One Semi-Automated Forces simulation. Using a broadband Internet connection, the computers had a 128 megabyte graphics card, 2 gigabytes of memory, and a 2.8 gigahertz Pentium 4 processor.

Usability

The accessibility of game based simulations is providing a new way to train soldiers, supplementing training at a home station, school houses, and live training centers. These simulations can prepare soldiers for the types of situations they will experience in theater. By training on various scenarios, soldiers can prepare for deployment by reinforcing basic skills, situational awareness, and decision-making techniques. While simulation based training may not replace live training, it can hone the skills needed to prepare soldiers for combat and enhance the live training they do receive.

One of the key features of the simulation was its flexibility and adaptability. Using the simulation as a virtual stage, a wide variety of training scenarios could be supported by changing the objects in the environment and the dialogue and actions of the live role-players. Before beginning a particular scenario, the trainer is given the opportunity to set up the scenario, using the scenario editor to place dynamic objects (IEDs, barriers, and so forth) in specific locations, allowing for variations in training scenarios. The flexibility of the system could enable the military to rapidly and effectively communicate and train new tactics to the unit level. While developed to support training for urban combat operations, the simulation also showed potential for Homeland Security training, supporting a force protection/antiterrorism exercise at Fort Riley, Kansas (Stahl, Long, & Grose, 2006).

Training Effectiveness Evaluation

The Simulation and Training Technology Center worked closely with the Army Research Institute to evaluate this new training technology. The Army Research Institute conducted formative evaluations with soldiers, consisting of an overview of the simulation, background questionnaires (experience, rank, training expertise, and computer familiarity), hands-on training on how to use the simulation, presentations of the features and functionality of the system, and questionnaires addressing the system aspects and features. After structured discussions on the system capabilities, tools, and features, the soldiers performed a specific preplanned mission. Following completion of the mission, the soldiers conducted an after action review and completed final questionnaires regarding the exercise and the system as a whole.

The Army Research Institute has conducted four evaluations of the simulation to date. The overall results of these studies showed that soldiers do benefit from the training provided. The soldiers recognized the simulation's ability to support and supplement situational awareness training exercises with the diverse environment provided by the system and the ability to model unpredictable behaviors. Another major training benefit was the after action review capability, namely, being able to replay a simulation exercise (Singer, Long, Stahl, & Kusumoto, 2007). As the popularity of these virtual worlds has grown, so has interest in our research program. Working with the Naval Air Warfare Center and our allies, we continue to explore how this technology can be leveraged to support training for the warfighter in joint and coalition warfare environments.

REFERENCES

Mayo, M., Singer, M. J., & Kusumoto, L. (2006, July). Massively Multi-player (MMP) environments for asymmetric warfare. *JDMS: The Journal of Defense Modeling and Simulation: Applications, Methodology, Technology 3*(3), 155–166. Retrieved April 19, 2008, from http://www.scs.org/pubs/jdms/vol3num3/JDMSIITSECvol3no3Mayo155-166.pdf

Singer, M., Long, R., Stahl, J., & Kusumoto, L. (2007, May). *Formative evaluation of a Massively Multi-player Persistent (MMP) environment for asymmetric warfare exercises* (Technical Rep. No. 1227). Arlington, VA: U.S. Army Research Institute for the Behavioral and Social Sciences.

Stahl, J., Long, R., & Grose, C. (2006, June). *The application and results of using MMOG technology for force protection/anti-terrorism training.* Paper presented at the European Simulation Interoperability Workshop, Stockholm, Sweden.

World of Warcraft reaches new milestone: 10 million subscribers. (2008, January). Retrieved April 19, 2008, from http://www.blizzard.com/us/press/080122.html

Part IV: International Training Examples

Chapter 16

A SURVEY OF INTERNATIONAL VIRTUAL ENVIRONMENT RESEARCH AND DEVELOPMENT CONTRIBUTIONS TO TRAINING

Robert Sottilare

The increased complexity of operational missions and environments has prompted researchers worldwide to evolve virtual environment (VE) technology to support more complex training missions and environments. Research and development in VEs has profound implications for all levels of training in the United States and around the world. These implications include, but are not limited to, improved representations of virtual humans and other intelligent agents to support training and training management, increased capabilities to support rapid construction of geospecific mission rehearsal environments, and increased capabilities to mix live and virtual environments to support more realistic training by incorporating real world data (for example, command and control information).

Other chapters have addressed existing VE technologies within the United States. This chapter focuses on international activities in the research, development, and application of "VEs for training" external to the United States.

It would take much more space than has been allocated here to do justice to all the excellent training research worldwide that utilizes VE technologies. This chapter attempts to (1) examine where VE research is taking place worldwide, (2) summarize a few of the current research vectors in VE for training, and (3) look ahead toward future research topics.

What follows is a sample of organizations conducting "training research involving the use of virtual environments." These organizations are reviewed in terms of their research thrusts, objectives, applications, and key publications. For the purposes of discussion, we will consider mission rehearsal to be "training for a specific mission" and therefore a subset of training. Specifically, the scope of Volume 3, Section 1 focuses on a sampling of research and technology development related to visual, aural, and haptic interaction in virtual environments. For this purpose, we consider any reality that contains a virtual component to be part of our discussions of virtual environments. This includes virtual reality and mixed reality (including augmented virtuality and augmented reality). Fundamental research in VEs, which has a potential impact on training, is also considered in this survey.

VE RESEARCH AT THE VIRTUAL REALITY LAB

The Virtual Reality Lab (VRLab) at the Swiss Federal Institute of Technology in Lausanne, Switzerland, was established in 1988 and is focused on the modeling and animation of "three-dimensional inhabited virtual worlds," including real time virtual humans, multimodal (visual, aural, and haptic) interaction, and immersive VE (Virtual Reality Lab, 2007). VRLab's research portfolio includes "JUST in Time Health Emergency Interventions" (Manganas et al., 2005), "Stable Real-Time AR Framework for Training and Planning in Industrial Environments" (Vacchetti et al., 2004), and "Immersive Vehicle Simulators for Prototyping, Training and Ergonomics" (Kallmann et al., 2003). VRLab is currently a research partner in seven European Union (EU) and four Swiss national projects involving virtual reality.

VE RESEARCH AT MIRALAB

MIRALab at the University of Geneva in Switzerland was founded in 1989 and presently includes about 30 researchers from various fields, including computer science, mathematics, medicine, telecommunications, architecture, fashion design, cognitive science, haptics, and augmented reality, who conduct research in computer graphics, computer animation, and virtual worlds. One MIRALab foundation project is the virtual life network (VLNET), a networked collaborative virtual environment that includes highly realistic virtual humans and allows users to meet in shared virtual worlds where they communicate and interact with each other and with the environment. The virtual human representations have appearance and behaviors similar to real humans in order to "enhance the sense of presence of the users in the environment as well as their sense of being together in a common virtual world" (Capin, Pandzic, Noser, Magnenat-Thalmann, & Thalmann, 1997). Extensions of this work can be seen in Joslin, Di Giacomo, and Magnenat-Thalmann's (2004) review of the creation and standardization of collaborative virtual environments.

VE RESEARCH AT THE VIRTUAL ENVIRONMENT LABORATORY

The Virtual Environment Laboratory (VEL) was established at Ryerson University in Toronto, Ontario, Canada, in 2003 to "advance the integration of geospatial, modeling, visualization and virtual reality technologies for use in urban environment applications" (Ryerson University, 2007). VEL's research and development goals include the automated detection, identification, correlation, and extraction of remotely sensed imagery; and the modeling of complex human-environmental interactions in urban environments. While initial operational applications of this technology include land use management, three-dimensional (3-D) urban modeling, landscape mapping, disaster management, and transportation planning, this cross-section of technologies could also be applied to urban environment training for first responders (that is, police, fire, and emergency medical personnel).

TNO DEFENSE SECURITY AND SAFETY

TNO in Soesterberg, the Netherlands, conducts extensive training research in the areas of human factors, behavioral representation, artificial intelligence, games research, and movement simulation. Two projects related to VE for training that highlight TNO's expertise are Ashley and Desdemona (DESoriëntatie DEMONstrator Amst).

Ashley is a virtual human whose roles include wingman, unmanned aerial vehicle pilot, and training mentor. Current versions of Ashley are being used to monitor trainee performance and provide feedback. Future versions will be able to perceive natural language and provide feedback when the trainee requests it (Clingendael Center for Strategic Studies, 2006).

Desdemona is a movement simulator that provides an extensive range of motion due to the combination of both the hexapod and the centrifuge. This motion base can be used for realistic virtual training for aircraft and automobiles or can be used for analysis of vehicle performance. Impressive accelerations and complex curves are possible where the additional realism of movement is required to support training objectives (TNO, 2008).

CHADWICK CARRETO COMPUTER RESEARCH CENTER

At the Chadwick Carreto Computer Research Center in Mexico City, Mexico, Menchaca, Balladares, Quintero, and Carreto (2005, p. 40) defined "a set of tools, based on software engineering, HCI techniques and Java technologies, to support the software development process of 3D Web based collaborative virtual [environments] (CVE) populated by non autonomous interactive entities." The thrust of this work is to define "a methodology supported by design, analysis and implementation tools that assist [in] the development of Web-based CVE" (Menchaca et al., p. 40). This research emphasizes collaboration and interaction of the VE entities.

The tools defined include a model of "social groups, a graph-based, high level notation to specify the interactions among the entities, and a Java-based software framework that gives support to the model and the interaction graph in order to facilitate the implementation of the CVE" (Menchaca et al., 2005, p. 40). This research has the potential to support a more flexible, tailor-made training VE based on the trainees and their relationships.

CENTER FOR ADVANCED STUDIES, RESEARCH, AND DEVELOPMENT IN SARDINIA

Center for Advanced Studies, Research, and Development in Sardinia (CRS4) is an applied research center "developing advanced simulation techniques and applying them, by means of High Performance Computing, to the solution of large scale computational problems" (CRS4, 2007). This research has significance to complex training in virtual environments where large numbers of trainees interact (that is, network centric warfare scenarios) and includes research on massive-model rendering techniques in which various output-sensitive rendering algorithms are used to overcome the challenge of rendering very large 3-D models in real time as required for interactive training (Dietrich, Gobbetti, & Yoon, 2007).

VE RESEARCH AT THE VIRTUAL REALITY AND VISUALIZATION RESEARCH CENTER

The Virtual Reality and Visualization (VRVis) Research Center in Vienna, Austria, has ongoing research in six key areas: real time rendering, virtual habitats, scientific visualization, medical visualization, virtual reality, and visual interactive analysis. Three of these research areas have a direct impact on training.

The basic research being conducted by VRVis in "virtual reality" includes goals to provide more realistic virtual environments and a more interactive 3-D representation and rapid display of realistic objects within a 3-D environment (Matkovic, Psik, Wagner, & Gracanin, 2005). The training benefit of this work is the ability to present more complex and realistic environments to the trainee. Additional benefits are faster and more cost-efficient content development for virtual environments.

Another research goal being pursued by VRVis is "interactive rendering," the real time response of complex, 3-D graphics to human actions for such applications as museum displays, one-to-one-training, and location based services.

A third research goal that is applicable to VE training is "medical visualization," which includes the development of extremely fast renderers that allow large, complex datasets to be viewed in real time for virtual training of surgeons. Endoscopy has recently been applied to pituitary surgery as a minimally invasive procedure for the removal of various kinds of pituitary tumors. Surgeons performing this procedure must be both familiar with the individual patient's anatomy and well trained. A VE for endoscopy training permits very precise training and pre-operative planning using a realistic representation of the patient's anatomy with no risk to the patient (Neubauer, et al., 2005).

REALISTIC BROADCASTING RESEARCH CENTER

Kim, Yoon, and Ho (2005) at the Gwangju Institute of Science and Technology (GIST), Gwangju, Korea, have defined a multimodal immersive media termed "realistic broadcasting" in which 3-D scenes are created by acquiring immersive media using a depth based camera or a multiview camera. According to Kim et al. (p. 164), "after converting the immersive media into broadcasting contents," the content is sent to users via high speed/high capacity transmission techniques where the user can experience realistic 3-D display, 3-D sound, and haptic interaction with the broadcast VE.

VE RESEARCH AT THE MARMARA RESEARCH CENTER OF TURKEY

The Marmara Research Center (MRC) in Kocaeli, Turkey, is a multidisciplinary organization that participates in fundamental and applied simulation research that includes partnerships in the European Simulation Network (ESN) project. ESN is developing procedures and software to overcome risks/obstacles to prepare for the creation of a Europe-wide military training and exercise simulation environment by performing a virtual environment analysis of the network requirements with the aid of artificial intelligence software (Marmara Research Center, 2007).

Additional research vectors at MRC include the development and use of command and control simulations to train decision makers and support the development of new concepts (Hocaoğlu & Firat, 2003) and the research, development, and evaluation of a common evaluation model and an intelligent evaluation system to simplify and minimize time (and cost) of training evaluations where training is conducted in a synthetic environment (Öztemel & Öztürk, 2003).

VE RESEARCH WITHIN THE NORTH ATLANTIC TREATY ORGANIZATION

The complex missions entrusted to the military of the North Atlantic Treaty Organization (NATO) countries are driving new performance requirements for military personnel (Alexander et al., 2005) and the need for new and improved (1) approaches to training, (2) methods to analyze changes to tactics, techniques, and procedures, and (3) virtual environments to support the transition and application of training research via experimentation.

NATO countries are researching and developing virtual environment technology to support the need for more realistic training and more natural/realistic human-systems interaction (HSI). Collaborative activities, topics, and VE research themes within NATO organizations include the following:

- *Human Factors and Medicine (HFM) Panel:* The mission of the HFM Panel is to provide the science and technology base for optimizing health, human protection, well-being, and performance of the human in operational environments with consideration of affordability. In 2005, the HFM Panel focused on the study of "Virtual Environments for Intuitive Human-System Interaction" (Alexander et al., 2005). This study and its associated workshop provided an overview of VE research in Canada, Denmark, Germany, Sweden, the Netherlands, the United Kingdom, and the United States. The application of VE technology encompassed vehicle operation training, individual skills and collective tactical training, and command and control training for maritime, ground combat, and aviation environments.

• *Simulated Mission and Rehearsal Training (SMART) initiative:* The mission of SMART is to "coordinate, develop, and implement the potential for interactive virtual military training amongst participant nations' armed forces" (NATO Research and Technology Organization, 2007). The "potential" includes emerging technologies developed by NATO panels and other research organizations.

VE RESEARCH IN THE EUROPEAN UNION

As training missions become more complex and the student's cognitive load increases, it is critical to provide students with tools to manage information within the VE. The management of information in VE training is being enhanced by the architecture and technologies for inspirational learning environments (ATELIER). ATELIER is a EU project within the programs for future and emerging technologies and includes participants from Italy, Sweden, Austria, and Finland. ATELIER focuses on learning environments for architecture and interaction design students, and its main purpose is to implement methods for students to manage large amounts of data collected in a VE during the learning process using intuitive criteria (Loregian, Matkovic, & Psik, 2006).

FINDINGS AND CONCLUSIONS

First, in surveying the VE literature, it is evident that significant contributions to the research and development of VE are worldwide. Collaborations between research centers are the norm with journal article by-lines that often include authors from multiple countries.

Next, the science involved in creating and interacting with VEs can be and is often broadly applied not only to training, but to other operational missions as well. The need for very complex VEs is driving innovative thinking resulting in the application of existing technologies in new ways and the creation of new technology to support large-scale VEs, highly realistic VEs, and more interactive VEs for planning, training/rehearsal, and evaluation. The research vectors consist of HSI design (including navigation, manipulation, and data management), rapid construction of VEs, large-scale simulation architectures, and the integration of real world data from sensors and other sources to form augmented and mixed reality environments.

Significant work remains to evaluate the effectiveness of evolving VE technologies for training applications. This future work will result in new standards, heuristics, and best practices. The integration of new VE technology with commercial products (that is, games) will continue to expand their capabilities and make them more affordable and available to a wider training audience (for example, Nintendo Wii includes a wireless game controller, which can be used as a handheld pointing device and can detect acceleration in three dimensions).

Finally, this chapter covers a very small percentage of the research being conducted worldwide. However, it can still serve as a guide to lead readers to pockets of expertise in VE, innovative methods and applications of VE, the identification of the strengths and limits of current VE technology, and the VE technology gaps that define future research and development programs.

REFERENCES

Alexander, T., Goldberg, S., Magee, L., Borgvall, J., Rasmussen, L., Lif, P., Gorzerino, P., Delleman, N., Mcintyre, H., Smith, E. & Cohn, J. (2005). *Virtual Environments for intuitive human-system interaction: National research activities in augmented, mixed and virtual environments* (RTO Tech. Rep. No. RTO-TR-HFM-121-Part-I). Neuilly-sur-Seine Cedex, France: NATO Research and Technology Organization.

Capin, T. K., Pandzic, I. S., Noser, H., Magnenat-Thalmann, N., & Thalmann, D. (1997). Virtual human representation and communication in VLNET networked virtual environments. *IEEE Computer Graphics and Applications, Special Issue on Multimedia Highways, 17*(2), 42–53.

Center for Advanced Studies, Research, and Development in Sardinia (CRS4). (2007). *Research and technological development activities.* Retrieved August 29, 2007, from http://www.crs4.it/

Clingendael Center for Strategic Studies. (2006). *Where humans count: Seventh of a series of nine essays on the future of the Air Force* (p. 19.).

Dietrich, A., Gobbetti, E., & Yoon, S. E. (2007). Massive-model rendering techniques: A tutorial. *IEEE Computer Graphics and Applications, 27*(6), 20–34.

Hocaoğlu, M. F., & Fırat, C. (2003, October). *Exploiting virtual C4ISR simulation in training decision makers and developing new concepts: TUBITAK's experience.* Paper presented at the C3I and Modeling and Simulation (M&S) Interoperability, NATO RTO Modeling and Simulation Conference, Antalya, Turkey.

Joslin, C., Di Giacomo, T., & Magnenat-Thalmann, N. (2004). Collaborative virtual environments, from Birth to Standardization. *IEEE Communications Magazine, Special Issue on Networked Virtual Environments, 42*(4), 65–74.

Kallmann, M., Lemoine, P., Thalmann, D., Cordier, F., Magnenat-Thalmann, N., Ruspa, C., & Quattrocolo, S. (2003, July). *Immersive vehicle simulators for prototyping, training and ergonomics.* Paper presented at Computer Graphics International CGI-03, Tokyo, Japan.

Kim, S. Y., Yoon, S. U., & Ho, Y. S. (2005). *Realistic broadcasting using multi-modal immersive media.* Gwangju, Korea: Gwangju Institute of Science and Technology (GIST).

Loregian, M., Matkovic, K., & Psik, T. (2006). Seamless browsing of visual contents in shared learning environments. *Proceedings of the Fourth IEEE International Conference on Pervasive Computing and Communications Workshops— PERCOMW'06* (pp. 235–239). Washington, DC: IEEE Computer Society.

Manganas, A., Tsiknakis, M., Leisch, E., Ponder, M., Molet, T., Herbelin, B., Magnenat-Thalmann, N., & Thalmann, D. (2005). JUST in time health emergency interventions: An innovative approach to training the citizen for emergency situations using virtual reality techniques and advanced IT tools (The VR Tool). *Journal on Information Technology in Healthcare, 2,* 399–412.

Marmara Research Center (MRC). (2007). *European simulation network.* Retrieved December 9, 2007, from http://www.mam.gov.tr/eng/

Matkovic, K., Psik, T., Wagner, I., & Gracanin, D. (2005). Dynamic texturing of real objects in an augmented reality system. *Proceedings of IEEE Virtual Reality Conference (VR 2005), 329,* 257–260.

Menchaca, R., Balladares, L., Quintero, R., & Carreto, C. (2005). Software engineering, HCI techniques and Java technologies joined to develop web-based 3D-collaborative virtual environments. *Proceedings of the 2005 Latin American conference on Human-computer interaction* (pp. 40–51). New York: Association for Computing Machinery.

NATO Research and Technology Organization. (2007). *Simulated Mission and Rehearsal Training (SMART).* Retrieved August 29, 2007 from http://www.rta.nato.int/panel.asp?panel=SMART

Neubauer, A., Wolfsberger, S., Forster, M., Mroz, L., Wegenkittl, R., & Buhler, K. (2005). Advanced virtual endoscopic pituitary surgery. *IEEE Transactions on Visualization and Computer Graphics, 11*(5), 497–507.

Öztemel, E., & Öztürk, V. (2003, April). *Intelligent evaluation definition of training systems in synthetic environments.* Paper presented at the International Training and Education Conference (ITEC 2003), London, United Kingdom.

Ryerson University. (2007). *Virtual environment laboratory.* Retrieved December 27, 2007, from http://www.ryerson.ca/civil/research/laboratories/

TNO. (2008). Desdemona: The next generation in movement simulation. Retrieved April 17, 2008, from http://www.tno.nl/downloads/veilig_training_desdemo-na_S080014_EN.pdf

Vacchetti, L., Lepetit, V., Papagiannakis, G., Ponder, M., Fua, P., Thalmann, D., & Magnenat-Thalmann, N. (2004). Stable real-time AR framework for training and planning in industrial environments. In S. K. Ong & A. Y. C. Nee (Eds.), *Virtual and augmented reality applications in manufacturing.* London: Springer-Verlag.

Virtual Reality Lab (VRLab) [website]. (2007). Retrieved December 7, 2007, from http://ligwww.epfl.ch/About/about_index.html

TRAINING EFFECTIVENESS AND EVALUATION

SECTION PERSPECTIVE

Eric Muth and Fred Switzer

Any training effectiveness evaluation (TEE) study that does not consider the training context would be useless. Effectiveness of a training system is always a joint function of the training system itself and the tasks, goals, and operational environment of the activities being trained. Training and determining the effectiveness of that training is an interdisciplinary problem that crosses numerous fields including the following: fields focused on human behavior, such as psychology and education; fields that focus on human-system interactions such as human factors and industrial engineering; and fields that focus on system development, including computer engineering and computer science. While some of the information presented below is covered in more depth in other sections of this book, we felt it necessary to review some basic principles of training to give the reader some broad context in which to place TEE. In addition, we discuss some of the training outcomes for which virtual environments (VEs) may be particularly suited and some of the outcomes for which VE may be inappropriate or ineffective. In any case, evaluating training effectiveness (and its economic dimension, utility of training) must always occur in context.

Finally, the evaluation of training effectiveness should begin before system development, else system development is not justifiable; that is, if current training is effective, why develop a new system? VE developers cannot simply develop a system and then wait until the system is operational to begin the evaluation process and justify the need and use of a system. This technology push is costly, not only in development dollars, but potentially in misappropriated training dollars to a good salesperson. Therefore, we discuss some context issues that should drive early informal and formal evaluation that should occur long before "full-up" operational evaluation.

A PRIMER ON HOW PEOPLE LEARN

There are a number of key principles in learning and skill acquisition—Levy (2006) referred to these principles as the "learning context" in training. These principles include active versus passive learning, massed (all the training is received at once) versus distributed (training is given over time) practice, whole versus part learning, positive versus negative transfer of training, feedback, and practice (Shultz & Shultz, 2002). These factors must be considered in designing a training program, especially when training novel and/or complex tasks.

Active learning (also known as active practice) refers to the doing that may be necessary for trainees to learn. While observational learning can be helpful, often active learning is necessary for true training effectiveness. For example, a trainee cannot just be told or shown how to go about clearing a building (see Hoover and Muth, Chapter 19, and Knerr and Goldberg, Chapter 23). He or she must actively experience it to acquire the understanding and skills necessary to be effective quickly. VE training systems can lend themselves to either approach, but VE offers unique opportunities for the trainee to actively participate in the learning process. For example, Darken and Banker (1998) found that VE training outperformed real world training in some conditions in an orienteering task. The design process should include at least some informal evaluation of the degree to which the training system allows the trainee to actively practice.

Massed or distributed practice refers to two different approaches. In one approach, massed practice, trainees are given only one or a few long practice sessions. In the other approach, trainees are given many short practice sessions. Military logistics usually dictate that a massed practice approach be taken. However, it is important to note that distributed practice typically results in more effective learning. Shoot houses, such as those discussed in Hoover and Muth (Chapter 19) afford massed practice. Such VEs as those described in Grant and Galanis (Chapter 21) Bachelder, Brickman, and Guibert (Chapter 22) and Knerr and Goldberg (Chapter 23) afford distributed practice, especially when they can be run without supervision and contain appropriate feedback and after action review functions. Also, the spacing of practice sessions may have important motivational effects (this is discussed in more detail below).

Whole and part learning refers to the portion of a task being learned. Whole learning refers to learning the entire task as a whole, for example, flying a plane. Part learning refers to breaking a task into its component parts, for example, emergency procedures, stick and rudder skills, communications, and so forth. The effectiveness of whole or part learning depends on the complexity of the task and the ability of the trainees. Typically, when a novice is confronted with a complex task that can be broken down into subskills, part learning will be more effective. The decision about how to subdivide the content of a training program has to be a joint function of an effective and accurate task analysis and an understanding of learning principles. Dorsey, Russell, and White (Chapter 20) and Goldiez and Liarokapis (Chapter 26) touch on aspects of this process in identifying common ("identical") elements in the training task and the real task and integrating virtual training and live training.

One of the biggest contributions VEs may make to the training process is the increased opportunity to practice. Obviously practice is a critical component in high level performance of complex tasks. But the practice must include feedback to avoid practicing *apparently* effective behaviors, that is, practicing bad habits instead of good ones. Also, practicing *beyond* the point of mastering the target skill (practicing to automaticity or "overlearning") has value, especially in the kinds of tasks that are likely to be taught in VEs.

Feedback (also called "knowledge of results" or "KR") refers to giving the trainees information on how they are doing in the training and how they are progressing. If trainees are not given feedback, they may continue to use inappropriate strategies and perform ineffective behaviors because they have no information on whether these strategies and behaviors are working. The feedback should be immediate and frequent, if possible, and both positive and negative feedback are effective when delivered correctly. Hoover and Muth (Chapter 19) discuss instrumentation for training and how data can be derived from training systems to give the training feedback. Lampton, Martin, Meliza, and Goldberg (Volume 2, Section 2, Chapter 14) discuss the use of after action review systems in VEs.

Transfer of training is, in many ways, the "bottom line" of training. Transfer of training refers to the application of knowledge and skills acquired during training to the actual work environment (Baldwin & Ford, 1988). The training program needs to give careful attention to the gap between the training and work environments to ensure that skills mastered during the training transition to performance in the work environment. For example, the relevance of any simulations, demonstrations, or lectures on such a topic as spatial disorientation, the feeling of being positioned in three-dimensional space in a position different from true position, to actual flight must be made obvious to trainees. Note that transfer of training can be both positive and negative. Negative transfer is acquiring bad habits, misinformation, and misperceptions during training and applying those negative behaviors in the actual work environment. Negative transfer can be reduced by ensuring that critical elements of the training environment match the critical elements of the operational environment. This is Edward L. Thorndike's famous "identical elements theory." Sullivan, Darken, and Becker (Chapter 27) and Becker, Burke, Sciarini, Milham, Carroll, Schaffer, and Wilbert (Chapter 28) discuss transfer of training; and Dorsey, Russell, and White (Chapter 20) discuss extensions of identical elements theory for use in VEs.

A PRIMER ON TRAINING EVALUATION

As in good training design, accurate training evaluation depends on having done the appropriate task, job, needs, and organizational analyses to determine the knowledge, skills, and abilities that the training is intended to deliver. Only then can the training outcome measures be chosen rationally and effectively. Those outcome measures have traditionally (Kirkpatrick, 1976) been divided into four categories, reaction measures, learning measures, behavioral measures, and results measures.

Reaction measures, that is, measures of the trainees' attitudes, their positive or negative evaluation of the training program, have long been used in training evaluation. However, it has also long been known that reaction measures have limited usefulness in evaluating training effectiveness (including the finding that there is often a positive bias in reaction measures). However, Alliger and Janak (1989) have proposed looking at trainees' "utility" reactions, that is, their perceptions of the amount of positive transfer of training they expect to occur. This type of measure may prove more useful than the traditional affective reactions measure. However, affective reactions to training should not be ignored. Participants' post-training attitudes (and especially belief and attitude changes caused by training—see below) may have effects on such motivation-critical variables as self-efficacy/confidence, motivation to engage in further training, and, in the case of teams, team efficacy, team communication, backup behavior, and so forth.

Learning measures, typically implemented by some form of testing at the conclusion of the training, have proven useful to a degree. Kraiger, Ford, and Salas (1993) proposed three categories of learning criteria:

1. Cognitive outcomes (knowledge and memory; this would likely include both semantic and procedural memory);
2. Skill based outcomes—skills, procedures, and performance; and
3. Affective outcomes—beliefs and attitude *changes* as a result of training.

In addition, other taxonomies of learning criteria have been proposed including Bloom's original taxonomy of educational objectives (Bloom, Englehart, Furst, Hill, & Krathwohl, 1956) and its later revision (Anderson et al., 2001). This revised version of Bloom's taxonomy categorizes learning criteria on two dimensions: knowledge (factual, conceptual, procedural, and metacognitive) and cognitive process (remembering, understanding, applying, analyzing, evaluating, and creating).

Behavioral measures are those that quantify actual changes in behavior in the operational environment that are a result of training. As Levy (2006, p. 244) noted, "An evaluation of a training intervention that didn't include measures of behavioral criteria would be seriously flawed." In some senses, behavioral measures are the "gold standard" of training evaluation. That said, there are typically serious design and measurement issues that often restrict or prevent the acquisition of accurate behavioral measures (see the discussion below).

Finally, the fourth type of outcome measure is "results." Results outcomes are the value of the training program to the organization and its goals. In other words, was the training worth the time, effort, expense, and so forth expended by the organization? While in many ways results are (at least theoretically) the ultimate measure of the effectiveness of a training program, there are two factors that often argue against the use of results measures (or worse, generate misleading data). One factor is criterion deficiency. This refers to situations in which outcome measures are not tapping into the full range of target behaviors and results. In other words, important aspects of performance are being left out of the measures such that they form an incomplete picture of actual training effectiveness. Another factor is criterion contamination. This refers to situations in which the outcome measure is tapping into *irrelevant* behaviors and results and therefore generating misleading data.

The other major issue in evaluating training effectiveness is the design of the evaluation process. This is the issue of research design—how should the evaluation study be configured and analyzed to give accurate and unconfounded answers to the question "did this training work?" Research design is such a broad topic that we cannot adequately even introduce it in this short primer. But good research design is critical to accurate evaluation of training effectiveness and positive transfer.

BUILDING EFFECTIVE VIRTUAL ENVIRONMENTS: SOME ISSUES (AND UNANSWERED QUESTIONS)

VEs are famous (or notorious) for allowing more extreme situations than are typically presented in most training situations. VEs allow the trainee to experience a variety of variants of the worst case scenario and its close relatives. VEs allow a degree of control and precision over the training process seldom seen in previous training methods. This is primarily because VEs allow control over a greater number of variables in the training process than has been possible previously. But this also creates some new problems. Control over variables means that choices must be made about the levels of those variables to implement in the VE. This means that the instructor has a new challenge—the requirement to specify the ranges and values (even if those values are fixed or made the default) for a potentially huge number of variables. Very often the current research literature offers little or no guidance about the proper choices. And there is a corollary to this issue—what variables can safely be ignored, minimized, or eliminated entirely? Further, as Salas, Wilson, Burke, and Bowers (2002, p. 21) point out, it is a myth that "everyone who has ever learned anything or has gone to training is a training expert and therefore can design it (the training)." Therefore, defaulting to subject matter experts may not be the best solution either. Salas and Burke (2002) offered a set of useful guidelines for use of simulations in training.

There is one hidden benefit to the development and use of VEs: we will be forced to simultaneously look at a broad range of variables in the training process, specify the relative importance of these variables to training outcomes, and specify the useful ranges and values of these variables to guide the trainer.

The development of VEs for training is spotlighting other issues that have either not been present before or have had minimal impact on the effectiveness of training: for example, issues like the physical interaction of the training equipment (that is, the actual training apparatus itself) with the trainee; questions such as, Does the weight or weight distribution of a head-mounted display adversely affect the training effectiveness of HMDs when the training goes on for long periods? Is the limited field of view (FOV) found in VEs, such as driving simulators, reducing training effectiveness (and how)? For that matter, can limited FOV be a *benefit* by reducing distractions that interfere with early skill or knowledge acquisition? The issue of simulator sickness (nausea brought on by the VE apparatus; see Drexler, Kennedy, and Malone (Volume 2, Section 1, Chapter 11) has spawned an entire research literature of its own.

Of course there is the classic issue of optimal level of fidelity. Note that we are characterizing the fidelity issue as seeking the optimal, not the highest, level of fidelity. As Salas et al. (2002, p. 23) point out, it is another myth that "the higher the fidelity of the simulation, the better one learns." There is an implicit assumption in much of the VE training literature (and which stems very naturally from the larger simulation literature in general and identical elements theory in particular) that higher fidelity is always better in simulations. Is this necessarily true? In classic training methods, simplified, "lower fidelity" situations are often used effectively in the early stages of learning. Think of the elementary school reading primer ("See Dick run.") as the classic example. Is there an analog in VE training? When is higher fidelity a liability rather than an asset? Again, the classic issue of psychological fidelity versus physical fidelity is something that must be considered if the goal is developing an effective VE.

One point should be made here. Unfortunately, VE training system designers and developers will be faced with design decisions that the general training literature has not yet answered. Another hidden benefit of VEs may be to force training researchers (such as ourselves) to answer questions that have not been adequately addressed previously. For example, task analysis has not been refined to the point where we can easily and accurately answer such questions as "what are the critical elements of a process-control task mental model such that the process control team has an adequate shared mental model?"

Another area in which VE may push training research is in the area of post-exercise learning (for example, debriefings, after action reviews, follow-up training sessions, remedial training sessions, and so forth; see also our comments on feedback above), An often-overlooked capability of VE systems is to make a complete recording of an entire training session. While this capability may be approached in a real world system (for example, by using multiple video cameras), VEs can literally record every variable presented to the trainees. This capability offers enormous possibilities for post-exercise reviews. The digital nature of the data can allow trainers to literally put trainees in someone else's (virtual) shoes. A trainee could see precisely how his or her actions affected a teammate's behavior by replaying the scenario from the teammate's point of view. For that matter, a combat simulator could put the trainee in the enemy's shoes (the "bad-guy cam").

This same high density recording capability also provides a wealth of new opportunities and challenges in training performance measurement (and in training system evaluation). The opportunities are largely in two forms: more fine-grained, even microscopic analysis of training performance is possible by examining very short time periods for very specific variables. At the other end of the level-of-analysis scale, sophisticated multiple weighted composite performance variables can be created that can capture the multiple-objective nature of most tasks (rather than oversimplifying task outcomes for convenience or to accommodate cruder performance measures). Note that this latter possibility demands a clear understanding of all of the task objectives, the performance measures associated with each, and what the relative importance of each objective is toward overall task performance—a tall (but desirable) order.

TRAINING MOTIVATION AND VIRTUAL ENVIRONMENTS

In many theories of training motivation and skill acquisition the pacing of the training is hypothesized to have a significant effect on trainees' motivation. One of the strengths of VE training is often the ability of VE systems to dynamically alter the timing and difficulty of the tasks being trained (as opposed to more traditional training methods that are often linear and operate at a fixed pace). This ability to dynamically change the task timing, duration, even introduce unplanned pauses in the training scenario may give VE training a unique advantage. This may yield a distinct advantage for the (common) training situation in which the trainees exhibit various initial levels of ability. For example, Kanfer and Ackerman (1990, as cited in Kanfer, 1996) found that, at the same stage of training, lower ability trainees were more likely to suffer from low self-efficacy and negative affect during skill acquisition, while higher ability trainees were subject to boredom because they had already achieved a level of proficiency with the task. This implies that task pacing and timing could be varied in order to keep the slower-acquisition trainees' self-efficacy up and the faster-acquisition trainees' boredom down. It should be noted that Kanfer and Ackerman (along with a number of other self-regulation and skill acquisition researchers; Horvath, 1999) used a VE (a simulated air traffic control task) to test their hypotheses. Kanfer, Ackerman, Murtha, Dugdale, and Nelson (1994) found that goals during training impaired performance during massed practice. However, they found that this effect could be ameliorated by providing short breaks during training, presumably because this allowed the trainees to self-regulate (process and assess the goals relative to their immediately preceding training performance). VEs are not only typically amenable to building in this kind of pause, but the timing of the pauses and breaks could potentially be tailored to the individual trainee's progress (or lack of it).

TRAINING CRITICAL TEAM VARIABLES IN VIRTUAL ENVIRONMENTS

Salas, Sims, and Burke (2005) have identified eight critical variables (five primary variables and three "coordinating mechanisms") critical to team functioning: team leadership, mutual performance monitoring, backup behavior, adaptability, and team orientation as the primary variables; and shared mental modes, closed-loop communication, and mutual trust as the coordinating mechanisms. If VEs are to be used effectively in team training, we must understand their capabilities and limitations relative to these critical variables. For example, VEs offer a wealth of opportunities to observe, measure, and test mutual performance monitoring. VEs can often monitor directly (or restrict for testing) the information channels by which team members monitor each other. Potentially, measures of both quantity and quality of mutual performance monitoring can be implemented in VE. Likewise, many forms of backup behavior, another critical team variable, can be observed, measured, and manipulated during team training (and observed and evaluated during debriefing replays and AARs).

At least two of Salas et al.'s (2005) coordinating mechanisms, shared mental models and closed-loop communication, have characteristics for which VEs likely offer unique and useful possibilities. VEs allow mental model assessments (for example, Endsley's SAGAT procedure, 1988), which can be used to probe the trainees' progress and identify problem areas, as well as allowing testing of shared (and individual) mental models by dynamically varying the training situation, introducing system failures and emergencies, simulating communication problems (forcing the trainees to rely on their shared models for tacit coordination), and so forth. As noted above, VEs also have the capability for innovative and unique training in team (closed-loop) communication. For example, VEs may be able to vary the quality and quantity of communication among team members. This could help develop trainees' skills in overcoming communication problems, make them aware of the necessity for communications, and train restraint in unnecessary communications.

Although speculative, it is not too much of a stretch to imagine VE training systems that can help train more esoteric team variables, such as mutual trust and team orientation. Mutual trust is typically developed over time in experienced teams, but VEs may allow new teams in training to experience a wide range of situations and challenges. Properly managed, that sequence of VE experiences might help new teams get to a higher level of mutual trust (and even develop higher levels of team orientation and team efficacy) than would be possible with other training methods.

VEs certainly offer a range of possibilities for training adaptability. If adaptability is a mix of experience and creativity (and creativity is "10 percent inspiration and 90 percent perspiration"), then VEs should allow the kinds of practice and the range of experience necessary to improve teams' adaptability. Also certainly VEs allow the testing of adaptability by allowing trainers to simulate failures and situations that put a premium on team adaptability and prohibit routine solutions.

Of course, VEs offer the opportunity for virtual teammates. Since virtual teammates are under the control of the trainer, many aspects of team communication and coordination, mutual performance monitoring, backup behavior, even team leadership can potentially be trained and tested.

SECTION OVERVIEW

The chapters in this section discuss how to assess and evaluate the effectiveness of training with a focus on training systems. Effectiveness is examined from a variety of perspectives including transfer of training analyses and cost-benefit trade-offs of various training approaches. A variety of training systems is discussed, including low tech and high tech VEs.

Technology innovations in the classroom are not new. Chalkboards, then overheads, then whiteboards, and now computers have, over time, changed the way traditional lectures are delivered. As computational power becomes less expensive and more pervasive, the types of technologies available for training will only increase in complexity and availability. Nonetheless, the same question persists, "Does it matter?" Those who develop the technology and many consumers of the technology assume it does or are forced to abandon previous training tools as old tools are replaced by newer, higher tech tools. Those who study training, as do the authors of these chapters, challenge the "develop it and it will train" philosophy. The main goal of this section is not to debate whether technology in training is good or bad; it is simply to give those interested in using technology in training an awareness and understanding that there are techniques to answer the question, "Does it matter?"

This section is organized into three subsections: factors for TEE, relevance of fidelity in TEE, and applications of TEE. The goals of the first subsection are to explain when, why, where, and how TEE should be completed. The goal of the second subsection is to examine the relationship between fidelity in training systems and effective training; that is, low tech versus high tech: does it matter, and if so, when? The goal of the third section is to present some illustrative examples of TEE in action to serve as templates for future studies of TEE.

REFERENCES

Alliger, G. M., & Janak, E. A. (1989). Kirkpatrick's levels of training criteria: Thirty years later. *Personnel Psychology, 42,* 331–342.

Anderson, L. W., Krathwohl, D. R., Airasian, P. W., Cruikshank, K. A., Mayer, R. E., Pintrich, P. R., Raths, J., & Wittrock, M. C. (2001). *A taxonomy for learning, teaching, and assessing: A revision of Bloom's taxonomy of educational objectives.* New York: Longman.

Baldwin, T. P., & Ford, J. K. (1988). Transfer of training: A review and directions for future research. *Personnel Psychology, 41,* 63–105.

Bloom, B. S., Englehart, M. D., Furst, E. J., Hill, W. H., & Krathwohl, D. R. (1956). *Taxonomy of educational objectives: The classification of educational goals. Handbook 1: Cognitive Domain.* New York: McKay.

Darken, R. P., & Banker, W. P. (1998). Navigating in natural environments: A virtual environment training transfer study. Virtual Reality Annual International Symposium. Proceedings of the IEEE 1998 National Conference, 12–19, Atlanta, GA.

Endsley, M. R. (1988). Situation awareness global assessment technique (SAGAT). *Proceedings of the IEEE 1988 National Conference, 3,* 789–795. Dayton, OH.

Horvath, M. (1999). *Self-regulation theories: A brief history, and analysis, and applications for the workplace.* Unpublished manuscript.

Kanfer, R. (1996). Self-regulating and other non-ability determinants of skill acquisition. In P. M. Gollwitzer & J. A. Bargh (Eds.), *The psychology of action: Linking cognition and motivation to behavior* (pp. 404–423). New York: The Guilford Press.

Kanfer, R., Ackerman, R. L., Murtha, T. C., Dugdale, B., & Nelson, L. (1994). Goal setting, conditions of practice, and task performance: A resource allocation perspective. *Journal of Applied Psychology, 79,* 826–835.

Kirkpatrick, D. L. (1976). Evaluation of training. In R. L. Craig (Ed.), *Training and development handbook.* New York: McGraw-Hill.

Kraiger, K., Ford, K., & Salas, E. (1993). Application of cognitive, skill-based, and affective theories of learning outcomes to new methods of training evaluation. *Journal of Applied Psychology, 78*(2), 311–328.

Levy, P. E. (2006). *Industrial/organizational psychology: Understanding the workplace.* Boston: Houghton Mifflin.

Salas, E., & Burke, C. S. (2002). Simulation for training is effective when . . . *Quality and Safety in Health Care, 11,* 119–120.

Salas, E., Sims, D. E., & Burke, C. S. (2005). Is there a "Big Five" in teamwork? *Small Group Research, 36*(5), 555–599.

Salas, E., Wilson, K. A., Burke, C. S., & Bowers, C. A. (2002). Myths about crew resource management training. *Ergonomics in Design: The Quarterly of Human Factors Applications, 10*(4), 20–24.

Shultz D., & Shultz, S. E. (2002). *Psychology and work today* (9th ed., pp. 167–170). Upper Saddle River, NJ: Prentice Hall.

Part V: Factors for Training Effectiveness and Evaluation

Chapter 17

TRAINING EFFECTIVENESS EVALUATION: FROM THEORY TO PRACTICE

Joseph Cohn, Kay Stanney, Laura Milham, Meredith Bell Carroll, David Jones, Joseph Sullivan, and Rudolph Darken

Training effectiveness evaluation (TEE) is a method of assessing the degree to which a system facilitates training on targeted objectives. Organizations can utilize TEEs to better understand the overall value of an existing or newly implemented training program by illustrating strengths and weaknesses of the training program that should be either maintained, further developed, or improved upon to benefit the organization's performance as a whole.

Although the inherent value of a TEE is undeniable, instantiating this practice has proven challenging. There are three primary difficulties that must be addressed when planning and performing a TEE. The first is difficulty in the development of meaningful and collectable measures of performance to indicate whether targeted knowledge, skills, and attitudes have been acquired. Traditional TEEs have primarily been designed based on Kirkpatrick's (1959) four level model of training evaluation: (1) reactions, (2) learning, (3) behavior, and (4) results. While Kirkpatrick's approach does provide a structural framework and an approach for its implementation across these four levels, there is a cost/time trade-off associated with developing each level of metric. The more advanced levels (3) and (4) often require larger data collection efforts, over greater time horizons, and, consequently, many TEEs have often fallen short of evaluating beyond trainee reactions and declarative learning, levels (1) and (2), failing to capture training transfer behaviors in the operational environment and the overall organizational impact of such training. This traditional approach to TEE is difficult to justify since it will not likely provide any diagnostic value or insight into why a particular training intervention was more or less effective. Additionally TEE methods that rely on performance evaluation generally assume continuous learning curves that reflect steady performance improvements over time. Clark (2006) provides examples, such as language acquisition skills, where improvements in underlying cognitive ability (transitioning from rule based to schema based performance) initially results in poorer performance. These under-lying cognitive effects are of much greater interest, but are much harder to measure and apply. The second difficulty associated with operational TEEs is the need to have an operational system or prototype to facilitate empirical evaluation. Traditionally, TEEs are performed after a training system is fully developed and instantiated in the training curriculum. As such, it has potentially already replaced a legacy trainer, and comparisons of comparative value cannot be evaluated. Although post-instantiation TEEs provide data regarding the utility of a training system or approach, many times these results cannot be leveraged to modify the system in order to increase the utility of the overall training program due to financial and time constraints. Thus, there is a need to take a more proactive approach by addressing how to build training effectiveness into the system/program from the conceptual design. The third difficulty is with the multiple logistical constraints that many times surround evaluation of a system in the operational field. For instance, in evaluating military training there are several factors inherent to the domain (for example, lack of experimental control in operational environments and lack of access to fielded systems), which threaten the validity of inferences made based upon training results (Boldovici, Bessemer, & Bolton, 2002). Other limiting logistical factors include limited numbers of participants and rigid scheduling issues, both of which require that evaluations are designed to be performed during trainees' traditional training courses. Sullivan, Darken, and Becker (Volume 3, Section 2, Chapter 27) provide an example of the problems and benefits of operational field tests of novel VE training systems.

LIFECYCLE APPROACH TO TEE

This chapter aims to bridge the gap between theory and practice by detailing a feasible approach to TEE, which allows the difficulties detailed above to be addressed during system development and evaluation. The solution lies in a life-cycle approach to TEE. A lifecycle approach to TEEs is designed to follow the development lifecycle in order to provide high level input as the training system is initially being developed and more precise guidance as training system releases are made. Driven by a detailed task analysis, training needs can be identified up front to not only allow training effectiveness to be built in during the development process, but also to facilitate the development of metrics to assess if the training is meeting intended training goals. The second and most evident advantage to this approach is that the input provided by training needs analysis can be used to shape future development of the training system, as well as theoretical evaluation of the system's ability to support targeted training objectives, prior to development. This spiral approach also provides an opportunity for multiple system evaluations, which allows training system designers and evaluators the opportunity to assess whether or not any instantiated changes are effectively increasing the utility of the training system that is under development and the training program as a whole. Third, the approach leverages quasi-experimental methods to allow evaluations of the system to be made with some level of certainty, grounded in research methods theory. This spiral TEE approach (compare Gagne, Wager, Gola, & Keller, 2005) thus incorporates an iterative cycle that encompasses training needs analysis, training systems user-centered design, and training effectiveness and training transfer evaluations to help mold a system that meets targeted training objectives (see Figure 17.1). This method can be used throughout the iterative design cycle: at conceptual design, with prototypes, and with fully functional systems pre- and post-fielding.

Training Needs Analysis

To ensure a training system is targeting the intended objectives, it is first necessary to identify training goals early in the training system design cycle. Training goals provide a foundation for all future stages of the TEE; if a system is not focusing training on appropriate goals and objectives, it will not have the desired effect on trainee performance. Hence, the first step in training evaluation should be the same as that of training development: determining who and what should be trained. For training courses that do not have defined training goals, these must be derived by performing a training needs analysis (TNA) (see Milham, Carroll, Stanney, and Becker, Volume 1, Section 2, Chapter 9). A TNA is the process of collecting data to determine what training needs exist to allow development of training that facilitates accomplishment of an organization's goals (Brown, 2002). TNAs are accomplished through doctrine review, instructor interviews, and direct observation. As such, in order to complete a thorough TNA it is necessary to have (1) access to documentation that adequately explicates the targeted tasks, (2) access to and dedicated time with instructors or subject matter experts (SMEs) with detailed knowledge of the tasks to be trained, and (3) admittance to training exercises or operational performance in order to observe task performance (see Milham, Carroll, Stanney, and Becker, Volume 1, Section 2, Chapter 9). These sources of information provide the background knowledge necessary to form a solid base of domain expertise to direct training objectives and training content design. Training objectives can be defined at a very high level (for example, mission level), which may lack the granularity to derive meaningful objectives to target, or at a more specific level (for example, task level) facilitating the identification of precise training objectives. In cases where the training goals are defined at a high level, it is crucial to eventually drill down into the task to operationalize the high level goals into specific task training objectives.

Task Analysis

From identifying high level training goals to drilling down to detailed task training objectives, it is necessary to perform a task analysis, focused on the identification of the specific tasks and subtasks necessary to complete a mission, task flow, frequency, timing, operators, and task performance requirements. As a result, training objectives identified during the TNA can be realized in the

Training System Design & Evaluation

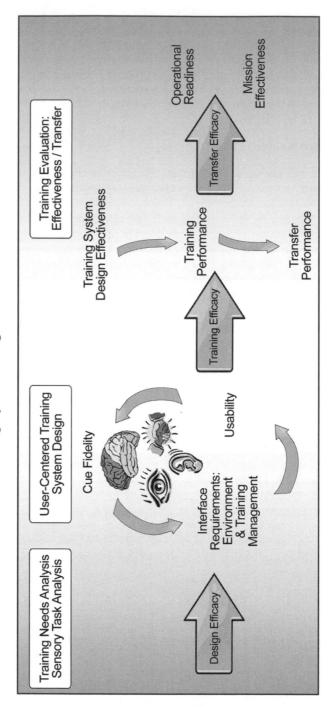

Figure 17.1.   Lifecycle Approach to Training System Design and Evaluation

training content design, and metrics to gauge task performance can be developed. With respect to the user, it is important that the task analysis identify the target trainee expertise level and the associated knowledge, skills, and attitudes (KSAs) currently possessed by trainees (pre-training), as well as those KSAs that are intended to be acquired through training. In order to identify scenario elements to be incorporated in the training content design, it is necessary to understand the subtasks with respect to how trainees gather information (what cues are required and through what mode—visual, auditory, or haptic) and how they act upon the environment in the real world. This form of task analysis, referred to as a sensory task analysis (STA), enables identification of multimodal cues and functionalities experienced in real world performance.

The outcome from the task analysis is deep, contextually rich data that can be used to inform and direct user-centered training system design (UCD), thereby leading to design efficacy (see Soloway, Guzdial, & Hay, 1994) for an early example of learner-centered design).

Training Systems Design

Based on the results of the TNA and task analysis, a mixed or "blended" fidelity training solution can be designed, which identifies how best to meet learning objectives that will close the target performance gap. The solution should integrate the optimal mix of classroom instruction training technologies and live events throughout a given course of training to ensure a desired level of readiness is achieved (Carter & Trollip, 1980). The composition of the material to be trained, characteristics of the target training community, cost of delivery, and operational requirements all contribute to the determination of the optimal blend of training delivery methods (Cohn et al., 2007). Thus, in addition to identifying the training goals and objectives, these must be translated into system requirements, including the following: cue and interface fidelity, overall training management approach, and underlying metrics. Once a design is agreed to, iterative usability analyses can provide insight into system potential during the build phases, allowing developers time to make high impact and critical modifications if they are identified.

From the composition design perspective, training systems design can be supported through UCD input. UCD ensures that a given training solution affords training by identifying the interactive, egocentric, and affective cues that best support training objectives, as well as functionalities that allow trainees to practice key competencies. UCD is driven by results from the task analysis, including the STA, which extends traditional task analysis beyond task breakdown, flow, sequence, and so forth to include identification of the critical multimodal cues (visual, auditory, and haptic) and functionality required to complete a task as defined by the operational context. UCD translates this contextual information into interface requirements by determining how the identified cues should be presented to the trainee to afford training of the task. This is done by coupling task information with training objectives to determine whether physical, psychological, or functional fidelity is required for successful task training. For instance, if a task requires one to detect aircraft orientation and project its response to changing environmental conditions, then the fidelity requirements demand high resolution visuals and realistic aircraft dynamic modeling, not just a low fidelity representation of an aircraft. On the other hand, given a task that primarily focuses on decision making in response to low granularity cues, low fidelity visuals, and a keyboard interface should suffice.

In addition to multimodal cue fidelity requirements, a training system must also support operational functionalities and coordination requirements. To support transfer, systems should facilitate performance of the actions and procedures trainees are required to execute in the field. These functionalities are identified through the task analysis, which facilitates systematic identification of the system functionalities/capabilities required for an operator to successfully complete each task (for example, system requirements for buttonology, temporal realism, and so forth).

Cue fidelity and functionality requirements are defined to facilitate interface design of an effective training environment that allows practice of targeted tasks. Practice alone, however, does not necessarily result in effective training, regardless of the quality of the training environment. In addition to the environment, a training management component is necessary to ensure effective training. Training management describes the process of understanding training objectives within a domain, creating training events to allow practice of targeted objectives, measuring and diagnosing performance, and providing feedback to trainees (Oser, Cannon-Bowers, Salas, & Dwyer, 1999). The training management component relies on both contextual data from the TNA and task analysis (for example, training objectives, metrics, scenario events, and scenario manipulation variables) to facilitate targeting training objectives at varying levels of difficulty, as well as theoretical and empirical data from the training science community to ensure effective training strategies (for example, feedback methods) are incorporated.

Usability analyses, which use this information, should be conducted to examine the degree to which the human-system interaction is optimized within the training system. First, a heuristic evaluation consisting of exploratory and scripted interaction with the system can be performed to identify any existing usability issues that could result in user perception and interaction errors. Second, SMEs conduct cognitive walkthroughs of the system employing a "thinking aloud" protocol, from which cognitive rationale while performing tasks with the system's interface is identified. Third, user testing is conducted to validate and extend findings from the heuristic evaluation and cognitive walkthroughs, and by examining the degree to which interface requirements identified in previous steps are instantiated within the system. The results of the usability evaluation are problem/solution tables describing interaction issues with the interface and redesign recommendations to enhance the training system design.

The outcome from the UCD stage is (1) a training system environment designed to achieve physical, psychological, and functional fidelity through the specification of contextually derived cue fidelity and interface requirements and (2) a training management component designed to ensure training objectives are effectively targeted and to diagnose trainees' performance and provide feedback or mitigations that enhance training outcomes, which should in turn lead to training efficacy.

Training Systems Evaluation

Training system evaluations are performed through iterative theoretical and empirical evaluation. This provides a framework for identifying problems in the simulation design, and for evaluating how well the training system affords performance enhancement. The most well-known model for evaluating the effectiveness of training systems is the aforementioned Kirkpatrick (1959, 1998) four-tier model. Level 1 (reaction) assesses a trainee's qualitative response (that is, how relevant the trainee felt the learning experience was) to the training system via interviews, questionnaires, informal comments, and focus group sessions. Level 2 (learning) assesses the trainee's increase in KSAs based on interaction with the training system, typically via instructor evaluations or more formally via pre- and post-test scores and team assessments. Level 3 (behavior) assesses the applied learning (transfer) to the target task via on-the-job observation over time or self-assessment questionnaires. Level 4 (results) assesses the organizational impact of the training (for example, return on investment, increased production, decreased cost, retention, quality ratings, and turnover) via post-training surveys, interviews with trainees and managers, ongoing appraisals, quality inspections, or financial reports. Thus, a well-planned TEE will utilize an integrated suite of methods to assess the degree to which a system facilitates training on targeted objectives, as defined at each level. In general, each level can be assessed via either process or outcome measures. *Process measures* examine the *manner* in which a task is accomplished (Cannon-Bowers & Salas, 1997; Salas, Milham, & Bowers, 2003), whereas *outcome measures* focus on how well a trainee accomplishes the overall task mission. Taken together, process and outcome measures can be useful diagnostic tools, in that they provide information about *what* happened and *why* (Fowlkes, Dwyer, Milham, Burns, & Pierce, 1999). Using such measures, TEEs provide training developers with data to better understand the overall value of an existing or newly implemented training program by illustrating strengths and weaknesses that should be either maintained, further developed, or improved upon to ensure that training goals are met. We have developed a three-tier TEE framework through which to gather evaluative data at all four levels of Kirkpatrick's model, which include (1) theoretical TEE, (2) trainee performance evaluation, and (3) transfer performance evaluation.

Theoretical TEE: Determining Training System Design Effectiveness

By performing a theoretical TEE, practitioners can answer the question of whether the training system design itself is effective in affording learning and whether learning can occur *before* the system is ever developed. Two "theoretical" TEE methods are used to answer these questions early in the development lifecycle, before trainees are ever brought to the system. *Cue fidelity evaluations* examine the degree to which the system includes the cue fidelity requirements identified in the UCD stage as necessary to address goal accomplishment (Herbert & Doverspike, 1990). This analysis identifies gaps between requirements driven by the task analysis and actual training system specifications. Next, *required capabilities analyses* are conducted to determine if the training system supports operational functionalities and coordination requirements necessary for a trainee to effectively perform targeted tasks. For example, given the task of using a laser designator to mark a target, in order to facilitate practice of the skills required to perform this task, some representation of the tool functionality must be present, the required fidelity of which is dependent on whether the goal is to train the cognitive or physical aspects of the task. The capabilities analysis identifies gaps with respect to functionalities required to perform such tasks and system capabilities. These theoretical evaluations can be complimented by user testing to support Kirkpatrick's level 1 (reaction) evaluation, which provides data on how relevant the training is to the trainee. Trainees can be asked to provide subjective evaluations via survey or interview after exposure to the training system. The reactions assessed vary based on the training program but may include the following:

- The relevance of the training to the trainee's job,
- The usability of the training system (for example, performance measures can include efficiency and intuitiveness, while outcome measures can include effectiveness and satisfaction), and
- The adaptability of the training system (for example, ability to personalize the learning both in terms of content and assessment).

Trainee Performance Evaluation: Did Learning Occur?

Trainee performance evaluation (that is, evaluating the learning that has taken place due to exposure to a training system) focuses on determining the following: What knowledge was acquired? What skills were developed or enhanced? What attitudes were changed? Kraiger, Ford, and Salas (1993) delineate a number of training outcomes related to the development of specific KSAs that can be used to structure an overall assessment of Kirkpatrick's level 2 (learning) metrics. Bloom's taxonomy, a widely used and accepted framework, can be used to assess learning behaviors along three dimensions—cognitive, psychomotor, and affective (Bloom, Englehart, Furst, Hill, & Krathwohl, 1956). Many VE training systems developed to date have targeted improved cognitive and/or psychomotor performance and are thought to be ideal for the development of complex cognitive task behaviors, such as situational awareness, cognitive expertise, and adaptive experience (Stanney et al., in press). While both cognitive and psychomotor behaviors are important to task learning and performance, the third component of learning identified by Bloom et al. (1956), affective behavior, encompasses the attitudes that students have toward the learned content. The attitudes of trainees have a significant effect on their learning potential. Thus, it is important to address all three learning outcomes. *Cognitive outcomes* involve the development of declarative, procedural, and strategic knowledge, that latter of which supports distinguishing between optimal and nonoptimal task strategies, as well as mental models, situation awareness, and self-regulation. *Psychomotor, skill based outcomes* assess how well skills associated with a given task or process (for example, perceptual, response selection, motor, and problem solving skills) have been developed and capture proficiency in domain-specific procedural knowledge (Kraiger et al., 1993; Proctor & Dutta, 1995). *Attitudinal outcomes* describe how attitudes are changed through training. These measures should assess such factors as affective behaviors that are critical to task persistence when faced with difficult operational objectives (for example, motivation, self-efficacy, and goal commitment; Kraiger et al.) and physiological responses associated with task demands (for example, stress). Targeted emotional responses (for example, remaining calm under pressure) may be achieved when trainees are given the opportunity to practice in affectively valid environments (that is, those that produce emotional responses similar to the operational environment).

A multifaceted approach to performance measurement, capturing each of these training metrics, is critical if training effectiveness is to be successfully interpreted. Specifically, competent performance in complex operational environments requires not only the basic knowledge of how to perform various tasks, but also a higher level conceptual and strategic understanding of how this knowledge is applied in order to optimally select the appropriate strategies and actions to meet task objectives (Fiore, Cuevas, Scielzo, & Salas, 2002; Smith, Ford, & Kozlowski, 1997). Moreover, it is also critical that trainees possess both well-defined, highly organized knowledge structures as well as the necessary self-regulatory skills to monitor their learning processes (Mayer, 1999). Thus, training evaluation should measure changes in the trainee (for example, post-training self-efficacy, cognitive learning, and training performance). Self-efficacy, or confidence with respect to trained tasks, which is an affective outcome of training, has been shown through several studies to be correlated with performance (Alvarez, Salas, & Garofano, 2004). Self-efficacy can be assessed via self-efficacy questionnaires (compare Scott, 2000), which are modified specifically for the tasks(s) being trained.

Given the logistical constraints imposed by factors inherent to the domain (for example, lack of experimental control in operational environments or limited participants), which threaten the validity of inferences made based upon the training results (Boldovici et al., 2002), quasi-experimental methods can be leveraged to provide some level of control regarding whether trainees have learned targeted skills. For example, often training cannot be withheld from a group of soldiers. In these cases, soldiers in pre-deployment training programs will receive all possible training intervention opportunities available to ensure their safety. As such, experiments comparing a control group (either with or without alternative training) to an experimental group may be infeasible. Given that inferences made from performance results cannot be unequivocally attributed to the training itself without a baseline control group against which to compare, evaluation techniques must be extended. In these cases, Sackett and Mullen (1993) suggest several pre-experimental designs: pre-test–post-test no-control-group designs or post-test-only nonequivalent control group designs. Although the pre-test and post-test design provides change data, Sackett and Mullen (1993) point out that training cannot be accountable as the sole reason for change. Haccoun and Hamtiaux (1994) proposed using an internal referencing strategy (IRS) within a pre-test and post-test design without a control group, which introduces both trained, relevant material and untrained, nonrelevant material. Training effectiveness is measured by showing that positive changes in trained material are significantly greater than for untrained materials. By carefully selecting skills that are orthogonal in training relevance, some implications of trainer effectiveness can be reported with confidence using this technique that allows trainees to serve as their own control.

Training Transfer Evaluation: Did Learning Impact Mission Success?

Training transfer evaluation (that is, Kirkpatrick's level 3 [behavior—transfer]) involves evaluating the trainees' capabilities to perform learned skills while in the target operational environment (that is, transfer of acquired KSAs). This can be accomplished by establishing a relationship between training objectives and operational requirements (that is, combat capability, operational readiness, and business competitiveness) as derived by a well-defined process and outcome measures. Specifically, level 3 metrics can link training objectives to operational requirements. This linkage can be established in the form of a taxonomy based on the outcome of the documentation review (for example, military standards, instruction and training manuals, and universal task lists) and task analysis and then refined, structured, and prioritized via interviews with SMEs and trainers. Resulting training metrics should be directly related to operational effectiveness; be qualitatively or quantitatively defined to allow for measurement and evaluation; have associated targeted performance levels, allowing for gradations (for example, trained, partially trained, and untrained); and have relative weights associated with impact on operational readiness. The objective is to achieve a set of precisely defined training metrics that enable trainers to predict how effective a given training solution can be in achieving the highest levels of operational readiness and mission effective performance.

To examine how training affects mission success, training transfer studies can be conducted to determine the impact of VE pre-training on performance in operational environments. Transfer performance (that is, the degree that training results in behavioral change on the job) is the gold standard of training (Alvarez et al., 2004). In the military domain, this could be expanded to include live fire exercises, given that live training is a rare and costly commodity. Yet, given logistical constraints, it is challenging to determine how best to use training technologies to optimize live fire training to reduce or better use the time spent in live fire. Within the aviation community, numerous studies have demonstrated the training efficacy and transfer of training from ground based flight simulators to actual flight operations (Finnegan, 1977; Flexman, Roscoe, Williams, & Williges, 1972; Jacobs & Roscoe, 1975; Lintern, Roscoe, & Sivier, 1989). To guide such studies, early on Roscoe and Williges (1980) provided a systematic approach for establishing the amount of live training time that could be saved through substitution of simulator training. One of the downfalls of these approaches, however, is the need for a large number of participants, a particular challenge in operational environments. To address this, methods have been developed to adapt the aviation transfer of training methods so as to deal with operational constraints (Champney, Milham, Bell-Carroll, Stanney, & Cohn, 2006) and provide an experimental paradigm to answer the questions of (1) how much VE pre-training should be provided to indicate stable and significant learning? and (2) for a set number of VE pre-training trials, how much live fire training can be saved? Ultimately, to evaluate transfer of training for a training system, the performance of a baseline group (trained to criterion in the live fire environment) is compared to trainees with VE pre-training (for a succession of training increments [trials/time intervals]) and then trained to criterion in the live environment

Transfer Efficacy

If the TNA and task analysis have led to design efficacy, which has in turn through UCD led to training efficacy, then transfer efficacy should be achieved thereby leading to operational readiness and mission effectiveness. To determine if these have been achieved, training results need to be translated to and compared to the organization's overall goals. Kirkpatrick's level 4 (results) evaluation involves determining the organizational impact (for example, transfer effectiveness and operational readiness), financial impact (for example, training throughput and return on investment), and internal impact (for example, achieving excellence; supporting organizational change or growth of trainees) of the training.

To examine organizational impact, the transfer effectiveness ratio (TER; Roscoe, 1971) can be used to specify the trials/time saved in the live environment as a function of prior trials/time in the training platform. The incremental transfer effectiveness ratio (ITER; Flexman ct al., 1972) can also be used to determine the transfer effectiveness of successive increments of training in the training platform, with successive increments of training predicted to decrease the average TER and ITER to a point where additional training is no longer effective. By examining the ITER, a training instructor can perform a trade-off analysis between live and simulator training time and prescribe the number of trials/time increments that should be run for a trainee to reduce the amount of live training needed to meet performance criterion (for example, spend 10 hours in live training versus spending 2 hours in simulator training and 7.5 hours in live training to obtain the equivalent performance criterion, thereby saving 2.5 hours of live training time). However, to examine the ITER, a large number of participants are required (that is, an experimental group is needed for each increment of time). As such, a method has been developed to identify a single point at which to compare the transfer effectiveness of VE training (Champney et al., 2006), allowing the calculation of TER and percent transfer despite logistical limitations.

In terms of operational readiness, the U.S. Armed Forces primarily use the status of resources and training system (SORTS) to report readiness in four critical areas: personnel, equipment on hand, equipment serviceability, and training

(Knapp, 2001). A "C rating" is produced for each of the four areas, as well as an overall rating assessment:

- C-1 indicates the unit has the requisite resources and can support the full wartime mission(s) for which it is assigned.
- C-2 indicates the unit can do most of the missions assigned.
- C-3 indicates the unit can do many, but not all, of portions of the missions assigned.
- C-4 indicates the unit needs more resources (people, parts, or training) before it can do its assigned missions.

Knapp (2001) recommends the addition of two reports that address higher levels of training and readiness: (1) a training report from the training event coordinator and (2) a concurrent readiness report. The training report assesses the level of training planned and the training goal and reports how well the training program was accomplished. The concurrent readiness report ties in current capabilities as depicted in the training report with current resources (for example, manning, equipment, and parts) to accomplish the mission.

As for the financial impact, the ITER and the cumulative transfer effectiveness ratio (CTER) allow for an examination of training-cost efficiency. Specifically, ITER describes the time saved in one training situation due to successive increments of training in another (Roscoe & Williges, 1980). As opposed to percent transfer methods, this method accounts for the time spent in a trainer and considers the savings on one training platform (for example, live training) that is realized by each training interval in another (generally less costly) training platform. Theoretical and empirical work suggests that ITER and CTER are negatively decelerated functions; even if the percent transfer is still increasing, the value of each additional interval in the original trainer is adding only a fraction of value added. Hypothetically speaking, 1 hour in a trainer may save more than an hour and a half in flight, whereas 15 hours in a trainer may only save 7.75 hours

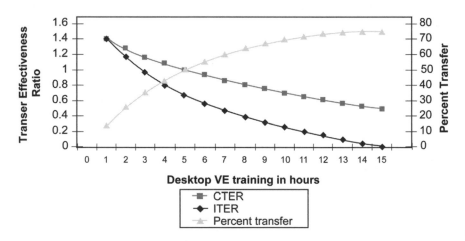

Figure 17.2.   Training Transfer Effectiveness. Cumulative Transfer Effectiveness Ratio (CTER); Incremental Transfer Effectiveness Ratio (ITER)

of flight. The ITER and CTER can be graphed to illustrate this function and the point of diminishing returns (see Figure 17.2). In our work, we have applied the Roscoe and Williges (1980) methodology to evaluate the ITER and CTER of several types of virtual environments (VEs), in an effort to examine the value of different levels of VE fidelity.

Table 17.1 presents examples and descriptions of the Level 1–Level 4 metrics used in an operational TEE performed on the multiplatform operational team training immersive virtual environment system.

Table 17.1.   TEE Evaluation Metrics

Kirkpatrick's Level	Metric	Metric Description	Example Item
Level 1— Reaction	Self-efficacy	Questionnaire assessing self-confidence in performing targeted task	How confident are you that you can determine correction from visual marks PROFICIENTLY during live fire training? 1 (certain cannot do it)–10 (certain can do it)
Level 2— Learning	Knowledge test	Test that targets procedural knowledge	You and your team are performing a close air support mission. The call for fire (including suppression of enemy air defense) and 9 line have been called in and time on target is approaching. What will you and each of your team members be doing at this time?
Level 3— Behavior	Event based checklist	A checklist to be filled out by observers that queries which tasks were/were not performed by the trainee and which tasks were instructor assisted	Corrections from mark communicated to pilot (occurred, instructor assisted, instructor corrected)
	Instructor evaluation form	A form in which the instructor rates trainee performance on targeted training objectives	Multiple effective marks on deck and simultaneous marks differentiated (poor, average, or excellent)
Level 4— Results*	Throughput	Number of trainees trained per unit time?	In 1 hour, how many trainees can complete the training scenario?

*Notional metric.

CONCLUSIONS

Developing, planning, and performing a training effectiveness evaluation is a deceptively complex exercise. From the basic knowledge elicitation efforts through metric development and ending with the actual assessment, there are many different factors that must be accounted for when attempting to quantify the impact of a training system. Oftentimes, significant trade-offs must be made, such as conducting an assessment that identifies benefits to the individual user at the cost of conducting a more holistic assessment to identify the return on investment to the organization. An enduring challenge for training system developers, procurers, and end users is ensuring that a complete lifecycle approach to TEEs is laid out as early as possible so that the costs, benefits, and risks to all stakeholders can be clearly identified and trade-offs made proactively, not reactively.

This chapter provides one such approach, starting with analyses to identify specific training goals and objectives, and then defining the types of scenario elements necessary to realize these goals. The results from these early analyses can then be used to inform system design, support iterative usability analyses that should be interspersed through the development cycle, and guide the development of a training management component that ties together the other elements into a cohesive training system. Last, training system effectiveness may be determined by quantifying system utility across several levels: reaction, which focuses on training usability; learning, which quantifies how well the system provides the necessary information; behavior, which provides insight into the degree to which performance is improved; and results, which demonstrate long-term impact to the organization through the use of the developed system. Ideally, all four levels would be determined through a comprehensive TEE; in practice, however, constraints may force some levels to be emphasized over, or to the exclusion of, others. Done well, though, these trade-offs may be made, and the assessments performed, in a way that still allows all those involved to understand the overall utility of the training system

REFERENCES

Alvarez, K., Salas, E., & Garofano, C. M. (2004). An integrated model of training evaluation and effectiveness. *Human Resource Development Review, 3*(4), 385–416.

Bloom, B., Englehart, M., Furst, E., Hill, W., & Krathwohl, D. (1956). *Taxonomy of educational objectives: The classification of educational goals. Handbook I: Cognitive domain.* New York: Longmans, Green.

Boldovici, J. A., Bessemer, D. W., & Bolton, A. E. (2002). *The elements of training effectiveness* (Rep. No. BK2002-01). Alexandria, VA: The U.S. Army Research Institute.

Brown, J. (2002). Training needs assessment: A must for developing an effective training program. *Public Personnel Management, 31*(4), 569–579.

Cannon-Bowers, J. A., & Salas, E. (1997). A framework for developing team performance measures in training. In M. T. Brannick, E. Salas, & C. Prince (Eds.), *Team performance assessment and measurement: Theory, methods, and applications. Series in applied psychology* (pp. 45–62). Mahwah, NJ: Lawrence Erlbaum.

Carter, G., & Trollip, S. R. (1980). A constrained maximization extension to incremental transfer effectiveness, or, how to mix your training technologies. *Human Factors, 22,* 141–152.

Champney, R., Milham, L., Bell-Carroll, M., Stanney, K., & Cohn, J. (2006). A method to determine optimal simulator training time: Examining performance improvement across the learning curve. *Proceedings of the Human Factors and Ergonomics Society 50th Annual Meeting* (pp. 2654–2658). Santa Monica, CA: Human Factors and Ergonomics Society.

Clark, E. V. (2006). Color, reference, and expertise in language acquisition. *Journal of Experimental Child Psychology, 94,* 339–343.

Cohn, J. V., Stanney, K. M., Milham, L. M., Jones, D. L., Hale, K. S., Darken, R. P., & Sullivan, J. A. (2007). Training evaluation of virtual environments. In E. L. Baker, J. Dickieson, W. Wulfeck, & H. O'Neil (Eds.), *Assessment of problem solving using simulations* (pp. 81–105). Mahwah, NJ: Lawrence Erlbaum.

Finnegan, J. P. (1977). *Evaluation of the transfer and cost effectiveness of a complex computer-assisted flight procedures trainer* (Tech. Rep. No. ARL-77-07/AFOSR-77-6). Savoy: University of Illinois, Aviation Research Lab, Institute of Aviation.

Fiore, S. M., Cuevas, H. M., Scielzo, S., & Salas, E. (2002). Training individuals for distributed teams: Problem solving assessment for distributed mission research. *Computers in Human Behavior, 18,* 729–744.

Flexman, R. E., Roscoe, S. N., Williams, A. C., Jr., & Williges, B. H. (1972, June). *Studies in pilot training* (Aviation Research Monographs, Vol. 2, No. 1). Savoy: University of Illinois, Aviation Research Lab, Institute of Aviation.

Fowlkes, J. E., Dwyer, D. J., Milham, L. M., Burns, J. J., & Pierce, L. G. (1999). Team skills assessment: A test and evaluation component for emerging weapons systems. *Proceedings of the 1999 Interservice/Industry Training, Simulation, and Education Conference* (pp. 994–1004). Arlington, VA: National Training Systems Association.

Gagne, R. M., Wager, W. W., Gola, K. C., & Keller, J. M. (2005). Principles of instructional design (5th ed.). Belmont, CA: Wadsworth/Thompson Learning.

Haccoun, R. R., & Hamtiaux, T. (1994). Optimizing knowledge tests for inferring learning acquisition levels in single group training evaluation designs: The internal referencing strategy. *Personnel Psychology, 47,* 593–604.

Herbert, G. R., & Doverspike, D. (1990). Performance appraisal in the training needs analysis process: A review and critique. *Public Personnel Management, 19*(3), 253–270.

Jacobs, R. S., & Roscoe, S. N. (1975). *Simulator cockpit motion and the transfer of initial flight training* (Tech. Rep. No. ARL-75-18/AFOSR-75-8). Savoy: University of Illinois Aviation Research Lab, Institute of Aviation.

Kirkpatrick, D. L. (1959). *Evaluating training programs* (2nd ed.). San Francisco: Berrett Koehler.

Kirkpatrick, D. L. (1998). *Another look at evaluating training programs.* Alexandria, VA: American Society for Training and Development.

Knapp, J. R. (2001). *Measuring operational readiness in today's inter-deployment training cycle.* Newport, RI: Naval War College. Retrieved June 3, 2007, from: http://stinet.dtic.mil/cgi-bin/GetTRDoc?AD=ADA393512&Location=U2&doc=GetTRDoc.pdf

Kraiger, K., Ford, J. K., & Salas, E. (1993). Application of cognitive, skill-based, and affective theories of learning outcomes to new methods of training evaluation. *Journal of Applied Psychology, 78,* 311–328.

Lintern, G., Roscoe, S. N., & Sivier, J. E. (1989). *Display principles, control dynamics, and environmental factors in pilot performance and transfer of training* (Tech. Rep. No. ARL-89-03/ONR-89-1). Savoy: University of Illinois, Aviation Research Lab, Institute of Aviation.

Mayer, R. E. (1999). Instructional technology. In F. T. Durso, R. S. Nickerson, R. W. Schvaneveldt, S. T. Dumais, D. S. Lindsay, & M. T. H. Chi (Eds.), *Handbook of applied cognition* (pp. 551–569). Chichester, England: John Wiley & Sons.

Oser, R. L., Cannon-Bowers, J. A., Salas, E., & Dwyer, D. J. (1999). Enhancing human performance in technology-rich environments: Guidelines for scenario-based training. In E. Salas (Ed.), *Human/technology interaction in complex systems* (Vol. 9, pp. 175–202). Stamford, CT: JAI Press.

Proctor, R. W., & Dutta, A. (1995). *Skill acquisition and human performance.* London: Sage.

Roscoe, S. (1971). Incremental transfer effectiveness. *Human Factors, 13,* 561–567.

Roscoe, S. N., & Williges, B. H. (1980). Measurement of transfer of training. In S. N. Roscoe (Ed.), *Aviation psychology* (pp. 182–193). Ames: Iowa State University Press.

Sackett, P., & Mullen, E. (1993). Beyond formal experimental design: Towards an expanded view of the training evaluation process. *Personnel Psychology, 46,* 613–627.

Salas, E., Milham, L., & Bowers, C. (2003). Training evaluation in the military: Misconceptions, opportunities, and challenges. *Military Psychology, 15*(1), 3–16.

Scott, W. (2000). *Training effectiveness of the VE-UNREP: Development of an UNREP self-efficacy scale (URSE).* Orlando, FL: Naval Air Warfare Center, Training Systems Division.

Smith, E. M., Ford, J. K., & Kozlowski, S. W. J. (1997). Building adaptive expertise: Implications for training design strategies. In M. A. Quinones & A. Ehrenstein (Eds.), *Training for a rapidly changing workplace: Applications of psychological research* (pp. 89–118). Washington, DC: American Psychological Association.

Soloway, E. Guzdial, M., & Hay, K. (1994). Learner-centered design: The challenge for HCI in the 21st century. *Interactions, 1*(2), 36–48.

Stanney, K. M., Cohn, J., Milham, L., Hale, K., Darken, R., & Sullivan, J. (in press). Deriving training strategies for spatial knowledge acquisition from behavioral, cognitive, and neural foundations. *Military Psychology.*

Chapter 18

TRANSFER UTILITY— QUANTIFYING UTILITY

Robert C. Kennedy and Robert S. Kennedy

Transfer effectiveness can be evaluated from numerous perspectives and can be quantified using a variety of metrics depending on the practical interests of the evaluator. This chapter provides several approaches that explain the challenges of transfer utility assessment and that can be implemented or modified for use in evaluating a virtual environment (VE) training system of interest. Such metrics as hours of practice, training days, simulator time, and flight time each quantify the training process in units of contact or learning time. Because of the large economic costs associated with administering VE systems, it is also useful to quantify such programs in fiscal units, such as budgetary dollars, capital outlay, or dollars saved over equivalent field training. Flight simulator training does not have any direct fuel costs compared to actual flight time, and with U.S. jet fuel consumption of over 600 million barrels (Energy Information Administration, 2007), simulator training administrators have an interest in looking at the marginal utility of the VE device in terms of fuel-cost savings. For example, Orlansky and String (1977) examined VE systems considering multiple cost variables and found that the investment of a multimillion dollar flight simulator could be amortized over about two years using fuel savings alone as an effectiveness criterion. Flight simulator administrators could also find value in quantifying the environmental impact of reduced fuel consumption, considering the fact that jet engines release over 20 pounds of carbon dioxide per gallon of fuel (Energy Information Administration).

Clearly VE training evaluation draws interest from various and often independent factions and, thus, the process of assessment may require a variety of quantitative indices. The process begins with criterion definition. In many cases, the outcomes of interest are well specified in terms of proximal performance. Less common are those that result on higher levels of the organization. (See Goldstein & Ford, 2002, on criterion development.) The remainder of Volume 3, Section 2 discusses several practical approaches to evaluating VE transfer effectiveness.

TRADE-OFF MODELS

VE training systems provide opportunities to the trainee for development of important knowledge and skills with the expectation that virtual training can alleviate some of the risks and costs associated with on-watch training. It is not always clear as to what those benefits are and, more importantly, what the relative utilities are to multiple benefits associated with implementing VE devices over their real world counterparts. Some have attempted to make comparisons of the relative utility using a trade-off approach. For example, early research on incremental training benefits of various fidelity qualities was done by Miller (1954), who suggested that while engineering fidelity development was important, its utility would tend to show diminishing returns at high levels.

One approach is the incremental transfer effectiveness ratio (ITER; Roscoe, 1971, 1980; Roscoe & Williges, 1980). Their work looked at the relationship between time spent in a flight simulator and its impact on savings using ITER [see Eq. (1)].

$$\text{ITER} = \frac{Y_{x-\Delta x} - Y_x}{\Delta x}. \tag{1}$$

X represents hours of simulator time, and *Y* indicates the subsequent flight time necessary to reach criterion performance requirements. Using tabular data presented in Roscoe and Williges (1980), a graph may be plotted showing the relationship between ITER and simulator hours (see Figure 18.1). The plot supports

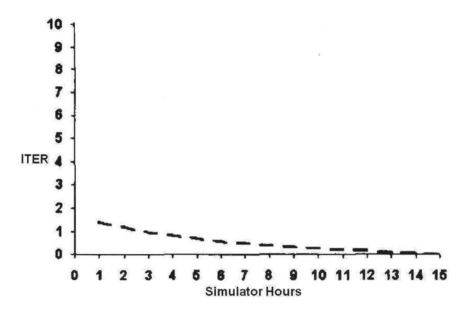

Figure 18.1.   Using tabular data presented in Roscoe and Williges (1980), a graph was plotted showing the relationship between ITER and simulator hours.

their contention that the first hour of training in a flight simulator saves more than an hour of training in an aircraft, with each subsequent hour saving less time until after 14 hours there is no significant savings in flight training time per unit VE training.

This approach illustrates that when evaluating cost of training to some criterion level, the costs associated with a VE system can be compared with that of flight training to determine the more cost-effective approach.

A related trade-off model called isoperformance derives contour lines or surfaces from all possible combinations of two or more determinants for a given performance (Jones, Kennedy, Kuntz, & Baltzley, 1987; Kennedy, Jones, & Baltzley, 1988). Like the ITER approach, isoperformance defines effectiveness at some desired outcome level, while the determinants act as boundaries in an iterative maximization analysis. The tabular data used in the previous example can also be used to construct an isoperformance curve. Figure 18.2 includes the ITER plot from Figure 18.1 with the addition of an isoperformance curve developed from the same tabulated flight and simulator training times for a given performance criterion.

Thus, a comparison is provided of the resultant ITER and an isoperformance curve illustrating that the ITER slope approximates the derivative of the isoperformance curve relative to simulator time. The isoperformance approach assumes a value for the performance criteria and provides a means for trading off various predictor values, such as simulator and flight time as with this case. Other variables could be incorporated using the same approach, such as classroom training, previous flight experience, or previous VE experience just to name a few.

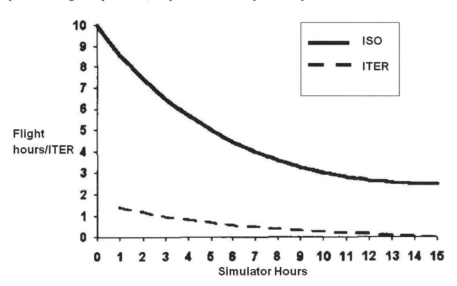

Figure 18.2.   This plot shows Figure 18.1 with an isoperformance curve developed from the same tabulated flight and simulator training times for a given performance criterion.

The need for objective data on fidelity payoffs was advanced through programmatic investigations such as that of Collyer and Chambers (1978), in which the architecture of the U.S. Navy aviation wide-angle visual system (AWAVS) program was defined. They noted the cost and complexity associated with unnecessarily high levels of physical or engineering fidelity and presented a program for acquiring behavioral research data designed to empirically investigate these questions. Consider the following four related elements: physical fidelity, realism, training effectiveness, and cost. If a seasoned aviator were asked to pick a good simulator, he or she would probably prefer one that has a high degree of realism. An engineer asked the same question might select one that possesses technological advances and that depicts faithfully the physical characteristics of the environment. Alternatively, a human resources manager, whose job it is to place the operator in an operational setting and demonstrate satisfactory performance, may consider training effectiveness as most important. The fourth element is cost, which is likely to increase along with the other three.

We readily admit that all four elements are likely to be correlated with each other, but to the extent that they are not isomorphic, focus on one does not guarantee that the other criteria are also met. A realistic scene can have high fidelity, which would be expected to have a high training transfer, and so on, but these are empirical questions and when they have been studied, the correspondence among them is far from perfect. It is important to remember that these four elements are not likely to be perfectly correlated, and all too often we can adhere to one element when another is intended, which can result in erroneous or unreliable assessment results. Furthermore, there may be overriding reasons (time, resource constraints, and user acceptance) to emphasize one of these elements over another in making managerial decisions related to a particular VE configuration or alternative training approach. From the standpoint of best return on investment, however, it is critically important to keep a clear focus on the ultimate goal of empirically demonstrated effective performance in the operational setting. Evaluation plans should be based on the fullest possible understanding of each of these elements and how they can be traded off, and experimental questions should be framed that will allow such trade-offs to occur.

TIME-COST MODELS

The evaluation of VE and other training systems involves several factors that include cognitive, attitudinal, affective, and skill improvement (Kraiger, Ford, & Salas, 1993). One of the most robust findings from the behavioral sciences is that which gives rise to learning theory: practice improves performance and a key element in practice is repetition (Ladd & Woodworth, 1911). This is particularly challenging for military applications when considering the significant level of turnover of its personnel. By design, officer and enlisted training programs assume that a large proportion of their graduates will complete an initial tour of duty and then separate from the armed services. There are certainly those who are career focused, but the rule has long been observed both in truth and in jest,

"Half the military spends half its time training the other half" (R. Ambler, personal communication, ca. 1960). Thus, the extremely costly military training model necessitates aggressive actions for identifying the most equitable training programs.

VE training evolved, in part, from the need to provide a high level of practice trials while minimizing the impact of variables, such as raw training dollars and risk/danger to the servicemember or equipment, and the need for specialized practice that is not possible in operational settings, as well as a need to consider secondary and tertiary consequences of training, such as the environmental impact of fossil fuel consumption.

Many VE systems have directly replaced or supplemented existing programs at reduced labor, fuel, operations, and maintenance costs while improving on safety and environmental impact. Each of these variables and many more could potentially describe the effectiveness of a VE training system. In other words, training effectiveness, which is often viewed in terms of efficacy of facilitating learning and practice, can also be evaluated in terms of business factors, such as return on investment, environmental impact, or even scrap-metal inventory.

The most basic time-cost evaluation involves the comparison of early performance levels with those subsequent to a VE training program. Hays and Vincenzi (2000) conducted a training effectiveness evaluation (TEE) on the virtual environment submarine VE device, which provides simulated training for such shipboard tasks as determining position, ship handling, and emergency operations. They conducted routine analyses of variance on pre-measures and post-measures, which showed positive learning effects. This approach may be ideal when post-training test performance is assumed to be a valid proxy for at-sea performance and can provide important data for use in cost analysis.

At this level, a VE system can be compared to other programs of training. Even the most complex programs consist of a series of more simple training exercises, measurable using a time-cost metric. Applying the time-cost metric in laboratory tasks involves little more than specifying a level of performance (for example, 85 percent of the items correct or latencies less than a specified level) and then determining how much time, or how much practice, it takes subjects drawn from a specified population to meet those standards. Real world training environments are not so straightforward. As a model for the process of real world training where sufficient data are available for analysis, a Red Cross sponsored swimming program was evaluated (Kennedy et al., 2005). Though this example does not specifically evaluate VE systems, the model could prove quite useful when attempting to model VE transfer utility.

The swimming lessons program was evaluated in which children first begin their swimming classes with a goal of achieving the level of "Beginner." Analogous to flight training, upon successful completion of the Beginner level, the next level was the "Advanced Beginner," followed by "Intermediate," and finally "Swimmer." Each level requires well-defined criteria that must be successfully achieved in order to move to the next level of qualification. Beginners are required to tread water for 30 seconds, swim the elementary backstroke for

Table 18.1. The Data from the Swimmers as They Progressed through Training

Level	$H(N, z\|z)$	q_z
Nonswimmer	0.00	1.00
Beginner	14.84	.72
Advanced Beginner	23.58	.54

25 yards, swim crawl for 25 yards, swim underwater for 15 feet, and so on. Table 18.1 presents data from a sample of 287 swimmers as they progress through the various levels of training. $H(N, z\|z)$ is the average number of hours taken to reach level z beginning at Nonswimmer (N). Q_z is the proportion of swimmers reaching level z. As with military flight training, the results show that each level retains fewer and fewer participants as they complete their requirements. This sequence indicates that the set of participants at each consecutive level is a subset of those in the previous level. As a group, moreover, they are faster learners than the group of all children who reached Beginner.

Children (also analogous to pilot students) who advance to higher levels reach lower levels faster than all children who reach those lower levels. Advanced Beginners who averaged 23.58 hours to reach that level are not the same children as the Beginners who averaged 14.84 hours to reach the Beginner level. The Intermediates took more time to reach Intermediate (30.42 hours) than the Swimmers. When analyzing training data, it is important to control for the potential confound that trainees who do not continue to a subsequent level may not be a random subset of those who reach a level. In the case of the Swimmers, they are slower to reach the level than others who get there and the further the Swimmers advance the stronger the company they keep becomes, until finally they are progressing no faster than average. If the Swimmers still continue, they will reach a level where they are among the late arrivals.

VE training programs are likely to involve substantially more complex sets of tasks, which require more complex consideration of the interactions between skill levels, longer training times, and restricted range in trainees at higher training levels when conducting this type of analysis. However, the parallels with the learning processes involved with swimming training suggest we have much to learn from their association, including potential approaches to evaluating the transfer effectiveness of various VE systems.

COST JUSTIFICATION

Orlansky and String (1977, 1979) approach effectiveness evaluation in terms of an economic efficiency criterion, which they suggest must satisfy one of the following:

1. Minimize economic cost for a given level of performance effectiveness or production;
2. Maximize performance effectiveness for a given cost.

Often the latter is of interest when there is a fixed budget and the transfer or performance is manipulable. In their example of flight training, criteria are required for flight training as well as simulator time. Cost/flight hour is specified through a process that distinguishes between relevant and irrelevant cost elements. Simulator effectiveness varies simultaneously based not only on training time, but also on how available the unit is for simulator time (Orlansky & String, 1977).

Orlansky and String (1979) evaluated multiple simulators in terms of cost and performance effectiveness using existing data from a coast guard program (Isley, Corley, & Caro, 1974) and a navy program (Browning, Ryan, Scott, & Smode, 1977). They make the assumption that transfer of performance is constant, focusing on the economic efficiency of the programs after the implementation of a new VE (flight simulator) system. In the coast guard study, they note that flight training without the simulator costs $3.1 million per year compared to $1.6 million per year with flight training and simulator training combined. Using these figures, their simulator is amortized over 2.1 years. The navy study showed similar results. Their P-3c simulator costs were $4.2 million and its implementation reduced the required flight training time from 15 hours to 9 hours. Based on an average of 200 pilots per year, their approach showed a $2.5 million per year savings. An amortization of the simulator costs resulted in less than two years. They further considered that this system also reduced the number of aircraft required for the training program, which would result in a 10-year savings of over $44 million.

UTILITY ANALYSIS

Training systems that use VE and other sophisticated training devices are quite expensive to administer and can run in the millions of budgetary dollars. There may be a desire to index training program effectiveness on economic scales using macroanalyses methods, such as utility analysis (UA; Brogden, 1949; Brogden & Taylor, 1950; Cronbach & Gleser, 1965; Taylor & Russel, 1939). UA is a general term used to describe a number of methods used in human resources program evaluation, especially in employment selection. Brogden (1949) and Cronbach and Gleser (1965) developed procedures that provided measures of selection program effectiveness in terms of dollars of profit. Iterations of these models are still in use, one of which is appropriately referred to as the Brogden-Cronbach-Gleser (BCG) model. The BCG model assumes a linear relationship between a predictor and a performance criterion, and employs the following formula:

$$\Delta U = N_s SD_Y(r_{xy})(\mu_{Xs}) - NC. \tag{2}$$

Here, ΔU is the incremental utility gain from a selection instrument or program, N_s represents the number of applicants hired, SD_Y is the standard deviation of job performance in dollar terms, r_{xy} is the validity coefficient of the measure, μ_{Xs} is the mean predictor score of the selectees, N is the total number of applicants, and C is the average cost of administering the instrument for an applicant.

This model also assumes that the predictor score is standardized (Z), that the organization is selecting based on a top-down hiring policy, and that those offered a position all accept the offer (as discussed in Gatewood & Field, 2001; Cabrera & Raju, 2001). Others have conducted utility analyses, with one of the substantive differences across these studies being how the authors decided to measure SD_Y (for example, Bobko, Karren, & Kerkar, 1987; Bobko, Karren, & Parkington, 1983; Boudreau, 1983; Cascio & Ramos, 1986; Raju, Cabrera, & Lezotte, 1996; Reilly & Smither, 1985; Schmidt, Hunter, McKenzie, & Muldrow, 1979; Schmidt, Hunter, & Perlman, 1982; Weekly, Frank, O'Connor, & Peters, 1985).

The principles of UA laid the foundation for various economic models for training program evaluation (for example, Barati & Tziner, 1999; Cascio, 1989, 1992; Cohen, 1985). Honeycutt, Karande, Attia, and Maurer (2001) adapted this approach in conducting a utility analysis of a sales training program using the following formula:

$$U = (T')(N')\,(d_1)(SD_y)\,(1{+}V)\,B\,(N)(C). \qquad (3)$$

The analysis defined utility (U) is the financial impact of the sales training program on the organization as a whole. Time (T) is defined as the period that the training continues to impact the firm; N is the number of employees that remained with the firm; d_1 represents performance changes resulting from the training; SD_y represents the pooled standard deviation of employee performance. Their calculations show that the sales training program resulted in a profit (U) of over \$45,000, which equated to \$2.63 of revenue (\$1.63 of profit) for each dollar spent on the training program. There are certainly limitations to this and each of the other methods. As with any attempt at data interpretation, it is important to understand the underlying assumptions and to test them accordingly. Given the assumptions stipulated by the Honeycutt et al. (2001) method, this method provides a potentially useful implementation of utility analysis models for use in training transfer. The extremely high costs associated with many VE training systems suggest a high likelihood for this or similar types of macroanalyses.

SUMMARY

As with any training program, VE programs seek to provide some treatment that will result in learning or practice ultimately intended for application on the job. The process of transfer effectiveness evaluation is intended to provide valid metrics that present training outcomes in terms of departmental and organizational objectives. Trade-off models, such as effectiveness ratios and isoperformance curves, are available and can be adapted for analyses where the evaluators seek to identify optimal configurations of training factors, such as fidelity and training time. Operational costs, such as maintenance, electrical/mechanical costs, trainer salaries, and trainee salaries, each have the potential to impact the overall effectiveness of a VE system, as well as corresponding departments. A careful consideration and definition of the effectiveness metrics are critical in prescribing a transfer effectiveness evaluation approach that will produce meaningful data and subsequent inferences.

REFERENCES

Barati, A., & Tziner, A. (1999). Economic utility of training programs. *Journal of Business and Psychology, 14,* 155–164.

Bobko, P., Karren, R., & Kerkar, S. P. (1987). Systematic research needs for understanding supervisor-based estimates of SD_Y in utility analysis. *Organizational Behavior and Human Decision Processes, 40,* 69–95.

Bobko, P., Karren, R., & Parkington, J. J. (1983). Estimation of standard deviations in utility analyses: An empirical test. *Journal of Applied Psychology, 68,* 170–176.

Boudreau, J. (1983). Effects of employee flows on utility analysis of human resource productivity improvement programs. *Journal of Applied Psychology, 68,* 396–406.

Brogden, H. E. (1949). When testing pays off. *Personnel Psychology, 2,* 171–185.

Brogden, H. E., & Taylor, E. K. (1950). The dollar criterion—Applying the cost accounting concept to criterion construction. *Personnel Psychology, 3,* 133–154.

Browning, R. F., Ryan, L. E., Scott, P. G., & Smode, A. F. (1977). *Training effectiveness evaluation of devices 2F87F, P-3C Operational Flight Trainer* (TAEG Report No. 42). Orlando, FL: Training Analysis and Evaluation Group.

Cabrera, E. F., & Raju, J. S. (2001). Utility analysis: Current trends and future directions. *International Journal of Selection and Assessment, 9,* 92–102.

Cascio, W. F. (1989). Using utility analysis to assess training outcomes. In I. L. Goldstein, (Ed.), *Training and development in organizations* (pp. 63–88). San Francisco, CA: Jossey-Bass.

Cascio, W. F. (1992). *Managing human resources: Productivity, quality of work life, profits.* New York: McGraw-Hill.

Cascio, W. F., & Ramos, R. A. (1986). Development and application of a new method for assessing job performance in behavioral/economic terms. *Journal of Applied Psychology, 71,* 20–28.

Cohen, S. I. (1985). A cost-benefit analysis of industrial training. *Economics of Education Review, 4,* 327–339.

Collyer, S. C., & Chambers, W. S. (1978). AWAVS, a research facility for defining flight trainer visual requirements. *Proceedings of the Human Factors Society 22nd Annual Meeting.* Santa Monica, CA: Human Factors Society.

Cronbach, L. J., & Gleser, G. C. (1965). *Psychological tests and personnel decisions.* Urbana: University of Illinois Press.

Energy Information Administration. (2007). Aviation gasoline and jet fuel consumption, price, and expenditure estimates by sector, 2005. Retrieved August 15, 2007, from http://www.eia.doe.gov/emeu/states/sep_fuel/html/fuel_av_jf.html

Gatewood, R. D., & Field, H. S. (2001). Human resource selection (5th ed.). Fort Worth, TX: Harcourt.

Goldstein, I. L., & Ford, J. K. (2002). A review of training in organizations: Needs assessment, development, and evaluation (4th ed.). Boston: Wadsworth/Thomson.

Hays, R. T., & Vincenzi, D. A. (2000). Fleet assessments of a virtual reality training system. *Military Psychology, 12,* 161–186.

Honeycutt, E., Karande, K., Attia, A., & Maurer, S. (2001). An utility based framework for evaluating the financial impact of sales force training programs. *Journal of Personal Selling & Sales Management, 21,* 229–238

Isley, R. N., Corley, W. E., & Caro, P. W. (1974). *The development of U. S. Coast Guard aviation synthetic training equipment and training programs* (Final Rep. No. FR-D6-74-4). Fort Rucker, AL: Human Resources Research Organization.

Jones, M. B., Kennedy, R. S., Kuntz, L. A., & Baltzley, D. R. (1987, August). *Isoperformance: Integrating personnel and training factors into equipment design.* Paper presented at the Second International Conference on Human-Computer Interaction, Honolulu, HI.

Kennedy, R. S., Drexler, J. M., Jones, M. B., Compton, D. E., & Ordy, J. M. (2005). Quantifying human information processing (QHIP): Can practice effects alleviate bottlenecks? In D. K. McBride & D. Schmorrow (Eds.), *Quantifying human information processing* (pp. 63–122). Lanham, MD: Lexington Books.

Kennedy, R. S., Jones, M. B., & Baltzley, D. R. (1988). Optimal solutions for complex design problems: Using isoperformance software for human factors trade-offs. *Proceedings of the Operations Automation and Robotics Workshop: Space Application of Artificial Intelligence, Human Factors, and Robotics* (pp. 313–319).

Kraiger, J., Ford, K., & Salas, E. (1993). Application of cognitive, skill-based, and affective theories of learning outcomes to new methods of training evaluation. *Journal of Applied Psychology, 78,* 311–328.

Ladd, G. T., & Woodworth, R. S. (1911). *Elements of physiological psychology: A treatise of the activities and nature of the mind, from the physical and experimental points of view.* New York: C. Scribner's Sons.

Miller, R. B. (1954). *Psychological considerations in the design of training equipment* (Tech. Rep. No. TR-54-563). Wright-Patterson Air Force Base, OH: Wright-Patterson Development Center.

Orlansky, J., & String, J. (1977). *Cost-effectiveness of flight simulators for military training: Volume I. Use and effectiveness of flight simulators* (IDA Paper No. P-1275). Arlington, VA: Institute for Defense Analyses.

Orlansky, J., & String, J. (1979). Cost effectiveness of computer-based instruction in military training (IDA Paper No. P-1375). Arlington, VA: Institute for Defense Analyses.

Raju, N. S., Cabrera, E. F., & Lezotte, D. V. (1996, April). *Utility analysis when employee performance is classified into two categories: An application of three utility models.* Paper presented at the Annual Meeting of the Society for Industrial and Organizational Psychology, San Diego, CA.

Reilly, R. R., & Smither, J. W. (1985). An examination of two alternative techniques to estimate the standard deviation of job performance in dollars. *Journal of Applied Psychology, 70,* 651–661.

Roscoe, S. N. (1971). Incremental transfer effectiveness. *Human Factors, 13*(6), 561–567.

Roscoe, S. N. (1980). Transfer and cost-effectiveness of ground based trainers. In S. N. Roscoe (Ed.), *Aviation psychology* (pp. 194–203). Ames: Iowa State University Press.

Roscoe, S. N., & Williges, B. H. (1980). Measurement of transfer of training. In S. N. Roscoe (Ed.), *Aviation Psychology* (pp. 182–193). Ames: Iowa State University Press.

Schmidt, F. L., Hunter, J. E., McKenzie, R. C., & Muldrow, T. W. (1979). Impact of valid selection procedures on work force productivity. *Journal of Applied Psychology, 64,* 609–626.

Schmidt, F. L., Hunter, J. E., & Pearlman, K. (1982). Assessing the economic impact of personnel programs on workforce productivity. *Personnel Psychology, 35,* 333–347.

Taylor, H. C., & Russel, J. T. (1939). The relationship of validity coefficients to the practical effectiveness of tests in selection. *Journal of Applied Psychology, 23,* 565–578.

Weekly, J. A., Frank, B., O'Connor E. J., & Peters, L. H. (1985). A comparison of three methods of estimating the standard deviation of performance in dollars. *Journal of Applied Psychology, 70,* 122–126.

Chapter 19

INSTRUMENTING FOR MEASURING

Adam Hoover and Eric Muth

This chapter considers the problem of instrumentation for the recording of live building clearing exercises. The recording of live building clearing exercises can help training in a number of ways. The recorded data can assist an after action review by allowing an instructor or group of trainees to replay an exercise to look for weaknesses and errors. It also allows for the development of automated methods to analyze performance. Such automation could potentially make training more objective, and more readily available, where dedicated facilities and instructors are lacking. Over time, a database of millions of exercises could be built, facilitating deeper studies of variability of team performance, as well as the evolution of performance of individual teams as training progresses.

In order to record live exercises, some amount of instrumentation is necessary. The instrumentation can be built into a dedicated training facility, for example, by placing sensors throughout the buildings used to conduct exercises. Instrumentation can also be placed on the trainees in order to record individual actions. In either case, there is some trade-off with regard to the variability and face validity of the allowed exercises: in general, the more data that are to be recorded, the more restricted the exercises. Instrumentation is by nature somewhat fragile. If trainees are required to wear or carry instrumentation, then their actions must be somewhat limited to ensure the correct operation of the instrumentation and to prevent its destruction. The same holds true for the facility and any instrumentation permanently deployed in the infrastructure. For example, trainees must not physically break through walls or fire rounds in the direction of cameras or other instruments recording the exercise. In addition, instrumentation fixed into the infrastructure of a facility is not easily redeployed; therefore, all the exercises must take place using the same buildings. This limits the variability in that the same structural layout and floor plans must be used every time.

Training for urban operations is not a new idea; it is the instrumentation that is changing. Many military bases are already using facilities in which building clearing exercises can be conducted. These facilities range in size from a single building to a multiple-block area consisting of tens of buildings. The buildings are usually made of concrete and have no glass windows or other easily broken materials. Soldiers practice against mock enemy forces using either simunitions, generally made either from paint or rubber, or "laser-tag" instrumented weaponry. It is important to note that while these facilities provide practice in urban operations, they already give up some realism due to the costs associated with simulating actual urban warfare. For example, the buildings cannot be harmed through breached entry without making it prohibitively costly to repair for the next group of trainees. Thus, adding instrumentation to augment and learn from training exercises is merely an extension of existing practices.

Some existing training facilities have already been instrumented to allow the recording of exercises. Examples include the concrete towns at Fort Benning, Georgia; Fort Polk, Louisiana; Quantico, Virginia; and Twentynine Palms, California. These facilities are in general focused on platoon-sized operations involving multiple buildings. Cameras may be placed throughout the facility to record video from a variety of angles. However, the video is not used to track trainees and is correlated manually (if at all) for playback. Some facilities use the global positioning system (GPS) to track participants. This limits position tracking to outdoors and limits accuracy to several meters, at best. Weapons can be instrumented with equipment that tracks the number of shots fired. MILES (multiple integrated laser engagement system) gear is the most widely known type of training gear; it operates like laser-tag equipment. This type of gear can track when participants were shot and by whom. However, it is intended for outdoor use and is not suited for the shorter distances involved in indoor engagements.

In contrast to these facilities, we are interested in the action that occurs inside a single building. While such actions can involve platoon- and larger-sized forces, we are interested in how a single fire team (four to five men) cooperates during building clearing. All the action we are interested in takes place indoors; therefore, GPS cannot be used to track the locations of participants. In addition, we require location accuracy on the order of 10 centimeters (cm) so that it is possible to identify what position a person occupies inside a room, not just what room that person is in. We also desire instrumentation that tracks where trainees are aiming weapons, and where they are looking, at all times. Monitoring the coverage of weapons and line of sight should allow for a deeper analysis of team performance.

The rest of this chapter describes a facility we built to meet these needs. We call our facility the Clemson Shoot House. It consists of reconfigurable walls so that the floor plan layout can be changed. It uses a network of cameras to automatically track trainee locations. We constructed custom laser-tag-style weapons and helmets to track shots, hits, and the orientations of weapons and heads. We also constructed heart-rate monitors to provide some physiological monitoring of the trainees. All the tracking information is gathered at a central recording station where it can be stored and replayed. While describing our facility, we break down the options currently available and discuss the lessons learned during the construction of this facility. To our knowledge, the Clemson Shoot House represents the current cutting edge of this type of facility.

There is almost no literature published regarding the construction of a shoot house or instrumentation for the recording of building clearing exercises. Even the relatively well-known MILES gear is barely discussed in the research literature. Therefore many of the lessons learned must be reported without reference to published literature; we hope that by documenting our facility and experiences this trend will change.

WALLS AND FACILITY INFRASTRUCTURE

We constructed our facility at the 263rd Army National Guard, Air and Missile Defense Command site in Anderson, South Carolina. This site was chosen because of its proximity to Clemson University (about 15 kilometers), and the large area available. The Army National Guard provided space within a large warehouse that has a 6.1 meter (m) (20 foot) high ceiling. Our facility covers approximately 200 square meters (sq m), the size of a single-floor house, and is constructed entirely inside the warehouse. It consists of a shoot house and an instructor operator station. The shoot house is approximately 180 sq m of reconfigurable rooms and hallways. The instructor operator station houses equipment and provides for centralized training observation and after action review. By constructing the shoot house entirely inside an existing warehouse, we were able to leave off a ceiling or roof, and yet still provide protection from the environment for the instrumentation in the facility.

Figure 19.1 shows a computer-aided design (CAD) diagram of the facility, where the instructor operator station is on the left side. The configuration of the shoot house can be changed by inserting walls at hallway junctions (creating various L, T, and Z hallways) and by removing entire walls between rooms (creating larger rectangular or L-shaped rooms). There are several external entrances to the shoot house so that various exercises can be scripted.

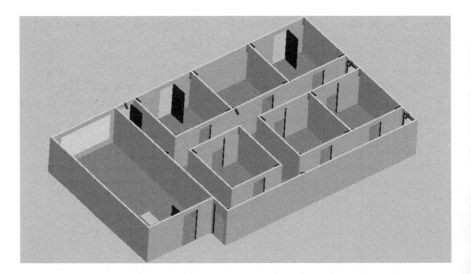

Figure 19.1.   Computer-Aided Design Diagram of the Clemson Shoot House

Within the shoot house, a small amount of furniture is placed into fixed positions. The furniture can be moved between exercises, but is expected to remain stationary during a single building clearing run. This is necessary to preclude the confusion that could be caused by tracking moving furniture and mistaking it for people. Figure 19.2 shows a picture overlooking a portion of the shoot house containing furniture. Note that in this picture, the shoot house is configured differently than shown in the CAD diagram (the walls between three of the small rooms have been removed, creating a large L-shaped room).

The materials used for the walls are similar to those used for office partitions. They consist of thick Styrofoam sandwiched between two pieces of paneling. The framing is metal and bolted into the concrete floor. Support at the top is provided by additional framing spanning across halls and other open areas. These materials withstand simple collisions and pressure from people leaning on them, but could be damaged by strong actions or point loads. Compared to using concrete blocks, the benefit is that it takes less than one hour to reconfigure the floor plan.

Lessons Learned

- Partition walls are sufficient (2.4 m [8 feet] high, movable, sturdy enough) when constructed inside a warehouse.

Figure 19.2.   An Overhead View of Part of the Shoot House

- The facility should have varying room sizes and shapes and door placements. Opposing symmetric doorways in a hallway are more challenging than offset doorways or single entries.

- If live observation is desired, an observation platform or tower near the center of the facility is useful to avoid having to go too high to see inside.

POSITION TRACKING

Our shoot house is equipped with 36 cameras (visible on top of the walls in Figure 19.2) wired to a rack of seven computers. The cameras are calibrated offline to a common coordinate system (Olsen & Hoover, 2001). The computers record the video feeds and process them in real time to track the spatial locations of subjects (Hoover & Olsen, 1999). Tracks are updated 20 times per second and are accurate to approximately 10 cm in two dimensions. This provides for a detailed analysis of team motion, such as how a team moves through a doorway and what positions are taken during clearing of a room. The raw video is also recorded and can be viewed during playback along with the tracking data.

Figure 19.3 shows a screenshot from a replay of a four-man exercise. The location of each tracked person is displayed as a colored circle, shown in the floor plan of the shoot house. Walls are displayed as white lines, and stationary furniture is displayed as white rectangles. The lines coming out of each circle represent the weapon orientation of each tracked person (this equipment is discussed in more detail in a following section). The orientations of helmets (not shown in this screenshot) can be displayed similarly. The video from a nearby camera is displayed at the right side of the screenshot, showing a live view of some of the

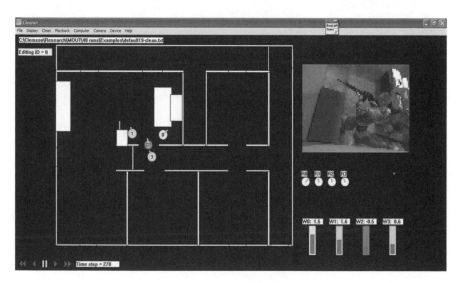

Figure 19.3.   A Screenshot of a Replay of an Exercise

action. The camera view can be manually selected to any of the 36 cameras, or it can be automated to select the camera closest to a particular track or to select the camera closest to groups of people. At the bottom right of the screenshot are data from the heart-rate monitors worn by each trainee (this equipment is also discussed more in a later section). The replay of data can be controlled through standard video controls, including pause, play, fast forward, and rewind. The replay can also be controlled according to shots fired, moving forward or backward, to successive shots fired. This mode of control lets those watching the replay quickly find moments of weapons action and examine the outcome.

The deployment of the camera network and computers to process the camera feeds can take up to a full day. Most of this time is spent on running wires and could be reduced by using wireless cameras. Changing the floor plan of the shoot house can require repositioning some cameras. Moving a few cameras does not take too much time (less than one hour), including the time for calibration. The position tracking software is made aware of wall and furniture locations in order to help maintain continuity. For example, knowing where a wall is prevents a track from inadvertently being associated with another track on the opposite side. This similarly helps with tracking trainees as they move around furniture, preventing the track from inadvertently locking onto the furniture. The floor plan of the walls and furniture is stored in a file that is loaded when position tracking starts and can be changed very quickly.

The lighting in the shoot house is provided by overhead spotlights of the variety commonly used in warehouses. The quality of this lighting is poor and causes a great deal of shadows. A single person standing in the middle of a room may cast three to four shadows of varying depth, each cast by a different overhead spotlight. These lighting conditions are among the worst possible for automated image processing because the shadows are difficult to differentiate from actual people. Although our tracking system is designed to be resilient to this problem, there are cases where multiple tracks within a small area are not properly differentiated. In such cases, a track may inadvertently lock onto a shadow, or two tracks may get mutually confused as shadows cross each other.

The position tracking also tends to suffer in hallways and in doorways. If several people bunch up in a hallway, then the cameras typically do not have enough vantage to see the correct positions of each person. This is because the cameras are placed at opposing ends of the hallway, looking toward each other. If four people stand in a line between the cameras, then neither camera can see the two people in the middle. In a room this problem does not occur because cameras are placed in all four corners and generally provide complete coverage. In a doorway, tracks must be "handed off" from one camera to another as a person moves through. Our tracking system tackles this problem by maintaining a global position for each track (as seen in Figure 19.3) and using that information to assist in camera hand-off. However, when multiple people quickly go through the same doorway, there can be some momentary confusion while the system performs the hand-offs of the multiple tracks.

Overall, we estimate that our position tracking performs correctly roughly 70 percent to 80 percent of the time, with no operator intervention. In order to create "clean" tracking data, a recording is manually reviewed by a human operator using the replay tool. The human uses the video to compare against the automatically recorded tracks. If an error is observed, the human can override the automatically recorded position and manually fill in a corrected position. Depending on the number of people tracked throughout the exercise (four to eight people), the length of the exercise (0:30–6:00 minutes), and the quality of the data, the position cleaning process can take anywhere from 15 minutes to 1 hour.

In addition to cleaning the position tracking data, the recording is also post-processed to correlate rifle, helmet, and heart-rate data to individual tracks. This is accomplished by indicating which devices correspond to which position tracks. Finally, the shots fired and hits are correlated to identify kill shots. Typically, the entire system registers a hit roughly 50–100 ms after the corresponding shot (or about one to two time steps at our 20 Hz [hertz] sampling rate). These correlations are identified and saved in the final post-processed (or clean) data file in order to facilitate subsequent analysis.

During the construction of this facility, we had the opportunity to observe several tracking technologies, such as radio frequency identification (RFID) based systems. It is our opinion that no currently available indoor tracking technology works better than what we have developed and that none of them (ours included) have fully solved the problem. There is still a need for a reliable, fully automated indoor tracking system, which can provide accuracy on the order of 10 cm and an update rate of 30 Hz. Preferably, the tracking system should be easy to deploy in any infrastructure and require minimal instrumentation on the bodies of tracked subjects. Until such a technology is developed, it will continue to be difficult to obtain tracking data on indoor operations.

Lessons Learned

- Video cameras are probably still the best sensor option to track people indoors at high accuracy (for example, on the order of 10 cm), even though they will not work near 100 percent of the time.

- Based upon observations of the performance of other sensor types (such as RFID, sonar, and wireless signal strength based tracking), all currently available options still suffer in hallways and doorways. A completely automated, hands-off solution for tracking inside a building remains unknown.

- Position data recorded at 20 Hz and at 10 cm accuracy allow for the visualization of the motion of trainees at a level heretofore unseen. Based upon watching over 1,000 exercises, we believe this fidelity of data should allow for new types of analysis of team performance; studies to validate this hypothesis are still ongoing.

WEAPONS AND BODY INSTRUMENTATION

The primary purpose of instrumentation on the weapons and bodies of trainees is to track who shot whom. However, the instrumentation can also be used to track the actions of trainees during periods when no firing is taking place. For example, it may be useful to track the orientations that trainees hold weapons, collectively and individually. It may be useful to track where trainees are looking at all times. Keeping weapons oriented properly, covering the blind areas of fellow team members, and watching weapons coverage are all likely related to team performance. Therefore, we desired instrumentation to track the orientations of the heads and weapons of trainees.

We investigated the availability of MILES gear and the suitability of gear produced by the laser-tag gaming industry. Neither was found to meet our needs. While MILES is a familiar term to many involved in this field, we found surprisingly little information on vendors. We also were unable to find any literature detailing the MILES standard. In the gaming industry, we had more success identifying and communicating with vendors. However, all the systems we found were sold as closed products, making it difficult to modify to suit our needs. The vendors tend to be small companies, with small markets, and are not geared toward custom solutions. Facing these obstacles, we decided to design and construct our own custom laser-tag gear.

We constructed several embedded devices, including weapons, helmets, and heart monitors. All of the devices are wireless and completely untethered. They use a chip built on the 802.11b networking standard to communicate data to the rack of computers in the instructor operator station. This allows us to use a commercial off-the-shelf 802.11 router to communicate with all our embedded devices, with all its advantages of throughput and error recovery. All data are updated at between 5–20 times per second, depending on the update rate of the individual sensors.

Our weapons are plastic M16 replicas (see Figure 19.4) gutted and fitted with electronics to facilitate tracking. An orientation sensor fitted in the barrel of the weapon (see Figure 19.5) measures the three-dimensional orientation of the weapon relative to Earth's magnetic field. The weapon emits a custom infrared signal upon firing, designed to avoid interference from ambient signals. Range is good to over 50 m, well beyond the size of the shoot house. The weapon is instrumented with a detector for the infrared signal for determining hits. All electronics are wired to a custom circuit board (see Figure 19.6). The circuitry details of our system can be found in Waller, Luck, Hoover, and Muth (2006).

Our helmets (see Figure 19.7) are constructed using many of the same parts. The electronic compass is stored in the top of the helmet and allows us to roughly track where a subject is looking. Four infrared detectors are used, one on each side of the helmet, to determine when a subject has been shot. The same circuit board used in the weapon is used in the helmet to control all the parts and communicate with the instructor operator station.

Our heart monitors (see Figure 19.8) use a standard electrocardiogram to measure heart activity. Individual heartbeats are detected onboard the device. The time between heartbeats is then used to compute heart-rate variability (Hoover & Muth, 2004). Heart-rate variability gives a longer-term measure related to the state of autonomic arousal of the subject, while heart rate gives the more familiar

Figure 19.4.   Custom Laser-Tag-Type Weapon

Figure 19.5.   Orientation Sensor in Weapon Barrel

Figure 19.6.   Custom Circuit Board in Weapon Stock

Figure 19.7.   Custom Helmet for the Laser-Tag System

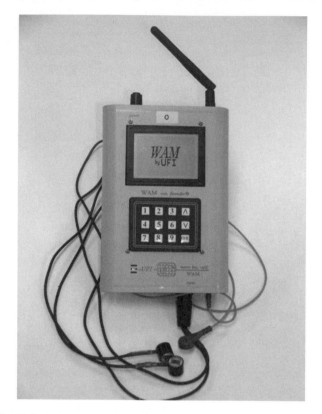

Figure 19.8.   Wireless Heart-Rate Monitor

shorter-term measure related to physical activity. We expect the data measured by this device to be noisy, as the subject will be mobile and active. Therefore the monitor includes methods to overcome errors in heartbeat detection, while still being able to accurately measure heart-rate variability (Rand et al., 2007).

Lessons Learned

- MILES gear is not well published; neither is laser-tag equipment in general. It is difficult to find schematics (ours is now in the literature; Waller et al., 2006). The laser-tag gaming industry does not produce equipment suitable for research and military training.

- Wireless connections should be handled by a standard, such as 802.11. Off-the-shelf chips are available for use in embedded systems without the need to design custom radio frequency methods. Essentially, wireless connections have become a commodity. For building-sized distances, 802.11 is ideal.

- Orientation sensing for small embedded systems is a relatively new endeavor and is not standardized yet (Waller, Hoover, & Muth, 2007). Until there is a standard method for performance evaluation, engineers and scientists should be wary of commercially published specifications for orientation tracking,

- Instrumentation should be worn on a helmet, a vest, and a weapon. Although the torso does not necessarily need to be tracked independently of other body parts, it is the most commonly fired-upon body part. It is useful to track head and weapon orientations of trainees.

CONCLUSION

In an ideal situation, it would be possible to track the actions of soldiers as they clear buildings in active combat. Of course, the conditions of the task leave no time or concern for instrumentation for data collection. It has already become standard practice for trainees to practice building clearing exercises in mock towns and buildings. This practice is now being extended by instrumenting these facilities in order to record the exercises. To date, there is very little literature published concerning the construction and instrumentation of shoot houses, or facilities where building clearing exercises are conducted. Our goal with this chapter was to document our facility and the lessons we learned in the hope of changing this trend.

REFERENCES

Hoover, A., & Olsen, B. (1999, May). A real-time occupancy map from multiple video streams. *Proceedings of IEEE International Conference on Robotics and Automation* (pp. 2261–2266). Washington, DC: IEEE.

Hoover, A., & Muth, E. (2004). A real-time index of vagal activity. *International Journal of Human-Computer Interaction, 17*(2), 197–209.

Olsen, B., & Hoover, A. (2001). Calibrating a camera network using a domino grid. *Pattern Recognition, 34*(5), 1105–1117.

Rand, J., Hoover, A., Fishel, S., Moss, J., Pappas, J., & Muth, E. (2007). Real-time correction of heart interbeat intervals. *IEEE Transactions on Biomedical Engineering, 54*, 946–950.

Waller, K., Hoover, A., & Muth, E. (2007). Methods for the evaluation of orientation sensors. *Proceedings of the 2006 World Congress in Computer Science Computer Engineering and Applied Computing* (pp. 141–146).

Waller, K., Luck, J., Hoover, A., & Muth, E. (2006). A trackable laser tag system. *Proceedings of the 2006 World Congress in Computer Science Computer Engineering, and Applied Computing* (pp. 416–422).

Part VI: Relevance of Fidelity in Training Effectiveness and Evaluation

Chapter 20

IDENTICAL ELEMENTS THEORY: EXTENSIONS AND IMPLICATIONS FOR TRAINING AND TRANSFER

David Dorsey, Steven Russell, and Susan White

In this chapter, we consider the issue of training transfer from an identical elements theory perspective. When organizations invest time and money in training, they do it for a specific reason—to realize a return on investment in terms of increased performance. However, even assuming task- or job-relevant training, there are no guarantees that what trainees learn in a training context will help them perform better on the job. This is the "transfer of training" problem that remains unresolved after 100 years of discussion in the academic literature (see Barnett & Cici, 2002; Cox, 1997). One of the most fundamental topics in this debate, and the one of central interest in this chapter, is how to construct a training environment for maximum transfer. In particular, we consider the implications of this debate on training that occurs in virtual environments (VEs).

Our starting point is the seminal work of Thorndike and Woodworth (1901a, 1901b, 1901c) and the idea of identical elements. The basic tenet of this theory is that transfer is facilitated to the extent that the training environment matches the performance environment in terms of physical features, cognitive requirements, and the like. The more dissimilar the features, the more transfer will be degraded. The theory has great intuitive appeal, and its basic approach guides many modern theories of instructional design.

From its conception, identical elements theory was much different from the theories that prevailed at the time of Thorndike and Woodworth's research. For example, their work appeared at a time when the "formal" or "mental discipline" model was popular, arguing that people can strengthen their mental capacities (that is, general cognitive capabilities) by exercising their brains. By learning difficult subject matter, such as Latin, one's capacity for doing other difficult things—performing difficult jobs—would presumably be improved. Despite its intuitive appeal, little research evidence supports the idea that performance on a task can be improved by improving on an unrelated one.

Identical elements theory was also a marked departure from the perspectives of the Gestalists. The idea of dividing a training environment into "elements" is inconsistent with the view that the whole is greater than the sum of its parts. Of course, putting identical elements theory in a Gestalt context does remind us to consider that it may be combinations/groups of features that need to be consistent between training and performance environments—and it may be that some elements are not separable from others or that some elements may cue others spontaneously (Cormier, 1984).

When we describe the training environment and its essential features, we should clarify that we are referring to the features that contribute to creating a particular psychological or perceptual state within learners. Note that in Thorndike's original works (for example, Thorndike & Woodworth, 1901a), the similarities of interest between training and performance environments were those between tasks or "mental habits"—not just the physical environment. Thus, in identifying the elements that should be matched, we focus on those that create a similarity in cognitive functioning between the two environments, as well as those that re-create the physical environment. Thinking of identical elements theory only in terms of matching features of the physical environment, and failing to distinguish between the physical and psychological features of the environment, substantially limits the depth and applicability of the theory. A wide variety of environmental dimensions impact the degree of transfer between learning and performance environments (see Barnett & Cici, 2002, for a comprehensive taxonomy of transfer dimensions).

MODERN EXTENSIONS OF IDENTICAL ELEMENTS THEORY

Although the difficulty in operationalizing identical elements has been recognized since their introduction, at a minimum, most modern accounts of training transfer invoke the tenets of identical elements theory as a jumping-off point. Fortunately, training research has progressed to a point of suggesting the transfer mechanisms necessary to overcome the absence of identical elements. In this chapter, we review these mechanisms in terms of how they extend identical elements theory, particularly as it applies to virtual environments.

Transfer mechanisms can be arranged into three general categories: person based, design based, and situation based. This type of classification scheme is consistent with previous efforts to organize the transfer-of-training literature (for example, Baldwin & Ford, 1988). Examples of person based mechanisms include cognitive abilities, noncognitive traits, training motivation, and skills. The most relevant design based mechanism for a discussion of virtual environments concerns VE fidelity. Last, situation based mechanisms include characteristics of the social/training environment and levels of analysis (for example, individual training versus team training versus teams-of-teams training). Each of the above mechanisms extends identical elements theory and will be discussed in turn.

Person Based Transfer Mechanisms

Person based transfer mechanisms are individual characteristics (knowledge, skills, and abilities) that promote transfer in the absence of identical elements between learning and transfer contexts. A review of these mechanisms will assist us in building a profile of trainees most likely to demonstrate transfer. Cognitive capabilities are among the most robust of these transfer mechanisms (Chen, Thomas, & Wallace, 2005; Day, Arthur, & Gettman, 2001; Colquitt, LePine, & Noe, 2000; Holladay & Quiñones, 2003; Singley & Anderson, 1989). For example, Colquitt et al. meta-analyzed studies from the training motivation literature and reported a corrected $r = .43$ relationship between cognitive ability and transfer. This suggests that high ability trainees can more easily overcome gaps between learning and transfer environments.

The manner in which trainees symbolically represent knowledge also plays an important role in transfer. John R. Anderson and his Carnegie Mellon University colleagues reframed Thorndike's identical elements to refer to units of procedural and (to a lesser extent) declarative knowledge across settings (Anderson, Corbett, Koedinger, & Pelletier, 1995; Singley & Anderson, 1989). Specifically, transfer occurs when students have learned the same production rules (that is, *if-then*) in one setting that are required to succeed in a second setting. Without knowledge of the exact production rules required in a transfer context, cognitive skill can be demonstrated only to the extent that the learner manages to convert or translate known productions or declarative knowledge into new actions.

Many cognitive based training interventions seek to influence the thought processes of novices by exposing them to experts' knowledge representations of training tasks. Anderson's laboratory has produced an impressive body of evidence supporting its approach to computer based tutoring (based on the adaptive control of thought–rational [ACT-R] theory of learning; Anderson et al., 1995), which tutors students by giving feedback based on discrepancies between student behaviors and a cognitive model of the problem. In a related vein, the authors of this chapter are presently conducting research investigating the effect of feedback on performance across different U.S. Navy air defense warfare (ADW) scenarios; the feedback we are using, not unlike the ACT-R tutors, is based on comparisons of student behaviors to those of an "expert" computational model. The positive link between trainee/expert knowledge structure similarities and transfer has also been demonstrated by Day et al. (2001) in a complex video game task.

A number of noncognitive characteristics can also facilitate transfer. Self-efficacy, the belief one has in his or her ability to successfully complete a given task (Bandura, 1997), is perhaps the most frequently studied of these mechanisms (for example, Ford, Smith, Weissbein, Gully, & Salas, 1998; Mathieu, Martineau, & Tannebaum, 1993; Mitchell, Hopper, Daniels, George-Falvy, & James, 1994; Stevens & Gist, 1997). In their meta-analysis of the training literature, Colquitt et al. (2000) reported moderate-sized relationships between self-efficacy and transfer, including corrected correlations of $r = .47$ for pre-training self-efficacy and $r = .50$ with post-training self-efficacy. Thus, in our efforts to profile trainees most likely to transfer new skills to the work environment, two critical components include ability and *belief* in one's ability.

Ford et al. (1998) found support for a model that linked both self-efficacy (directly) and an "identical elements" learning strategy (indirectly) to transfer in an ADW training simulator. In a novel extension of identical elements theory, learners in this study were asked to choose the task complexity of their practice trials, with the understanding that the most complex practice trial would be the most similar to the final transfer task. This allowed Ford et al. to operationalize identical elements between practice and transfer tasks as simply the number of self-selected practice trials of maximum difficulty; students choosing more complex practice trials minimized the practice-performance gap and performed better during the final task.

Other noncognitive moderators of transfer include goal orientation (for example, Fisher & Ford, 1998; Heimbeck, Frese, Sonnentag, & Keith, 2003), personality traits (for example, Herold, Davis, Fedor, & Parsons, 2002), motivation (for example, Mathieu, Tannenbaum, & Salas, 1992; Colquitt et al., 2000), and such work "involvement" variables as job involvement and organizational commitment (for example, Colquitt et al., 2000). However, as Baldwin and Ford (1988) noted, the utility of identifying individual difference transfer mechanisms is limited by the fact that in most circumstances *all* employees in a given organizational unit must undergo training, not merely the employees most likely to demonstrate transfer. Unfortunately, an aptitude-treatment approach to personnel training—whereby individual trainees are matched to optimal training interventions (for example, low fidelity simulations or high fidelity simulations)—has received little research attention and still represents a "next frontier" for the instructional sciences.

A final person based transfer mechanism worth noting is the acquisition of general learning skills that foster the acquisition of other skills. Metacognition and adaptive behavior are two such skills. Metacognition (sometimes used interchangeably with "self-regulation") refers to an individual's knowledge of, and control over, his or her own thoughts (Flavell, 1979; Ford et al., 1998). An individual who is trained to monitor his or her own learning progress, for example, can more easily identify trouble areas and adjust learning strategies than trainees with less developed metacognitive skills. The advantage of metacognitive strategies is their wide applicability, which does not depend upon specific content or contexts. Although the value of metacognitive procedures for transfer has been pointed out in educational research (Cox, 1997; Pressley, Borkowski, & O'Sullivan, 1984), metacognitive based instructional strategies have been slow to catch on in the classroom (Moley et al., 1992). Recent research involving computer based learning tasks (Ford et al., 1998; Keith & Frese, 2005) reinforces the notion that metacognition promotes transfer and is likely to generalize to virtual environments.

In addition, current approaches to personnel training assume that ever-changing situational elements can be overcome by learning "adaptive" behavioral skills (for example, Bell & Koslowski, 2002; Pulakos, Arad, Donovan, & Plamondon, 2000; Pulakos, Schmitt, Dorsey, Hedge, & Borman, 2002). Individuals who successfully modify their behaviors in response to changing work conditions are more adaptive than those who do not modify their behaviors (or those who do modify their behaviors, but choose unsuccessful behaviors). Recent research has documented that adaptability relates to training performance (Lievens, Harris, & Van Keer, 2003) and other job performance criteria (Johnson, 2001). Although such a proposition has not been tested directly in the training literature, we suspect that individuals who can demonstrate adaptive performance will be more likely to transfer new skills beyond original learning contexts.

Design Based Transfer Mechanisms

Design based transfer mechanisms represent instructional design characteristics of the virtual environment that can be altered to promote transfer. Of course, several instructional design features may impact transfer of training (for example, use of advanced organizers in training, encouraging self-reflection as a learning tool, and goal setting techniques), but we will maintain our identical elements focus here and discuss fidelity between the virtual learning and performance environments. Blade and Padgett (2002) defined fidelity as the "degree to which a VE . . . duplicates the appearance and feel of operational equipment (i.e., physical fidelity) and sensory stimulation (i.e., functional fidelity) of the simulated context" (p. 19). The transfer context is typically a real world setting (for example, the cockpit of an actual military aircraft); however, in certain instructional pipelines or sequences, VE training may be a precursor to training in a next-step, higher fidelity simulator.

According to identical elements theory, high fidelity training environments should facilitate transfer that is greater than low fidelity environments because the former more closely approximates the situational characteristics of the transfer environment (for example, ergonomic design, visual or auditory features, time pressure, and distractions). For example, there is considerable evidence in conventional training and educational research that physical differences between the learning and transfer contexts affect transfer in a negative way (for example, Ceci, 1996; Chen & Klahr, 1999; Rovee-Collier, 1993; Spencer & Weisberg, 1986). Despite widely held views to the contrary, however, there is little evidence available to support a relationship between the degree of the physical fidelity of VEs and transfer success (Koonce & Bramble, 1998; Lathan, Tracey, Sebrechts, Clawson, & Higgins, 2002; Salas, Bowers, & Rhodenizer, 1998). In fact, there is some evidence to suggest that distorting or augmenting the visual capabilities of VEs, in ways that no longer mirror real world parameters, can improve transfer (Dorsey, Campbell, & Russell, in press). Lathan et al. suggest that the primary benefit of high physical fidelity lies in motivating trainees: VEs with high tech appeal have greater "face" validity and therefore will be more likely to engage trainees.

Functional or psychological fidelity, however, does appear to be in play for training within virtual environments. In a convincing demonstration of the power of identical elements, Lathan et al. (2002) described a program of research demonstrating that participants who learn routes using VEs develop "orientation specificity." Participants tested on routes aligned with their original training were successful, but struggled to transfer route-learning skills when the required route was in a direction opposite (contra-aligned) to what had been learned in virtual training. Rizzo, Morie, Williams, Pair, and Buckwalter (2005) described an ongoing research program examining the impact of what might be termed "emotional" fidelity on training outcomes. Specifically, Rizzo et al. were applying state-dependent learning principles (Overton, 1964) to test whether stress induced during virtual reality training could improve transfer to similarly stressful military environments. More research is needed to determine the strength of the relationship between degree of functional fidelity and transfer success, but at least some research does suggest that overlap in the functional/psychological elements of VEs promotes transfer.

Situation Based Transfer Mechanisms

Situation based transfer mechanisms represent social characteristics of either the training or transfer environment that can be altered to promote transfer. In conventional training studies, a number of researchers have examined perceptions of the post-training work environment as facilitators of transfer, including organizational culture, transfer climate, and supervisory support (for example, Facteau, Dobbins, Russell, Ladd, & Kudisch, 1995; Tracey, Tannenbaum, & Kavanagh, 1995; Smith-Crowe, Burke, & Landis, 2003). The social context of the transfer environment can determine whether skills that are learned in training will be supported and maintained or will be extinguished. Rouiller and Goldstein (1993) outlined several facets of an organization's climate that promote transfer, including goal cues, social cues, task and structural cues, and feedback. Tracey et al. found empirical support for a model linking both transfer climate and an organizational culture of continuous learning (that is, a culture that places high value on social support, continuous innovation, and competitiveness) to transfer among a managerial population. Although post-training environment has not been systematically studied in the VE literature, we believe that an unsupportive training climate can undermine the effectiveness of any training intervention.

Levels of Analysis

Last, levels of analysis can impact how both person based and situation based mechanisms operate. Chen et al. (2005) conducted one of the few existing studies examining how individual characteristics and processes influence transfer at both the individual and team levels. Using a low fidelity flight simulator, they found that task knowledge and skill had a greater influence on transfer at the individual level than at the team level, whereas the impact of efficacy was greater at the team level. Thus, previously conflicting findings from individual and team training research may be better understood using a multilevel perspective. Regarding situation based mechanisms, levels of analysis might moderate the impact of organizational climate on transfer. Because transfer climates (and cultures of continuous learning) probably differ among work groups within organizations, as well as across organizations, Baldwin and Ford (1988) speculated that the same training program conducted in different work groups or organizations might result in different degrees of transfer. Until a greater number of multilevel studies of transfer accumulate, however, the influence of levels factors on transfer will remain largely speculative. Regardless, consideration of identical elements at multiple levels of analysis may be fruitful for both theory and practice. For example, a VE simulator may be quite similar to the actual performance environment at the level of individual performers, yet lack critical features at a team or group level that impact transfer.

CONCLUSION AND IMPLICATIONS FOR RESEARCH AND PRACTICE

Although some elements of Thorndike's original identical elements theory were not fully explicated in terms of understanding the range of cognitive, social, and emotional elements and mechanisms that can impact transfer, training research and practice are unlikely to escape the fundamental tenets of identical elements. Completely identical training and performance environments may be an unrealistic goal, but, as suggested by Cox (1997), the practice of decomposing training and transfer situations into their constituent elements is fundamental to both developing effective learning interventions and conducting experimental science. Understanding similarity in elements may be even more important in VE training domains, as little is understood about transfer in such settings. For example, the kinds of knowledge or skills that are best suited to VE training has received little, if any, research attention. Declarative knowledge (that is, facts and figures) does not seem to be as natural a fit to VE training as procedural knowledge (that is, how-to knowledge), but what about spatial or sensorimotor skills, vigilance, memory, or complex problem solving? Van Buskirk, Cornejo, Astwood, Russell, Dorsey, and Dalton (Volume 1, Section 1, Chapter 6) present a framework for beginning to address these questions by mapping a taxonomy of training interventions to a complementary taxonomy of learning objectives.

To facilitate future work on identical elements in virtual settings, we offer a few summary thoughts and ideas regarding future research. First, as discussed above, any simulated environment, including a virtual one, can be characterized along various dimensions of fidelity (for example, physical fidelity and functional fidelity). From a design perspective, the fidelity of various features and elements represents potential points of correspondence between training and performance environments. Second, various factors known to moderate training effectiveness —be they cognitive, skill, social, affective, or environmental—must be considered alongside issues of identical elements. Such factors may act to amplify or attenuate the effects of identical elements discrepancies on transfer. Third, there is currently a dearth of research on many of the issues highlighted here. Further theory development and empirical research is needed, including measurement models and approaches to assess fidelity and identical elements correspondence in a multifaceted/multilevel manner. Research to inform the choice of VE based instructional techniques and strategies, in order to maximize identical elements and transfer across a wide variety of domains, does not currently exist.

By specifying the theory of identical elements, Thorndike and Woodworth (1901a, 1901b, 1901c) provided an important and foundational perspective on training and performance environments and related issues of transfer. By continuing to reflect upon their ideas, while considering modern extensions of identical elements, researchers and practitioners have much to gain in designing learning interventions that optimize transfer and promote learning that is reflected in the real world.

REFERENCES

Anderson, J. R., Corbett, A. T., Koedinger, K. R., & Pelletier, R. (1995). Cognitive tutors: Lessons learned. *The Journal of the Learning Sciences, 4,* 167–207.

Baldwin, T. T., & Ford, J. K. (1988). Transfer of training: A review and directions for future research. *Personnel Psychology, 41,* 63–105.

Bandura, A. (1997). *Self-efficacy: The exercise of control.* New York: Freeman.

Barnett, S. M., & Cici, S. J. (2002). When and where do we apply what we learn? A taxonomy for far transfer. *Psychological Bulletin, 128*(4), 612–637.

Bell, B. S., & Koslowski, S. W. J. (2002). Adaptive guidance: Enhancing self-regulation, knowledge, and performance in technology-based training. *Personnel Psychology, 55,* 267–306.

Blade, R. A., & Padgett, M. L. (2002). Virtual environments standards and technology. In K. M. Stanney (Ed.), *Handbook of virtual environments: Design, implementation, and applications.* Mahwah, NJ: Lawrence Erlbaum.

Ceci, S. J. (1996). *On intelligence: A bioecological treatise on intellectual development.* Cambridge, MA: Harvard University Press.

Chen, Z., & Klahr, D. (1999). All other things being equal: Acquisition and transfer of the control of variables strategy. *Child Development, 70,* 1098–1120.

Chen, G., Thomas, B., & Wallace, J. C. (2005). A multilevel examination of the relationships among training outcomes. *Journal of Applied Psychology, 90,* 827–841.

Colquitt, J. A., LePine, J. A., & Noe, R. A. (2000). Toward an integrative theory of training motivation: A meta-analytic path analysis of 20 years of research. *Journal of Applied Psychology, 85,* 678–707.

Cormier, S. (1984). *Transfer of training: An interpretive review* (Technical Report No. 608). Alexandria, VA: Army Research Institute for the Behavioral and Social Sciences.

Cox, B. D. (1997). The rediscovery of the active learner in adaptive contexts: A developmental-historical analysis of transfer of training. *Educational Psychologist, 32*(1), 41–55.

Day, E. A., Arthur, W. Jr., & Gettman, D. (2001). Knowledge structures and the acquisition of a complex skill. *Journal of Applied Psychology, 86,* 1022–1033.

Dorsey, D., Campbell, G., & Russell, S. (in press). Adopting the instructional science paradigm to encompass training in virtual environments. *Theoretical Issues in Ergonomic Science.*

Facteau, J. D., Dobbins, G. H., Russell, J. E., Ladd, R. T., & Kudisch, J. D. (1995). The influence of general perceptions of the training environment on pretraining motivation and perceived training transfer. *Journal of Management, 21,* 1–25.

Fisher, S. L., & Ford, J. K. (1998). Differential effects of learner effort and goal orientation on two learning outcomes. *Personnel Psychology, 51,* 397–420.

Flavell, J. H. (1979). Metacognition and cognitive monitoring: A new area of cognitive-developmental inquiry. *American Psychologist, 34,* 906–911.

Ford, J. K., Smith, E. M., Weissbein, D. A., Gully, S. M., & Salas, E. (1998). Relationships of goal orientation, metacognitive activity, and practice strategies with learning outcomes and transfer. *Journal of Applied Psychology, 83,* 218–233.

Heimbeck, D., Frese, M., Sonnentag, S., & Keith, N. (2003). Integrating errors into the training process: The function of error management instructions and the role of goal. *Personnel Psychology, 56,* 333–361.

Herold, D. M., Davis, W., Fedor, D. B., & Parsons, C. K. (2002). Dispositional influences on transfer of learning in multistage training programs. *Personnel Psychology, 55,* 851–869.

Holladay, C. L., & Quiñones, M. A. (2003). Practice variability and transfer of training: The role of self-efficacy generality. *Journal of Applied Psychology, 88,* 1094–1103.

Johnson, J. W. (2001). The relative importance of task and contextual performance dimensions to supervisor judgments of overall performance. *Journal of Applied Psychology, 86,* 984–996.

Keith, N., & Frese, M. (2005). Self-regulation in error management training: Emotion control and metacognition as mediators of performance effects. *Journal of Applied Psychology, 90,* 677–691.

Koonce, J. M., & Bramble, W. J., Jr. (1998). Personal computer-based flight training devices. *International Journal of Aviation Psychology, 8,* 277–292.

Lathan, C. E., Tracey, M. E., Sebrechts, M. M., Clawson, D. M., & Higgins, G. A. (2002). Using virtual environments as training simulators: Measuring transfer. In K. M. Stanney (Ed.), *Handbook of virtual environments: Design, implementation, and applications.* Mahwah, NJ: Lawrence Erlbaum.

Lievens, F., Harris, M. M., Van Keer, E., & Bisqueret, C. (2003). Predicting cross-cultural training performance: The validity of personality, cognitive ability, and dimensions measured by an assessment center and a behavior description interview. *Journal of Applied Psychology, 88,* 476–489.

Mathieu, J. E., Martineau, J. W., & Tannebaum, S. I. (1993). Individual and situational influences on the development of self-efficacy: Implications for training effectiveness. *Personnel Psychology, 46,* 125–147.

Mathieu, J. E., Tannenbaum, S. I., & Salas, E. (1992). Influences of individual and situational characteristics on measures of training effectiveness. *Academy of Management Journal, 35,* 828–847.

Mitchell, T. R., Hopper, H., Daniels, D., George-Falvy, J., & James, L. R. (1994). Predicting self-efficacy and performance during skill acquisition. *Journal of Applied Psychology, 79,* 506–517.

Moley, B. E., et al. (1992). The teacher's role in facilitating memory and study strategy development in the elementary school classroom. *Child Development, 63,* 653–672.

Overton, D. A. (1964). State-dependent or "dissociated" learning produced with pentobarbital. *Journal of Comparative Physiological Psychology, 57,* 3–12.

Pressley, M., Borkowski, J. G., & O'Sullivan, J. T. (1984). Memory strategy instruction is made of this: Metamemory and durable strategy use. *Educational Psychologist, 19,* 94–107.

Pulakos, E. D., Arad, S., Donovan, M. A., & Plamondon, K. E., (2000). Adaptability in the workplace: Development of a taxonomy of adaptive performance. *Journal of Applied Psychology, 85,* 612–624.

Pulakos, E. D., Schmitt, N., Dorsey, D. W., Hedge, J. W., & Borman, W. C. (2002). Predicting adaptive performance: Further tests of a model of adaptability. *Human Performance, 15,* 299–324.

Rizzo, A., Morie, J. F., Williams, J., Pair, J., & Buckwalter, J. G. (2005). Human emotional state and its relevance for military VR training. *Proceedings of the 11th International Conference on Human Computer Interaction.*

Rouiller, J. Z., & Goldstein, I. L. (1993). The relationship between organizational transfer climate and positive transfer of training. *Human Resource Development Quarterly, 4,* 377–390.

Rovee-Collier, C. (1993). The capacity for long-term memory in infancy. *Current Directions in Psychological Science, 2,* 130–135.

Salas, E., Bowers, C. A., & Rhodenizer, L. (1998). It is not how much you have but how you use it: Toward a rational use of simulation to support aviation training. *International Journal of Aviation Psychology, 8,* 197–208.

Singley, M. K., & Anderson, J. R. (1989). *The transfer of cognitive skill.* Cambridge, MA: Harvard University Press.

Smith-Crowe, K., Burke, M. J., & Landis, R. S. (2003). Organizational climate as a moderator of safety knowledge-safety performance relationships. *Journal of Organizational Behavior, 24,* 861–876.

Spencer, R. M., & Weisberg, R. W. (1986). Context-dependent effects on analogical transfer. *Memory & Cognition, 14,* 442–449.

Stevens, C. K., & Gist, M. E. (1997). Effects of self-efficacy and goal-orientation training on negotiation skill maintenance: What are the mechanisms? *Personnel Psychology, 50,* 955–978.

Thorndike, E. L., & Woodworth, R. S. (1901a). The influence of improvement in one mental function upon the efficiency of other functions. *Psychological Review, 8,* 247–261.

Thorndike, E. L., & Woodworth, R. S. (1901b). The influence of improvement in one mental function upon the efficiency of other functions: The estimation of magnitudes. *Psychological Review, 8,* 384–395.

Thorndike, E. L., & Woodworth, R. S. (1901c). The influence of improvement in one mental function upon the efficiency of other functions: Functions involving attention, observation, and discrimination. *Psychological Review, 8,* 553–564.

Tracey, J. B., Tannenbaum, S. I., & Kavanagh, M. J. (1995). Applying trained skills on the job: The importance of the work environment. *Journal of Applied Psychology, 80,* 239–252.

Chapter 21

ASSESSMENT AND PREDICTION OF EFFECTIVENESS OF VIRTUAL ENVIRONMENTS: LESSONS LEARNED FROM SMALL ARMS SIMULATION[1]

Stuart Grant and George Galanis

Fielded small arms appear to be reaching the limits of development (Jane's Information Group, 2006/2007), and the operational environment facing Western militaries continues to increase in complexity. A greater likelihood of close quarters battle, the more difficult friend-versus-foe discriminations, and the wider presence of noncombatants on the battlefield increase performance demands. Training and training technologies are one avenue for meeting the challenge.

Virtual environments (VEs), as represented by the current generation of simulators for the training of marksmanship skills, are readily available as training solutions. Indeed, there are a number of commercially available marksmanship simulators in existence based on commercial off-the-shelf components. Such devices appear to offer significant cost savings compared to expensive-to-operate live ranges (English & Marsden, 1995). In addition, the simulators also offer unprecedented levels of safety given they do not employ live ammunition. Training is not subject to adverse weather conditions, and because simulators are instrumented extensively, there are possibilities for coaching and feedback to trainees that are not available in live ranges.

Compared to flight simulators, rifle-range simulators appear to be relatively simple environments. Rifle ranges do not have the complex systems of an aircraft. The simulation of a rifle and its ballistics is simpler than the complex systems of a modern aircraft and its flight dynamics. However, researchers evaluating small arms simulators have been perplexed at the difficulty they have encountered in finding quantitative evidence of transfer of training or significant levels of correlation between marksmanship skills in the live range to performance in the simulators. To find greater levels of transfer and closer relationships between live and simulated fire, this chapter argues that higher levels of marksmanship training and more knowledge of how humans employ live and simulated small arms are required.

ASSESSING TRANSFER OF TRAINING

In evaluating a simulator for marksmanship training, the transfer of training to live-fire performance is the principal criterion. Although criteria for assessing a training device are naturally influenced by the benefits stakeholders seek from the device (for example, increased safety, reduced environmental impact, lower operating cost, or smaller footprint), if it cannot be demonstrated that the device contributes to successful live-fire performance, the other criteria are moot. Kirkpatrick (1959) identified trainee reactions, knowledge obtained in training, subsequent performance in the operational environment, and the impact on overall organizational performance as possible criteria. However, meta-analysis (Alliger, Tannenbaum, Bennett, Traver, & Shotland, 1998) indicates that neither evaluation of the training by the trainees nor the amount of knowledge acquired during training correlated strongly with subsequent performance on the job ($r \leq 0.21$). If the goal is to determine the effect of training on performance in the operational setting, then it should be assessed directly.

In assessing transfer, the existing assessment tests employed by the target training organization are very valuable performance metrics because they permit comparison against historical data and have inherent meaning and validity to the training organization that will strongly influence their acceptance of the device. However, these tests may incorporate factors that, while certainly relevant to effectively employing a rifle, are not strictly marksmanship per se. For an accurate evaluation, careful consideration must then be given to the requirements of the training device and how it is used. For example, the Canadian Personal Weapons Test—Level 3, used for assessing infantry use of the C7A1 assault rifle, includes a "run down" serial (Department of National Defence, 1995) that requires the firer wearing fighting gear to run 100 meters between successive timed target exposures. The physical fitness of the firer will certainly affect the soldier's score. Whether and how a marksmanship training device should train physical fitness pertaining to marksmanship should be established first to frame the assessment. In addition, many military marksmanship tests count hits of targets as the measure of marksmanship. This provides a single binary data point for each round fired. Although that is a meaningful result for combat, it is a relatively impoverished way to score the result. For these reasons, the collection of additional measures is worthwhile. Among other possible measures, constant and variable error of impact point from the target's center have desirable properties (Johnson, 2001; Taylor, Dyer, & Osborne, 1986). As continuous variables, they provide more information than binary scoring, and being measured on a ratio scale, they can support various summary and inferential statistics. Finally, they correspond to zeroing and grouping aspects of marksmanship and so have inherent meaning to the subject matter.

Performance during skill acquisition is governed by the learner's evolving supporting knowledge base. Initially performance is based largely on declarative knowledge that is more readily communicated verbally, but further practice of the skill results in the chaining together of the initially separate components of performance, until ultimately a smooth, automatic level of performance is reached (Anderson, 1983; Fitts, 1964). In applying this model of skill acquisition to marksmanship, Chung, Delacruz, de Vries, Bewley, and Baker (2006) noted that different training devices could support different stages of marksmanship skill. This suggests not only that different types of training devices can support expert performance by supporting different stages of performance, but that estimates of the efficacy of those devices may differ depending on the stage of skill attained when the estimate is made.

PREDICTION

Accurate predictions by a simulation are desirable for the purpose of validation and effective employment. The correspondence between results people obtain in the simulated and live environments speak to the validity of the simulation. Furthermore, a training simulator that accurately predicts live performance allows the live training and testing to be scheduled only when the trainees are ready.

The ability to predict live-fire performance from results obtained in marksmanship simulators is limited (see Chung et al., 2006, for a good set of references). Correlations between performance in marksmanship simulators and live fire are typically in the $r = 0.3$ to 0.4 range, usually accounting for less than 20 percent of the variance in live-fire scores. This is typical across various types of simulators. Simulators with dedicated simulator weapons that use lasers have provided correlations with live-fire scores ranging from 0.01 and 0.45 (Filippidis & Puri, 1999); 0.02 and 0.28 (Filippidis & Puri, 1996); 0.4 (Gula, 1998); 0.41 (Yates, 2004); to 0.68 (Hagman, 1998). A training device employing a laser insert for regular service weapons has achieved similar correlations, ranging from 0.16, 0.24, 0.55 (Smith & Hagman, 2000, 2001) to 0.5 and 0.55 (Smith & Hagman, 2003).

A surprising outcome of research looking for correlations between marksmanship performance in simulators and the live range is that performance in simulators appears to be worse (Filippidis & Puri, 1996; Yates, 2004). This is surprising as anecdotal evidence based on face validity suggests that recoil in simulators is significantly less and that the lack of live ammunition in simulators should make the simulator a less stressful environment—hence the expectancy is that simulators should induce superior marksmanship performance. Further investigations into this effect indicate that pixilation of targets in the simulator degrade marksmanship performance. When targets on the live range were modified to exhibit pixilation similar to that present in a simulator, marksmanship performance degraded by the same amount found in simulated conditions (Temby, Ryder, Vozzo, & Galanis, 2005). This finding suggests that eye-limited resolution of targets is a necessary requirement for simulators that are to be used for prediction of live-fire performance.

Predictions that can account for substantial amounts of variance in live-fire scores (especially up to 46 percent) can have practical value in screening trainees for live-fire training or testing (Smith & Hagman, 2003). It is worth noting, however, that questionnaires regarding affect and knowledge have shown equivalent predictive power (Chung et al., 2006).

TRANSFER OF TRAINING

Researchers attempting to find transfer of skill acquired in marksmanship simulators to live firing have often focused their attention on measuring the ability of the devices to train for performance on defined rifle marksmanship qualification tests. This approach achieves direct relevance to the military client's training requirement and exploits the underlying validity of the qualification test. The typical control group is the standard method of instruction. Obtaining solid support for the training effect has proven surprisingly elusive.

Support is often based on the finding of equivalent live-fire test scores for those trained on the device of interest and those trained in the conventional manner. However, this approach is dependent on the statistical power available for the comparison. White, Carson, and Wilbourn (1991) substituted marksmanship simulator training for the dry fire and sighting exercises used for U.S. Air Force Security Police weapons training. Their results showed no overall difference between the simulator-trained group and the control group, although the trainees with less prior weapons experience achieved higher scores if they received simulator training. The relatively large sample size ($n = 247$) provided the basis for good statistical power, making the claim of training equivalence convincing. The treatment effect was weak, however. The experimental manipulation being 30 minutes of conventional training versus 10–20 minutes of simulator training did not appear to provide a great training effect. Both the control and experimental groups achieved low, failing scores on the post-training, live-fire practice test, but then more than doubled their scores when the test was immediately repeated for qualification purposes.

Hagman (2000) found significant benefits of using a laser insert device (Laser Marksmanship Training System) over a suite of other training devices in grouping, zeroing, and known range firing. These were the tasks actually trained on the devices. The experimental group's advantage was not repeated on other tasks that comprised the marksmanship course. Both the control and experimental groups performed well on the live record fire test, with no significant difference between them.

Yates (2004) also found equivalence between one platoon trained using a dedicated laser based simulator and another trained with dry fire. Both platoons achieved success on the final qualification test, although inclement weather experienced by the experimental group on the range could have suppressed a training benefit. Comparing with soldiers trained using live fire, English and Marsden (1995) detected no difference to scores of soldiers trained with a dedicated laser system. Testing similar simulator technology, Grant (2007) found soldiers trained entirely in simulation could obtain qualification results that were successful and indistinguishable from those trained entirely with live fire and that significantly superior results were found with an equivalent amount of training using a mix of live and simulated fire.

CHALLENGES IN ASSESSING SIMULATIONS FOR MARKSMANSHIP TRAINING

Marksmanship scores encountered in the assessment of training devices typically show a large amount of error variance (Torre, Maxey, & Piper, 1987). Attempts to attribute scores in a live-fire environment to prior experience in a simulation environment or to predict live-fire scores on the basis of performance on a training device must contend with the fact that subject performance is unstable, as one would expect from various skill acquisition theories (Anderson, 1983; Fitts, 1964). Studies using live-fire training for live-fire testing as a control group typically show only a weak consistency in scores. Torre et al. found correlations between live-fire sessions of 0.3 and 0.54. Over a one-year interval the correlation between successive live-fire tests has been found to be 0.37 (Smith & Hagman, 2000). Indeed, Hagman (2005) examined 180 trainees firing a 40-round test and found that the score achieved after 20 rounds had been fired accounted only for less than 70 percent of the variance of the final score.

Assessments of marksmanship training devices are frequently hampered by the limitations imposed by the subject matter (the use of deadly weapons) and the subject populations (transient military personnel). Ideally, transfer of training studies provide a pre-test prior to any training to provide assurance that there are no (or at least a basis for controlling for) preexisting differences among the experimental groups that could be mistaken for a differential training effect. This is not feasible if the training audience is without any prior experience with firearms. Limiting the control group's training to that required to safely discharge the weapon (Boyce, 1987) may be the condition closest to the experimentalist's ideal.

FIDELITY IN MARKSMANSHIP SIMULATION

Training technologies continue to increase in power. For example, Moore's law (Moore, 1965) predicts a doubling of computational power every two years; theoretical predictions appear to suggest that this increase should continue for the coming decade, and some futurists suggest that this trend may continue in new forms beyond that period (for example, Moore, 1995). The implication is that new technology may provide new possibilities and modes of training delivery—including new types of VEs.

One of the simplest ways to apply new VEs to training systems is to substitute existing live-training systems with new VEs. This approach to VE design has been common in the past and will arguably continue to be prevalent in the near future. In such a paradigm, the simulator designer's role is reduced to analyzing an apparently functional live training environment and replicating that functional environment with a more cost-effective VE. Instructional staff and trainees are already familiar with the existing live environment, and hence there is no requirement to make significant changes to instructional and learning techniques once the new VE is introduced into service. This evolutionary approach places the emphasis of the VE design on the analysis of the existing live training environment. As such, the main disciplines for the analysis of the live environment and synthesis of the VEs replicating the physical are from the physical sciences and engineering. This approach minimizes—but does not completely eliminate—the requirement for detailed costly studies of human learning and instructional techniques.

Determining Fidelity Requirements for Simulators

When considering VE fidelity requirements, a typical goal is to learn what elements of the environment must be simulated and to what degree of fidelity for the task to be trained. This question is framed by Lintern (1996) in the following form:

> For instruction of high-speed, low-level flight in a fixed wing aircraft, a simulated visual scene with a Field of View of $w°$ by $h°$, a frame rate of $f Hz$ and a scene content level of s *units* when placed on a simulator cockpit of c *type* can substitute for $n\%$ of the hours normally required in the aircraft to reach proficiency. The time required in the simulator to achieve proficiency is t *hours.*

Although Lintern's (1996) framing of the simulation fidelity problem is stated in terms of flight simulation and is limited to issues related to out-the-window scenery, such an approach can be translated into statements for other tasks and VEs. Given the apparently extensive knowledge of marksmanship, it would appear a relatively straightforward matter to list the issues identified in marksmanship, and, using the knowledge of one or more subject matter experts, produce a statement of performance requirements for a small arms trainer. For example, one could refer to such documents as an army marksmanship pamphlet and begin listing the major considerations in marksmanship (for example, Department of National Defence, 1994). Such factors as grouping requirements, target clarity, wind, and lighting requirements could be included. Similarly, design engineers could also refer to the design manuals for particular rifles (and other documents) to ascertain the operational characteristics of a rifle to determine the type of projectile, caliber of the rifle, characteristics of the firing mechanisms, and the functioning of the sights. Once a training system was designed, the evaluation process of the training devices could then include comparison of the training device to the specifications, as well as the subject assessment of the complete operation by expert marksmen.

However, it appears that there are problems with the design approach discussed above and that such problems persist even though there has been research for several decades in this area. It has been suggested by a number of researchers that the design process places an overreliance on subject matter experts for developing the statement of requirements, and evaluation of the VEs, and that both aspects continue to have an overreliance on face validity as a measure of suitability of the design (Salas, Milham, & Bowers, 2003). We are not disputing the requirement for subject matter experts in the design and evaluation of such systems, but we suggest that some of the shortfalls of current marksmanship simulators are occurring despite this design practice.

Limitations of Rational and Replication Approach

One of the reasons why overreliance on face validity and subjective assessment of a training system occurs is that there is a disconnect between the ability to verbalize how psychomotor skills are performed and the tendency of humans to confabulate explanations when asked how such skills are performed. Hence even the subject matter experts may not be aware that they are performing the task in a manner quite different from the way in which they verbalize the task. So, for example, when marksmen are asked how they perform the task, they may report recognizing the target, positioning and holding the weapon, pointing the weapon toward the target, and then carefully releasing the projectile toward the target. However, how exactly a target was "recognized," the nature of the rates of movement involved in "pointing," and the way the trigger was activated are difficult to articulate; research would suggest these are not actually available to verbal consciousness.

Consider the apparently simple task of picking up a coin that is lying on a table. The broad parameters (similar to marksmanship) might include recognizing the coin to be picked up, positioning oneself close enough to reach the coin, and then reaching out to grasp the coin. However, even this apparently simple act has all sorts of complications. Research conducted by Westwood and Goodale (2003) investigated what was involved in picking up various shaped objects. One example considered how the apparent size of a coin changes when surrounded by other coins. If the surrounding coins are smaller than the central coin, then the central coin "looks" larger than it really is, whereas if the surrounding coins are larger, the central coin looks smaller than it really is. However, Westwood and Goodale also found that although experimental subjects could verbalize the apparent change in size of the coin to be picked up, when subjects actually reached for the coin, the reaching and grasping behavior (the psychomotor component of the task) did not reflect the verbal descriptions. Goodale speculates that there are possibly two different regions of the brain involved in such tasks—one region for the verbalization and recognition skills and the other for the psychomotor skills.

The academic research in coin-reaching experiments of the 1990s is reflected in real applications involving VEs. In the late 1950s and early 1960s considerable research effort was carried out relating to airplane accidents in the approach to landing at night. During the investigations as to how pilots actually perform this task, Lane and Cumming (1956) administered a questionnaire asking experienced airline pilots to indicate the geometry they believed that they used in performing the slant perception task during the approach to landing. To their astonishment, just over 50 percent of the responses were geometrically implausible, while another 25 percent of the subjects stated they did not know how they performed the task. Only 25 percent of the respondents indicated geometrically plausible explanations. Lane and Cumming concluded that the only way to design an improved landing system was through a lengthy process of analysis and evaluation of final task performance—and that expert opinion was limited.

Later research in approach to landing investigated the effects of artifacts of simulators in the approach to landing. So, for example, simulator displays are pixilated, the scenery is not displayed at eye-limited resolution, and textures in synthetic scenery are not as rich or dense as those found in the real world. A series of rigorous evaluations conducted by Lintern and a number of co-researchers in the 1980s demonstrated that the simulator artifacts create biases in pilots' slant perception, so there is a danger that pilots training in simulators will be incorrectly calibrating their perceptions while training in simulators (Lintern & Walker, 1991). The work in the field of approach to landings would then appear to apply some weight to the research in the field of visual perception (such as Westwood & Goodale, 2003) and that for psychomotor skills learning, VEs must be validated by empirical experiments, since expert verbalization cannot be relied upon to reveal the principles underlying human performance in complex psychomotor tasks.

The implication for VE design then is that asking a subject matter expert how a psychomotor task is performed may not reveal the actual learning underlying performance of the manual control part of a task. So then, an analysis of the task to be performed (flying an aircraft at a low level or aiming a weapon) must be based on empirical data, not solely on simple verbalizations. It requires models and evaluations of performance of the complete task as it is actually performed. Such analyses and evaluations are often time consuming and expensive, but the scientific literature would indicate that analyses and experimental validation are critical.

CONCLUSION

Marksmanship simulators have been used as part of successful marksmanship training programs. Their contribution to trainee success is not always easy to estimate, however, and the relationship between performance in the VE and in live fire is weak.

As these systems are refined, effort should be invested in obtaining more and better data regarding how the task is performed and what is learned. In particular, data should be collected on skilled firers whose performance shows little variable error within the rifle-firer system. Although the highest levels of expertise can take years to achieve (for example, Crossman, 1959), thereby making true experts difficult to find, using subjects who can demonstrate a high level of consistency across repeated live-fire tests will provide greater precision to developers and trainers. If reliable discrepancies can be found between the power of simulator and live-fire data to predict performance on a live-fire test, then researchers will be in a position to understand and overcome a simulator's limiting factors. This does not assume that live fire will be the most reliable predictor or even that achieving a comparable level of prediction with a simulator demonstrates that all the underlying factors in marksmanship have been captured in the simulator, but simply that a reliable research tool is available for evaluating simulator design.

Additionally, extensive data collection on the acquisition of marksmanship skill should be sought. Theory-driven data collection on novices, experts, and people transitioning between those levels should be used to complement marksmanship subject matter experts. These will inform decisions regarding visual resolution requirements, acceptable transport delays, and the type and precision of data required of the scoring systems.

Finally, improving prediction and demonstrable transfer of training of current marksmanship simulators is a significant challenge, and the challenge may be increasing. There is an emerging call for marksmanship training to explicitly and thoroughly address the highly dynamic, close quarters and built-up situations that are characteristic of current operations (Ellison, 2005). Current marksmanship simulators face significant obstacles in presenting these situations (Muller, Cohn, & Nicholson, 2004). These situations can be created for soldiers using live simulation, but marksmanship training was not the driving force behind the technologies used to instrument the soldiers and simulate their weapons fire. Nevertheless, the knowledge gained in overcoming the challenges to existing marksmanship trainers will go a long way toward solving them in the live domain.

NOTE

1. This chapter was originally published by the Government of Canada, DRDC Toronto Publications, Jack P. Landdt, Ph.D., Editor.

REFERENCES

Alliger, G. M., Tannenbaum, S. I., Bennett, W., Traver, H., & Shotland, A. (1998). *A meta-analysis of the relations among training criteria* (Rep. No. AFRL-HE-BR-TR-1998-0130). Brooks Air Force Base, TX: Air Force Research Laboratory.

Anderson, J. R. (1983). *The architecture of cognition.* Cambridge, MA: Harvard University Press.

Boyce, B. A. (1987). Effect of two instructional strategies on acquisition of a shooting task. *Perceptual and Motor Skills, 65,* 1003–1010.

Chung, G. K., Delacruz, G. C., de Vries, L. F., Bewley, W. L., & Baker, E. L. (2006). New directions in rifle marksmanship research. *Military Psychology, 18*(2), 161–179.

Chung, G. K., Delacruz, G. C., de Vries, L. F., Kim, J., Bewley, W. L., de Souza e Silva, A. A., Sylvester, R. M., & Baker, E. L. (2004). *Determinants of rifle marksmanship performance: Predicting shooting performance with advanced distributed learning assessments* (Rep. No. A178354). Los Angeles: UCLA CSE/CRESST.

Crossman, E. R. F. W. (1959). A theory of the acquisition of speed-skill. *Ergonomics, 2*(2), 153–166.

Department of National Defence. (1994). *The rifle 5.56 mm C7 and the carbine 5.56 mm C8* (Report No. B-GL-317-018 / PT-001). Ottawa, Ontario, Canada: Department of National Defence.

Department of National Defence. (1995). *Shoot to live: Part 1—Policy* (Report No. B-GL-382-002/FP-001). Ottawa, Ontario, Canada: Department of National Defence.

Ellison, I. W. (2005). *Current inadequacy of small arms training for all military occupational specialties in the conventional army* (Master's thesis; Rep. No. A425634). Fort Leavenworth, KS: U.S. Army Command and General Staff College.

English, N., & Marsden, J. (1995). *An evaluation of the training and cost effectiveness of SAT for recruit training* (Report No. DRA/CHS/HS3/CR95039/01). Farnborough, Hampshire, United Kingdom: Defence Research Agency.

Filippidis, D., & Puri, V. P. (1996). An analysis of Fire Arms Training System (FATS) for small arms training. In *Annual Meeting of TTCP HUM Technical Panel 2*. Toronto, Ontario, Canada: The Technical Cooperation Program.

Filippidis, D., & Puri, V. (1999, November). *Development of training methodology for F-88 Austeyr using an optimum combination of sim/live training*. Paper presented at the Land Weapons System Conference, Salisbury, South Australia.

Fitts, P. M. (1964). Perceptual-motor learning. In A. W. Melton (Ed.), *Categories of human learning* (pp. 243–285). New York: Academic Press.

Grant, S. C. (2007). *Small arms trainer validation and transfer of training: C7 Rifle* (Rep. No. TR 2007-163). Toronto, ON: Defence Research and Development Canada.

Gula, C. A. (1998). *FATS III combat firing simulator validation study* (DCIEM Rep. No. 98-CR-26). North York, Ontario, Canada: Defence and Civil Institute of Environmental Medicine.

Hagman, J. D. (1998). Using the engagement skills trainer to predict rifle marksmanship performance. *Military Psychology, 10*(4), 215–224.

Hagman, J. D. (2000). *Basic rifle marksmanship training with the laser marksmanship training system* (Research Rep. No. 1761). Alexandria, VA: U.S. Army Research Institute for the Behavioral and Social Sciences.

Hagman, J. D. (2005). *More efficient live-fire rifle marksmanship evaluation* (Rep. No. A762144). Alexandria, VA: U.S. Army Research Institute for the Behavioral and Social Sciences.

Jane's Information Group. (2006/2007). Executive overview: Infantry weapons. *Jane's infantry weapons*. Alexandria, VA: Jane's Information Group.

Johnson, R. F. (2001). *Statistical measures of marksmanship* (Rep. No. TN-01/2). Natick, MA: US Army Institute of Environmental Medicine.

Kirkpatrick, D. L. (1959). Techniques for evaluating training programs. *Journal of ASTD, 13*(11), 3–9.

Lane, J. C., & Cumming, R. W. (1956). *The role of visual cues in final approach to landing (Human Engineering Note 1)*. Melbourne, Australia: Aeronautical Research Laboratories, Defence Science and Technology Organisation.

Lintern, G. (1996). Human performance research for virtual training environments. *Proceedings of the Simulation Technology and Training (SimTecT) Conference* (pp. 239–244). Melbourne, Australia: Simulation Industry Association of Australia.

Lintern, G., & Walker, M. B. (1991). Scene content and runway breadth effects on simulated landing approaches. *The International Journal of Aviation Psychology, 1*(2), 117–132.

Moore, G. E. (1965). Cramming more components onto integrated circuits. *Electronics, 38*(8), 114–117.

Moore, G. E. (1995). Lithography and the future of Moore's law. *Proceedings of SPIE—Volume 2437* (pp. 2–17). Santa Clara, CA: The International Society for Optical Engineering.

Muller, P., Cohn, J., & Nicholson, D. (2004). Immersing humans in virtual environments: Where's the Holodeck? *Proceedings of the Interservice/Industry Training, Simulation, and Education Conference* (pp. 1321–1329). Arlington, VA: National Training Systems Association.

Salas, E., Milham, L. M., & Bowers, C. (2003). Training evaluation in the military: Misconceptions, opportunities, and challenges. *Military Psychology, 15*(1), 3–16.

Smith, M. D., & Hagman, J. D. (2000). *Predicting rifle and pistol marksmanship performance with the laser marksmanship training system* (Tech. Rep. No. 1106). Alexandria, VA: U.S. Army Research Institute for the Behavioral and Social Sciences.

Smith, M. D., & Hagman, J. D. (2001). *A review of research on the laser marksmanship training system* (ARI Research Note No. 2001-05). Alexandria, VA: U.S. Army Research Institute for the Behavioral Science.

Smith, M. D., & Hagman, J. D. (2003). *Using the laser marksmanship training system to predict rifle marksmanship qualification* (Research Rep. No. 1804). Alexandria, VA: U.S. Army Research Institute of the Behavioral and Social Sciences.

Taylor, C. J., Dyer, F. N., & Osborne, A. (1986). *Effects of rifle zero and size of shot group on marksmanship scores* (ARI Research Note 86-15). Fort Benning, GA: U.S. Army Research Institute.

Temby, P., Ryder, C., Vozzo, A., & Galanis, G. (2005). Sharp shooting in fuzzy fields: Effects of image clarity on virtual environments. *Proceedings of the 10th Simulation Technology and Training (SimTecT) Conference.* Sydney, Australia: Simulation Industry Association of Australia.

Torre, J. P., Maxey, J. L., & Piper, S. (1987). *Live fire and simulator marksmanship performance with the M16A1 rifle. Study 1: A validation of the artificial intelligence direct fire weapons research test bed* (Vol. 1, Technical Memorandum No. 7-87). Aberdeen Proving Ground, MD: U.S. Army Human Engineering Laboratory.

Westwood, D. A., & Goodale, M. A. (2003). Perception illusion and the real-time control of action. *Spatial Vision, 16,* 243–254.

White, C. R., Carson, J. L., & Wilbourn, J. M. (1991). Training effectiveness of an M-16 rifle simulator. *Military Psychology, 3*(3), 177–184.

Yates, W. W. (2004). *A training transfer study of the indoor simulated marksmanship trainer.* Unpublished master's thesis, Naval Postgraduate School, Monterey, CA.

Chapter 22

SIMULATION TRAINING USING FUSED REALITY

Ed Bachelder, Noah Brickman, and Matt Guibert

This chapter describes a novel mixed reality technique for real time, color based video processing for training applications combining software and off-the-shelf hardware called "fused reality." This technique allows an operator, using a helmet-mounted display, to view and interact with the physical environment in real time while viewing the virtual environment through color-designated portals (that is, painted surfaces, such as window panels). Additionally, physical objects can be deployed in real time into the virtual scene (for example, a person gesturing outside the simulator cabin can be virtually moved relative to the vehicle). Fused reality's adaptive feature recognition allows for realistic set lighting, colors, and user movement and positioning. It also enables multiple keying colors to be used (versus just blue or green), which in turn allows "reverse chromakeying"—preserving only keyed colors and rendering all others transparent. This technology enables hands-on immersive training for a very wide range of environments and tasks.

FUSED REALITY

Due to physical constraints and fidelity limitations, current simulation designs often fail to provide both functional utility and immersive realism. Fused reality (Bachelder, 2006) is a mixed reality approach that employs three proven technologies—live video capture, real time video editing, and virtual environment simulation—offering a quantum jump in training realism and capability. Video from the trainee's perspective is sent to a processor that preserves pixels in the near-space environment (that is, the cockpit) and makes transparent the far-space environment (outside the cockpit windows) pixels using blue screen imaging techniques. This bitmap is overlaid on a virtual environment and is sent to the trainee's helmet-mounted display (HMD). Thus the user can directly view and interact with the physical environment, while the simulated outside world serves as a backdrop.

Fused reality is a technique conceived at Systems Technology, Inc. (STI). It is similar in certain respects to the blue screen technique that Hollywood is using (such as that employed by Alfred Hitchcock in *Vertigo*). However, Hollywood processes its blue screening offline—STI is conducting it in real time—and, in contrast with blue screening, the backdrop required by fused reality allows large variations in color aberrations and lighting intensity.

HISTORY OF CHROMAKEY

Well before the film industry employed modern computer-generated imagery (CGI) to create stunning visual effects, many directors relied on simpler techniques. One of the earliest methods developed was the static matte. This technique is referred to as a static matte because the same roll of film is used to create the effect—there is no need (and really no ability) to overlay different mattes. The most common application of mattes consists of exposing two different parts of film to light at different times, which is known as a double-exposure matte. For example, many directors used the double-exposure technique to combine a tranquil terrain scene with a turbulent sky that is seemingly in fast-forward (with dark, seething clouds). This was accomplished by filming the tranquil scene with a sheet of black paper over the upper portion of the camera lens to prevent the sky from exposing the film. Once the initial scene was captured, the film would then be rewound and the terrain side of the film would be masked with the black paper (to shield the film that has already been exposed). The cameraman then filmed the stormy sky with a slower film speed, so that when the final video was played, two very different elements are combined into the same scene. Another simple example of using static mattes is the creation of widescreen films. By simply placing thin black strips of paper on the top and bottom edges of a lens, the film is instantly converted to widescreen dimensions. This process is referred to as a hard matte (whereas a soft matte requires a film projectionist to mask the projector to create the widescreen effect). Unfortunately, static mattes are not very versatile and cannot be used with moving objects.

Following the invention of the static matte technique, the traveling matte was developed as a more complicated improvement that allows mattes to "follow" moving objects. In a traveling matte shot, multiple mattes are used to designate the exact shapes of different elements of a scene. For instance, if the scene consists of an actor falling from a building, one film would be created to film the actor simulating a fall in a studio and another film would simply capture the building on location. Two different mattes are then created: one with the actor's figure masked in black and one with the actor's background (the studio) in black. During each frame, a new matte is created to adjust to the actor's movement along the background (to account for arm/leg movements and so forth). Once the filtering is complete, the film consists of four different pieces: the two originals and the two mattes. Finally, the building image is combined with the actor's blackened figure (so that a dark "gap" appears in the building), then the film is rewound and reexposed to the matte with the actor in it. Although this process did provide more flexibility in filming, it was very difficult to accomplish and required a tremendous amount of time and effort.

As CGI technology improved in the 1950s, the film industry began to conduct research in order to create better techniques for more efficient use of traveling mattes. Two of the most prominent researchers in this field were Arthur Widmer and Petro Vlahos, who are widely credited with the development of the chromakey process (also referred to as blue screen or green screen). Widmer first began developing chromakey while working for Warner Bros., and it was soon used in the 1958 making of the *Old Man and the Sea*—the adaptation of Ernest Hemingway's novel. Petro Vlahos's work earned him an Oscar in 1964 for blue screen compositing technology. With chromakey, a predetermined key color (often blue or green) is rendered transparent in one image in order to reveal the image behind it. Chromakey typically refers to the use of analog methods, whereas the more modern processes rely on digital compositing techniques (henceforth, the process will be referred to as "blue screen").

The blue screen technique allows for the combination of multiple sets of film (or computer images) into one. The process begins by filming an actor in front of a blue background. When filming is complete, the images are put through a blue filter that will allow only the background to be exposed on black and white film. This new image, referred to as the female matte, now consists of a black background with a blank space where the actor stood. Next, the original blue screen shot is now processed through a green and red filter in order to capture the actor's figure. This time, the black and white film, referred to as the male matte, shows a black figure where the actor stood and a clear background. It is important to note that in both the male and female mattes, the areas that are not black are actually clear (not white) because they are unexposed.

With the actor's (inclusive and exclusive) mattes completed, it is now possible to start combining the background and foreground images. First, the background image is filmed using the male matte as a filter so that the male matte occludes the background image and prevents portions of it from being exposed (now the background image has an unexposed gap where the actor's figure can be placed). Afterward, the original blue screen film (with the actor and the screen) is refilmed using the female matte as a filter so that only the actor's figure is exposed on the film and not the background. Finally, the images are combined frame by frame using high powered computers or special film equipment (such as optical printers and so forth). This process can also be accomplished during production: rather than filming the entire scene and then compositing it afterward (post-production), computers can be used to break down each frame as it is filmed and generate the composite. The ability of computers to do this in real time has opened the doors for many modern applications, such as the weather screen in many TV stations. The weather screen is what TV viewers see, but not what the weather anchor sees—in order for the weather anchor to view the simulated environment from his or her eyepoint, a helmet-mounted display is required.

Blue and green are typically used as the key colors because these two colors are not noticeably visible in human skin tone. Furthermore, digital cameras preserve more detail from green channels and the color has a higher luminance value, so it needs less light. However, it is always important to consider the background image of the current scenario. If the background contains a lot of natural scenery (that is, grass or sky), it would be wise to use magenta as the key.

The Naval Postgraduate School has conducted research in flight training using chromakey, and its most recent project is called the Chromakey Augmented Virtual Environment (ChrAVE) 3.0 System (Hahn, 2005). This system is more hardware relative to fused reality, employing (1) a compositing device, (2) a video graphics array–to-digital scan converter, and (3) an analog-to-digital signal converter. Green light emitting diode ring lighting is used to illuminate highly reflective material to produce the keying color; however, the main disadvantage of this technique is that the user's viewing angle is limited to small deviations from head on. In other words, if the surface was viewed at an angle of 45° relative to the surface's normal, very little of the source's energy will return back to the eye's viewpoint, destroying the chromakey effect.

Oda (1996) at Carnegie Mellon University uses stereoscopic pixel-by-pixel depth information in the form of a depth map as a switch (z-keying), allowing space to be segmented based on the detailed shape and position of surfaces. Frame rate using the process is very low (15 frames per second), which makes it unsuitable for much real time simulation training. Another serious drawback of this technique is that it fails when the background surface is featureless (such as a uniformly painted wall).

CHROMAKEY PROCESS

The blue in blue screening was chosen, as mentioned above, because blue is not present in human skin tones. These backdrops preclude the use of similar colors in the physical environment. Similarly, fused reality uses magenta as the target color, since it is rarely encountered in simulation environments. The color recognition technique used in fused reality can accurately distinguish between skin tones and magenta. Figure 22.1 gives pixel scatter plots of the red, green, and blue (RGB) components comprising the magenta color target. Due to nonuniformities across the material surface, as well as sensor artifacts and lens artifacts (there are darker areas within the magenta screen), there is a wide variance in RGB values. In order to algorithmically define the target color, scatter plots of the pixel colors were created, as shown in Figure 22.2, with the areas of the scatter plots approximated by bounding polygons. This technique using polygon templates was initially used by fused reality.

Hue saturation value has been identified as a simpler technique to RGB for color decomposition and is now used by fused reality to define a surface's color. Figure 22.2 shows the pixel scatter plots for saturation and value corresponding to the image in Figure 22.1. Note that these coordinates produce scatter plots that can be defined via bands (instead of complex and relatively imprecise polygons) based on their probability densities, shown below the scatter plots. Thus it is possible to statistically define the color characteristics of an image simply through lower and upper boundaries—a much simpler process than the RGB mapping (which requires linear interpolation) shown in Figure 22.2.

The robustness of this technique is demonstrated in Figure 22.3, where a magenta surface (shown top) mounted on a placard serves as a virtual display. The bottom photos in Figure 22.3 show two very different lighting environments

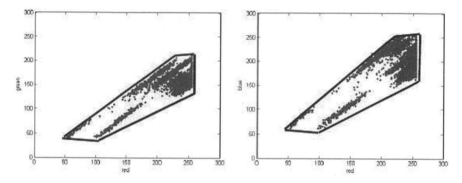

Figure 22.1.   RGB Components That Comprise the Magenta Color Target

(note the desk reflection brightness), but the magenta is correctly identified despite the variation in lighting.

The advantages of fused reality and its current recognition scheme thus include the following: (1) target color backdrops can be made from inexpensive and widely available cotton sheets, (2) chromakey is independent of the viewing angle, (3) it can use any lighting (incandescent or fluorescent) and brightness that makes the backdrop visually distinguishable from its surroundings, (4) more than

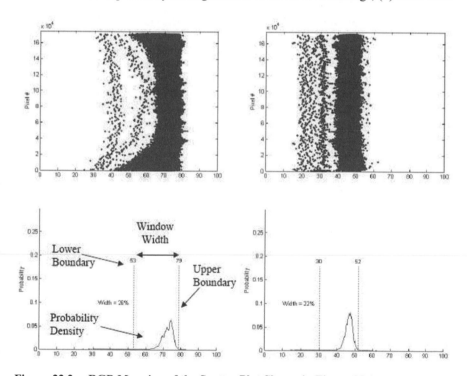

Figure 22.2.   RGB Mapping of the Scatter Plot Shown in Figure 22.1

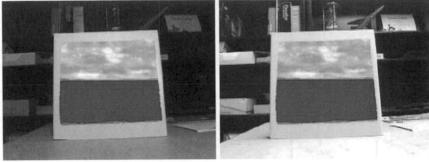

Figure 22.3.   Lighting Variations of the Magenta Color Target

one target color can be used, and (5) the lighting level does not have to be low to simulate low light visual environments or night vision as each pixel can be operated on to change the light level or color displayed to user. Fused feality divides space into the near tangible environment and the distant virtual environment and maintains high perceptual fidelity in both domains while minimizing computational expense. The user naturally encounters the high detail of the physical world through vision and touch, and excellent perception of the distant virtual world requires a low to medium level of detail.

VISUAL SYSTEM

A preliminary visual system is shown in Figure 22.4, where a Sumix camera has been mounted onto an eMagin HMD. A 12mm Computar lens is shown mounted on the camera, and an inertial head tracker made by Intersense (IC2) is attached. The camera is flush with the eye level. The HMD has a diagonal 40° field of view, with the resolution being 800 × 600 pixels. The system frame rate is approximately 70 hertz.

DEMONSTRATION

Fused reality was integrated with two of Systems Technology's simulation products: ParaSim (a parachute simulator) and STISIM Drive (a driving simulator).

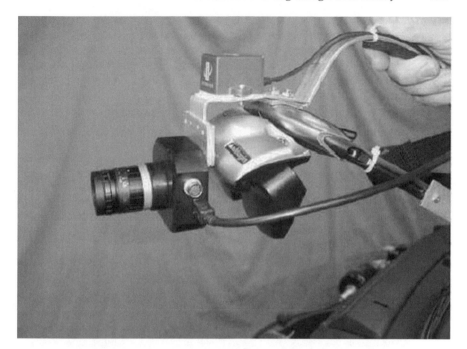

Figure 22.4.   Preliminary Visual System

ParaSim

ParaSim was first developed for the U.S. Forest Service to train firefighting smokejumpers. More recent versions are used to train navy and air force aircrews. A multiple concurrent jumper version has been developed for the Special Operations Command for mission planning and rehearsal. A typical configuration for ParaSim is shown in Figure 22.5, where a jumper is suspended in an actual parachute harness that is attached to a scaffolding. The user views a virtual jump scene using an HMD. The immersive effect is limited, however, due to the absence of visual cues corresponding to the physical tangible environment: harness, jump equipment, and limb location. The capability to view the virtual scene relative to one's boots would be especially helpful in assisting the jumper's spatial orientation.

In Figure 22.6 the scaffolding is shrouded with magenta cloth on all sides (including top and bottom) except the rear, which the jumper is not able to turn and view.

The trainee in the fused reality simulation sees the simulation display wherever the key color (in this case, magenta) exists. The monochrome drape over the simulation frame becomes an immersive display, completely surrounding the trainee, yet allowing him to see his own arm movements, body position, and the direction of his feet, as shown in Figure 22.7. The instructor's displays in Figure 22.8 show

Figure 22.5.   Typical ParaSim Configuration

the simulation controls, the simulation display, and the live video feed from the camera mounted on the HMD.

STISIM Drive

This simulator was originally developed for the Arizona Highway Patrol to evaluate the fitness for duty of long-haul truckers (Stein, Parseghian, Allen, & Rosenthal, 1991). Recent applications include use by medical research institutions to study, for example, the effects of new drugs, the cognitive impact of brain injuries, and the effects of HIV (human immunodeficiency virus) medications. Current STI research applications include programs for the National Institutes of Health to study the impact of simulator based training on novice drivers

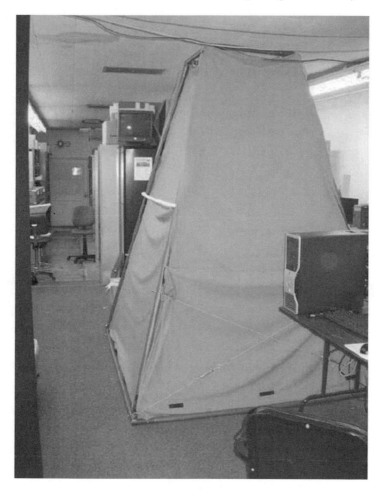

Figure 22.6.   Shrouded ParaSim Configuration

(Allen et al., 2000) and cognition of impaired drivers. The STISIM Drive device shown in Figure 22.9 has the following key system features that include off-the-shelf hardware components: Pentium processor; Windows 2000 Professional operating system; nVidia G-Force4 graphics processor; multiple screen display provides a 135° field of view, input/output interface processor for control inputs (that is, steering, braking, throttle, and optional clutch); steering force feedback (feel system); network for distributed processing; and optional motion base. Recent advances in commercial off-the-shelf laptop computer technology have enabled both these simulators to be operated with a laptop computer, greatly enhancing the portability of the simulators. Figure 22.10 shows a scene generated by STISIM Drive.

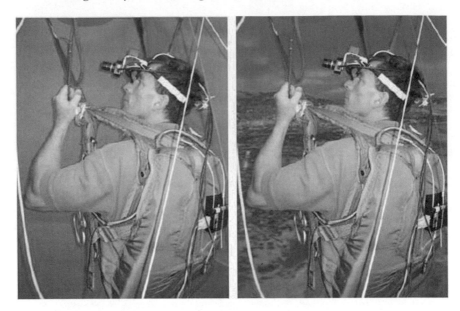

Figure 22.7. Immersive View of ParaSim Configuration

The STISIM Drive configuration employed with fused reality used a Honda car cab with a force-feedback steering wheel, a brake, and gas pedal as inputs to the simulation. Dashboard instruments, such as the speedometer and the tachometer, responded to the simulated car states. Magenta cloth was draped in front of the car cab as shown in Figure 22.11, so that the driver's field of regard was approximately 135°. A flat screen monitor, mounted outside the rear left window, displayed the same image the user was seeing in the HMD. Figures 22.11 and

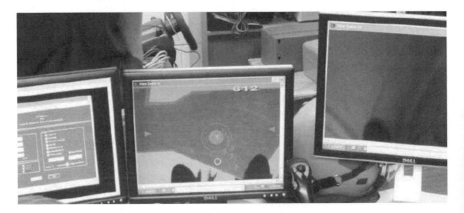

Figure 22.8. ParaSim Configuration Instructor's View

Figure 22.9.   STISIM Drive Device

22.12 show views of the user looking forward right and forward, respectively (note the driver's hands on the steering wheel in Figure 22.12).

Some key advantages that fused reality offers in the driving simulation include the following:

- Unlimited field of regard—every window can be covered with magenta to create a virtual portal;
- Enhanced experimental capability—drivers can interact with physical objects (such as maps and cell phones);
- Enhanced immersion realism—drivers can observe and operate real equipment while being framed in an authentic near-field environment, also known as embedded simulation; and

Figure 22.10.   Scene generated by STISIM Drive

Figure 22.11.   View of Driver Looking Forward Right

Figure 22.12.   View of Driver Looking Forward

- Lighting effects, such as blinding headlights, can be applied to both the virtual and video layers.

VIRTUAL DEPLOYMENT OF PHYSICAL OBJECTS

One of the most powerful aspects of fused reality is the capability to capture real time video and maneuver the video within the virtual scene, allowing the video to be occluded by virtual objects. This technique would significantly enhance realism and component recognition by users, since components of a virtual SolidWorks assembly could be textured with real time images of the actual assembly. These physical images would be extracted individually by the object recognition tool. Figure 22.13 shows an example of this in the current version of fused reality. Here the physical object, a hand, is identified by pixel brightness (the background is a black cloth). The pixels associated with the hand are maneuvered in 3-D virtual space via a joystick. Note that the strut of the water tower is occluding the hand that is behind it.

The following is an example of how this technique could be used in a driving simulation. An actual person standing in front of the driver could be gesturing as a policeman commanding traffic at an intersection (Figure 22.14). Although the policeman is physically fixed at some distance away from the car, the real time bitmap of the policeman that the driver sees can be made to move anywhere within the virtual scene. In this way the policeman's gesturing image would appear small in the distance and loom larger as the vehicle approaches. The person playing the part of the policeman could be viewing a screen mounted on top of the car so that he or she can respond to the simulated motion and position of the driver's car. It should be noted that this scenario does not require the policeman to be in the same physical location as the trainee. Allowing the policeman (or a specific person, such as a police chief) to be filmed at a remote location makes the technology even more useful. Thus two trainees in different locations could be set up to interact with each other in a virtual world, enabling teamwork training. A key feature of using live persons in fused reality vice models is that models perform according to a script, eliminating scenario flexibility and adaptability. With live actors, all participants can interact in a more natural flow of

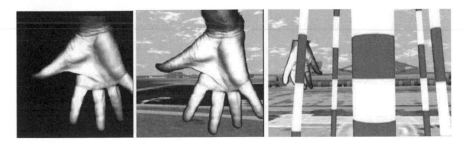

Figure 22.13.   Physical Image of a Hand in the Current Version of Fused Reality

Policeman   Flatscreen Monitor

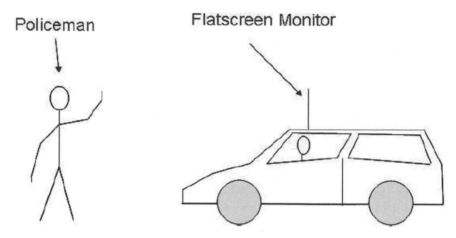

Figure 22.14.   Example of Fused Reality Technique in a Driving Simulation

events. As an example, if the driver does not initially notice the signals of the traffic policeman (and the policeman is virtually in danger of being run over), the officer can become decidedly more animated in his or her gestures—perhaps even dodging the vehicle as a last resort.

FUTURE FEATURES USING FUSED REALITY

Target areas (portals) could be designated by infrared (IR) and ultraviolet (UV) reflection, giving rise to virtual reality portal generation on command (that is, directing an IR or UV source toward a reflective surface). Dual reality portals could be made by coating glass with transparent material that reflects IR or UV light, so that the trainee sees a virtual environment while the naked-eye observer can view the actual environment that exists beyond the glass. This would allow training in actual vehicles, such as cars (driven in vacant parking lots) and aircraft, while a safety observer looks for potential conflicts or other hazards. Thus a trainee would experience the actual forces that he or she is effecting while operating in a virtual world.

CONCLUSION

This general description of fused reality, as applied to a parachute flight training system and a driving simulator, has demonstrated the difference between fused reality and traditional blue screen techniques. Fused reality offers a more cost-effective solution for real time image compositing for multiple color keys, wider ranges of color aberrations, and greater robustness to a variety of environmental lighting conditions. Virtual deployment of physical objects is an aspect of fused reality that further expands the capabilities of simulation and greatly enhances user immersion. The superiority of fused reality is clear in that the near tangible environment and the distant virtual environment are both accessible to the human operator and a high perceptual fidelity is maintained in both domains with minimal computational expense. The user experiences details of the physical world plus excellent perception of the generally less-detailed, distant virtual world.

REFERENCES

Allen, R. W., Cook, M. L., Rosenthal, T. J., Parseghian, Z., Aponso, B. L., Harmsen, A., et al. (2000). *A novice driver training experiment using low-cost PC simulation technology.* Paper presented at the Driving Simulator Conference (DSC) 2000, Paris, France.

Bachelder, E. N. (2006). Helicopter aircrew training using fused reality. In *Virtual Media for Military Applications* (RTO Meeting Proceedings No. MP-HFM-136, pp. 27-1–27-14). Neuilly-sur-Seine, France: Research Technology Organisation.

Hahn, M. E. (2005). *Implementation and Analysis of the Chromakey Augmented Virtual Environment (ChrAVE) Version 3.0 and Virtual Environment Helicopter (VEHELO) Version 2.0 in Simulated Helicopter Training.* Master of Science, Naval Postgraduate School, Monterey, CA.

Oda, K. (1996). *Z-Key: A New Method for Creating Virtual Reality.* Retrieved April 25, 2008, from http://www.cs.cmu.edu/afs/cs/project/stereo-machine/www/z-key.html

Stein, A. C., Parseghian, Z., Allen, R. W., & Rosenthal, T. J. (1991). *High risk driver project: validation of the truck operator proficiency system (TOPS)* (STI-TR-2406-01). Hawthorne, CA: Systems Technology, Inc.

Chapter 23

DISMOUNTED COMBATANT SIMULATION TRAINING SYSTEMS

Bruce Knerr and Stephen Goldberg

The term "dismounted combatant" may invoke a variety of colorful images, from a weather-beaten American Civil War cavalryman holding the reins of his horse to a medieval knight pinned to the ground by the weight of his armor. We will use it to describe contemporary army soldiers or marines who perform their missions in direct contact with the people, places, and objects in their environments rather than from inside a combat vehicle or via remote sensors or weapons. These are the soldiers whom we have traditionally described as "infantry."

Over the past decade two factors have conspired to make the job of the dismounted combatant more complex: changes in the variety and types of missions they perform and changes in the environment in which they perform those missions.

VARIETY AND TYPES OF MISSIONS

Dismounted combatants have traditionally been trained to conduct combat operations in an environment occupied almost exclusively by friendly and enemy forces. Today they are required to carry out a variety of activities ranging from food distribution and traffic checkpoint operation to combat, in an environment that includes a large number of people who are not clearly either friendly or enemy. Moreover, they must frequently transition rapidly from one type of activity to another. Their success is often dependent on the decisions and actions of relatively junior personnel (Krulak, 1999).

THE ENVIRONMENT

While we previously trained our infantrymen to operate in open terrain, there are two reasons why we must prepare them for urban combat as well. First, the world is becoming increasingly urban. Second, the two Gulf wars demonstrated U.S. superiority in open terrain. We can expect future enemies to attempt to fight us on urban terrain, which is more to their advantage. Urban areas are more complex than open terrain; buildings add a vertical dimension, limit visibility and communication, and are usually occupied by noncombatants, whose presence complicates decision making and limits options for the use of force.

Dismounted combatant simulations have different requirements than more traditional training simulations, such as flight or vehicle simulators. Their tasks frequently have the following characteristics.

DIRECT INTERACTION WITH THE SIMULATED PHYSICAL ENVIRONMENT

Unlike crew members of aircraft or armored vehicles, dismounted combatants interact directly with their weapons and the objects in their environment. They walk and run through streets, climb stairs, throw grenades, and drop behind barricades. They fire their weapons by lifting them into position and pulling a trigger. They obtain information about their environment directly through their basic senses (sight, hearing, touch, and smell), not an electronic display.

DIRECT PERSON-TO-PERSON INTERACTION UNMEDIATED BY EQUIPMENT

Communication with others is face-to-face and direct. They make eye contact and interpret posture and gestures.

EMPHASIS ON PHYSICAL ACTIVITY

The actions that dismounted combatants take are predominantly physical. While they make situation assessments, plan, evaluate alternative courses of action, and make decisions, the results of those cognitive activities are physical actions.

EARLY HISTORY—THE 1990s

The simulation networking program, begun in the early 1980s, and the close combat tactical trainer program, begun in 1992, established the feasibility of using networked simulators to train for combat in ground vehicles (Cosby, 1995). Interest in the use of immersive simulation for dismounted infantry training began in the early 1990s. Partly as a result of the efforts of Gorman (1990), a conference held in 1990 to discuss individual soldier systems and the role that an individual immersive simulator would play in their development provided the impetus for the initiation of research programs in the area (Goldberg & Knerr, 1997).

The navy was the first service to produce a prototype of a virtual individual combatant simulator. The team tactical engagement simulator program was begun in 1993. The team tactical engagement simulator consisted of an 8′ × 10′ rear projection display, demilitarized rifle tracker, head tracker, computer graphics generator, and system software. Trainees moved through the virtual world through the use of a foot pedal; pressure on the front of the pedal moved the trainee in the direction of gaze, while pressure on the back of the pedal moved the trainee in the opposite direction. J. H. Lind (Lind & Adams, 1994; Lind, 1995) used a structured process to obtain subject matter expert ratings that indicated high potential usefulness of the team tactical engagement simulator for training tactical situations, marksmanship, discretionary decisions, mission preview, and mission rehearsal. No empirical evaluations were conducted.

The army began its dismounted warrior network program in 1997. This program developed and evaluated a variety of different simulators and simulator technologies that could be used by dismounted combatants. The program did not evaluate training effectiveness, but did obtain data about task performance in the virtual simulators used (Lockheed Martin Corporation, 1997, 1998; Pleban, Dyer, Salter, & Brown, 1998; Salter, Eakin, & Knerr, 1999). It showed that soldiers could perform basic infantry tasks, such as engaging targets, simulating locomotion, and identifying people and objects, in the simulators. It also revealed limitations. Salter et al. identified four major areas for improvement following experiments conducted at the conclusion of the program. First, improved position and orientation tracking was necessary to improve weapons accuracy. Second, "walking" in the virtual world required conscious effort and may have impaired the performance of other tasks or interfered with training. Soldiers did not acquire full proficiency in the simulators in the time available and consequently moved more slowly in the virtual world than in the real world and frequently collided with walls. The third was providing a means of nonverbal communication, such as gestures and facial expressions. The fourth was increasing the field of view of visual displays.

RECENT HISTORY: 1999–2005

From 1999 to 2005, five army organizations conducted a series of related programs that shared a common assessment methodology and used nearly identical measures of task performance and training effectiveness. The organizations were the U.S. Army Research Institute Simulator Systems and Infantry Forces Research Units, the U.S. Army Simulation Training and Instrumentation Command, the U.S. Army Research Laboratory Human Research and Engineering Directorate, and the U.S. Army Research Laboratory Computational and Information Sciences Directorate. Each organization had a particular area of interest, but all worked together to explore concepts and systems. Evaluations conducted in 1999, 2001, 2002, 2004, and 2005 will be described. These were comprehensive assessments conducted with squads of soldiers, using as much of the developed technology as was feasible, in a realistic training exercise. The squads consisted of a squad leader (the primary trainee) and one or two four-man fire teams. Each assessment involved a different group of soldiers.

The assessments differed in their procedural details, but each involved three squads of six to nine soldiers. Each squad conducted a series of tactical scenarios in the simulators over one or, more commonly, two days. Each scenario consisted of a planning period, scenario execution in the simulators (lasting about 20 minutes), and an after action review. Only the 2004 assessment used a mix of immersive and desktop simulators. Otherwise, immersive simulators were used exclusively. More detail on these assessments can be found in Knerr (2007).

Typical hardware and software included the following networked components:

- SVS (Soldier Visualization Station) individual soldier simulators. The SVS is a personal-computer (PC) based dismounted infantry simulator developed by Advanced Interactive Systems, Inc. Immersive and desktop versions are functionally similar, but have different displays and controls. The immersive SVS uses a rear-screen projection system to present images (800 × 600 resolution) on a screen approximately 10 feet wide by 7.5 feet high. The soldier's head and weapon are tracked using an acoustic tracking system. The soldier navigates through the environment via a thumb switch located on the weapon. The desktop SVS is functionally similar to the immersive SVS, but the soldier sits at a PC and views the simulation on a monitor. Immersive and desktop visual displays presented the same information. A joystick is used to control view, movement, and weapon use. In these assessments, squad and fire team leaders always used the immersive SVS, role-players always used the desktop SVS, and fire team members usually used the immersive SVS. The simulators were typically equipped with radio headsets, which permitted verbal communication within the squad and between the squad leader and his higher headquarters.
- Dismounted Infantry Semi-Automated Forces Operator Station. An operator and the exercise controller used this station. Dismounted Infantry Semi-Automated Forces was developed by SAIC to provide a realistic representation of simulated entities.
- Dismounted Infantry Virtual After Action Review System. The Dismounted Infantry Virtual After Action Review System is a PC based system developed by the Army Research Institute and the University of Central Florida Institute for Simulation and Training specifically to meet the after action review requirements for dismounted infantry in urban combat (Knerr, Lampton, Martin, Washburn, & Cope, 2002). It was used in all evaluations beginning in 2001. The key capabilities are replay with synchronized audio and video, including the capability to jump to pre-designated segments or views and to produce tabular data summaries. It also includes the capability to display building interiors and to capture and replay voice communications.

All assessments used essentially the same list of 54 soldier activities to determine how well soldier tasks could be performed in the simulators (from Pleban, Eakin, & Salter, 2000). Soldiers rated their abilities to perform each activity as very poor, poor, good, or very good. The list of activities is shown in Table 23.1. Squad and fire team leaders completed a questionnaire that asked them to rate their improvement in 11 areas as a result of their training. Improvement was rated on a four-point scale from "no improvement" (0) to "vast improvement" (3). The 11 areas are shown in Table 23.3. Each assessment involved a different group of soldiers.

ASSESSMENT RESULTS

The capabilities of the SVS were fairly static during the 2002–2005 period, and for this reason their ratings of the 53 soldiers who completed simulator capability

Table 23.1. Combined Ratings of Simulator Capability (2002–2005)*

Task	Mean Rating	Task	Mean Rating
Move through open areas as a widely separated group.	2.49	Identify areas that mask supporting fires.	2.00
Identify civilians.	2.47	Maneuver below windows.	2.00
Execute planned route.	2.42	Use flash-bang grenades to help clear rooms.	2.00
Move in single file.	2.38	Take hasty defensive positions.	1.98
Fire weapon in short bursts.	2.35	Engage targets within a room.	1.98
Understand verbal commands.	2.31	Scan from side to side.	1.90
Locate assigned areas of observation, for example, across the street.	2.30	Look around corners.	1.89
Identify assigned sectors of observation.	2.29	Determine other team/squad members' positions.	1.84
Move according to directions.	2.29	Take position to one side of a doorway.	1.81
Identify noncombatants within a room.	2.27	Locate enemy soldiers inside buildings firing at your unit.	1.80
Identify sector of responsibility.	2.25	Move close to walls.	1.77
Execute the assault as planned.	2.25	Scan the room quickly for hostile combatants.	1.75
Communicate enemy location to team member.	2.22	Maneuver/move around obstacles.	1.75
Move quickly to the point of attack.	2.21	Use fragmentation grenades.	1.70
Aim weapon.	2.21	Estimate distances from self to a distant object/point.	1.67
Communicate spot reports to squad leader.	2.21	Maneuver close to others.	1.66
Fire weapon accurately.	2.19	Take a tactical position within a room.	1.66
Coordinate with other squad members.	2.17	Move past furniture in a room.	1.63

Identify covered and concealed routes.	2.16	Climb up or down stairs.	1.59
Identify safe and danger areas.	2.13	Maneuver around corners.	1.53
Assume defensive positions.	2.11	Visually locate the source of enemy fire.	1.52
Locate support team positions.	2.11	Move quickly through doorways.	1.40
Use handheld illumination (flares).	2.10	Distinguish between friendly and enemy fire.	1.39
Identify enemy soldiers.	2.08	Maneuver past other personnel within a room.	1.36
Locate buddy team firing positions.	2.06	Determine the direction from which enemy rounds are coming.	1.33
Employ tactical handheld smoke grenades.	2.03	Scan vertically.	1.19
Maintain position relative to other team members.	2.00	Determine the source of enemy fire by sound.	1.15

Note: N varies from 41 to 53. Ratings are on a scale from "very poor" (0) to "very good" (3).

ratings were combined to produce the summary rating shown in Table 23.1. Activities are ordered from best (very good, 3.0) to worst (very poor, 0.0).

Thirty of the 54 activities were rated good or better (2.00 and above), 18 were rated between the good/poor midpoint (1.50) and good (2.00), and 6 were rated poor (1.00 1.49). Activities that were rated highly included outdoor movement, identification of types of people (civilians, noncombatants within a room, and enemy soldiers), identification of tactically significant areas (sectors of observation and responsibility), and individual weapons use (but not grenades). Poorly rated items included maneuver indoors (close to others, past furniture, close to walls, around objects, past other personnel, around corners, through doorways, and up and down stairs), and identifying the source and type of fire (enemy or friendly), either by auditory or visual cues. Issues with maneuver and auditory cues will be addressed further in the discussion section.

Table 23.2 summarizes the overall results of the administration of the Training Effectiveness Questionnaire over the five assessments. It was completed by squad and fire team leaders in every assessment and by fire team members in the assessments conducted in 2004 and 2005. The pattern is consistent, with the mean ratings increasing consistently every year, from 0.82 (less than slight improvement) in 1999 to 2.06 (moderate improvement) in 2005.

Table 23.3 compares leader and soldier ratings on individual tasks. Leaders report the most improvement in controlling their units, assessing the tactical situation, and communication. Soldiers report the most improvement in

Table 23.2. Training Effectiveness Questionnaire Results*

Assessment Year	Leaders	Soldiers	Combined
	Mean (N)	Mean (N)	Mean (N)
1999	*0.82 (9)*	–	–
2001	1.24 (9)	–	–
2002	1.45 (9)	–	–
2004	1.74 (9)	1.82 (18)	1.79 (27)
2005	2.06 (9)	1.30 (17)	1.55 (26)
Combined 2002–2005	1.75 (27)	1.57 (35)	1.65 (62)

Note: Ratings are on a scale from "very poor" (0) to "very good" (3).

communication and planning a tactical operation. Overall, the general pattern is for more improvement to be reported for planning, coordination, and control tasks and less improvement for more rigidly described tasks and drills.

The 2002 assessment included an objective measure of unit performance, a 14-item checklist of unit behaviors that was scored independently by three raters for each scenario. Scores on similar scenarios improved with practice, providing rare evidence, beyond trainee opinion, that the training was effective.

Table 23.3. Combined Training Effectiveness Questionnaire Results (2002–2005)*

Question	Leaders	Soldiers	Combined
N	27	35	62
Assess the tactical situation.	2.09	1.68	1.86
Control of squad/fire team movement during the assault.	2.17	1.61	1.85
Communicate with members of your team or squad.	1.80	1.83	1.82
Coordinate activities with your chain of command.	1.80	1.71	1.75
Plan a tactical operation.	1.72	1.76	1.74
Control squad or fire team movement while not in contact with the enemy.	1.78	1.38	1.56
React to Contact Battle Drill.	1.55	1.53	1.54
Control your squad or fire team.	1.83	1.30	1.53
Clear a building.	1.50	1.54	1.53
Locate known or suspected enemy positions.	1.54	1.47	1.50
Clear a room.	1.48	1.42	1.45
Mean	1.75	1.57	1.65

Note: Ratings are on a scale from "no improvement" (0) to "vast improvement" (3).

DISCUSSION

Training Effectiveness

Soldiers and small unit leaders have reported that their skills improved as a result of training in dismounted combatant simulations, with the most improvement in controlling, coordinating, communicating, and planning and less improvement in the mechanics of tasks. Their reports have generally, if informally, been confirmed by observers. Objective measures of performance, obtained only in the 2002 assessment, indicated improvement in performance over the course of the training. There have been no successful attempts to measure the transfer of this training to a more advanced phase of training, such as a live simulation.

These results were obtained despite a number of limitations. The total amount of time that trainees spent conducting training scenarios was actually relatively short (no more than three hours and usually considerably less). In most cases, the software was still under development, and not fully problem-free. The scenarios were usually not tailored to the skill level of the leader or unit. The trainer/after action review leader was frequently unfamiliar with the technology and did not necessarily use it to best advantage.

In terms of an overall training approach, dismounted infantry simulation appears well suited to provide "walk" level training in a "crawl, walk, run" sequence of instruction. In the assessments it was used in that way, although it was not usually possible to include the "run" phase. The crawl level consists of individual skill training and initial demonstration and practice of the collective tasks. The run level consists of exercises conducted in an instrumented facility using military versions of laser tag or paintball. Both the walk and run (virtual and live) training consist of the planning and execution phases of tactical scenarios, with each exercise followed immediately by feedback. In accordance with standard army practice, this feedback takes the form of an after action review during which the trainees seek to discover what happened, why it happened, and how to improve.

Soldier Task Performance

Fine or precision movement in confined areas, such as movement indoors, has consistently emerged as the most important improvement required. The difficulties of indoor maneuver are likely a result of several contributing factors: the size of the bounding box that detects collisions between soldiers and other objects and between soldiers, the limited texture and shading cues on the walls inside buildings, the relatively narrow field of view of both the desktop and the immersive visual displays, the problem of representing objects in the immersive simulators that are located between the soldier and the rear-projection screen, and the linkage between the direction of gaze and the direction of movement.

While soldiers reported difficulty localizing auditory cues in the simulators, subject matter experts reported that it is difficult to localize auditory cues in real urban combat situations. It is not clear, therefore, whether this difficulty reflects limitations in the simulators or the complexity of the real world.

Simulators may differ in the specifics of the physical actions that they permit, particularly with regard to locomotion and touch. For example, trainees in the SVS could not physically "throw" grenades. They selected a type of grenade from a menu, chose direction and launch angle by pointing their mock weapons, adjusted the force desired by moving a slider with a pair of buttons on the weapon, and pulled the triggers to launch the grenades. While this bears little resemblance to the physical act of throwing a real grenade, it does provide practice in deciding which type of grenade to use and where and when to employ it. Such constraints as these limit training effectiveness only if those actions that cannot be performed in the simulator are not trained by other means. It is important that dismounted combatant simulations be used as part of a planned sequence of instruction that provides trainees with the prerequisite skills prior to the simulation and provides subsequent live training to improve the physical skills that cannot be trained in the virtual simulation.

Even though a simulator may not allow soldiers to perform many of the physical actions they need to perform in the way they need to perform them, it may allow them to practice and learn the cognitive skills they need. These include planning prior to the start of missions, maintaining situational awareness, and making appropriate decisions in complex situations.

RECOMMENDATIONS

Conduct a Large-Scale Evaluation Using Current Technology

The evidence that immersive simulation can provide effective training is not sufficient to justify the immediate acquisition of such systems on a large scale. However, it is sufficient to justify a rigorous training effectiveness evaluation of a prototype system that would permit quantification of the effectiveness of the training and comparison with alternatives. While this would be costly, in terms of both dollars and soldier time, it would lead to a more informed acquisition decision. This evaluation should consider objective measurement of both skill improvement on the simulator and transfer of those skills to a live simulation environment. It should involve a sufficient number of units that any meaningful real differences in the training effectiveness can be detected. The evaluation should be embedded in the unit's normal training progression, and unit personnel should be involved in its development and delivery.

Consider Cost-Effectiveness of Fully Immersive and Desktop or Laptop Systems

The cost differential between fully immersive simulators using large projection displays or head-mounted displays and simulators using desktop or laptop computers to provide the same functionality can be enormous. Knerr (2006) estimated the difference at over $75,000 per individual simulator. While the cost difference is likely to decrease as the cost of large visual displays decreases, the major cost drivers for fully immersive simulators are the interface devices, particularly the position and orientation trackers and visual display systems. The computers are relatively cheap. The interface devices also increase space and support requirements.

In contrast, the research evidence indicates that any difference in training effectiveness between immersive and desktop systems is likely to be small. Loftin et al. (2004) found small differences in effectiveness between immersive and desktop simulators and questioned the cost-effectiveness of the immersive simulator. A comparison of soldier (trained in the desktop simulators) and leader (trained in the immersive simulators) ratings of training effectiveness in the 2004 assessment raises the same question.

Other research addressing the question of interface fidelity on training effectiveness is limited. The basic concept is that immersive systems, as compared to desktop systems, provide the trainee with more information about their orientation in and movement through physical space. It appears that head-tracked visual displays, body-controlled movement, or a combination of the two can improve the performance of spatially oriented tasks and acquisition of spatial knowledge (for example, Grant & Magee, 1998; Lathrop & Kaiser, 2005; Singer, Allen, McDonald, & Gildea, 1997; Waller, Hunt, & Knapp, 1998), but this difference does not appear to be large.

Whether immersive simulators are also better when training squads to conduct urban or counterinsurgency operations and, if so, whether the difference is large enough to justify the increased cost, are unknown, but should be investigated, perhaps in conjunction with the large-scale evaluation recommended above.

REFERENCES

Cosby, L. N. (1995). *SIMNET: An insider's perspective* (IDA Document D-1661). Alexandria, VA: Institute for Defense Analyses. (ADA294786)

Goldberg, S. L., & Knerr, B. W. (1997). Collective training in virtual environments: Exploring performance requirements for dismounted soldier simulation. In R. J. Seidel & P. R. Chatelier (Eds.), *Virtual reality, training's future?* (pp. 41–52). New York: Plenum Press.

Gorman, P. F. (1990). *Supertroop via I-Port: Distributed simulation technology for combat development and training development* (IDA Paper No. P-2374). Alexandria, VA: Institute for Defense Analyses. (ADA229037)

Grant, S. C., & Magee, L. E. (1998). Contributions of proprioception to navigation in virtual environments. *Human Factors, 40(3)*, pp. 489–497.

Knerr, B. W. (2006). Current issues in the use of virtual simulations for dismounted soldier training. In *Virtual Media for Military Applications* (RTO Meeting Proceedings No. RTO-MP-HFM-136, pp. 21-1–27-11). Neuilly-sur-Seine, France: Research and Technology Organization.

Knerr, B. W. (2007). *Immersive Simulation Training for the Dismounted Soldier* (Study Rep. No. 2007-01). Alexandria, VA: U.S. Army Research Institute for the Behavioral and Social Sciences. (ADA464022)

Knerr, B. W., Lampton, D. R., Martin, G. A., Washburn, D. A., & Cope, D. (2002). Developing an after action review system for virtual dismounted infantry simulations. *Proceedings of the 2002 Interservice/Industry Training, Simulation and Education Conference.* Arlington, VA: National Training Systems Association.

Krulak, C. C. (1999, January). The Strategic Corporal: Leadership in the three block war. *Marines Magazine.* Retrieved April 15, 2008, from http://www.au.af.mil/au/awc/awcgate/usmc/strategic_corporal.htm

Lathrop, W. B., & Kaiser, M. K. (2005). Acquiring spatial knowledge while traveling simple and complex paths with immersive and non-immersive interfaces. *Presence, 14*(3), 249–263.

Lind, J. H. (1995). *Perceived usefulness of the Team Tactical Engagement Simulator (TTES): A second look* (Rep. No. NPS-OR-95-005). Monterey, CA: Naval Postgraduate School.

Lind, J. H., & Adams, S. R. (1994). *Team Tactical Engagement Simulator (TTES): Perceived training value* (Rep. No. NAWCWPNS TM 7724). China Lake, CA: Naval Air Warfare Center Weapons Division.

Lockheed Martin Corporation. (1997). *Dismounted warrior network front end analysis experiments* (Advanced Distributed Simulation Technology II, Dismounted Warrior Network DO #0020, CDRL AB06, ADST-II-CDRL-DWN-9700392A). Orlando, FL: U.S. Army Simulation, Training and Instrumentation Command. (ADA344365)

Lockheed Martin Corporation. (1998). *Dismounted warrior network enhancements for restricted terrain* (Advanced Distributed Simulation Technology II, Dismounted Warrior Network DO #0055, CDRL AB01, ADST-II-CDRL-DWN-9800258A). Orlando, FL: U.S. Army Simulation, Training and Instrumentation Command. (ADA370504)

Loftin, R. B., Scerbo, M. W., McKenzie, R., Catanzaro, J. M., Bailey, N. R., Phillips, M. A., & Perry, G. (2004, October). *Training in peacekeeping operations using virtual environments.* Paper presented at the RTO HFM Symposium on Advanced Technologies for Military Training, Genoa, Italy. (ADA428142)

Pleban, R. J., Dyer, J. L., Salter, M. S., & Brown, J. B. (1998). *Functional capabilities of four virtual individual combatant (VIC) simulator technologies: An independent assessment* (Tech. Rep. No. 1078). Alexandria, VA: U.S. Army Research Institute for the Behavioral and Social Sciences (ADA343575)

Pleban, R. J., Eakin, D. E., & Salter, M. S. (2000). *Analysis of mission-based scenarios for training soldiers and small unit leaders in virtual environments* (Rep. No. RR 1754). Alexandria, VA: U.S. Army Research Institute for the Behavioral and Social Sciences.

Salter, M. S., Eakin, D. E., & Knerr, B. W. (1999). *Dismounted warrior network enhancements for restricted terrain (DWN ERT): An independent assessment* (Research Rep. No. 1742). Alexandria, VA: U.S. Army Research Institute for the Behavioral and Social Sciences. (ADA364607)

Singer, M. J., Allen, R. C., McDonald, D. P., & Gildea, J. P. (1997). *Terrain appreciation in virtual environments: Spatial knowledge acquisition* (Tech. Rep. No. 1056). Alexandria, VA: U.S. Army Research Institute for the Behavioral and Social Sciences. (ADA325520)

Waller, D., Hunt, E., & Knapp, D. (1998). The transfer of spatial knowledge in virtual environment training. *Presence: Teleoperators and Virtual Environments, 7*(2), 129–143.

Part VII: Training Effectiveness and Evaluation Applications

Chapter 24

CONDUCTING TRAINING TRANSFER STUDIES IN COMPLEX OPERATIONAL ENVIRONMENTS

Roberto Champney, Laura Milham, Meredith Bell Carroll, Ali Ahmad, Kay Stanney, Joseph Cohn, and Eric Muth

Training effectiveness evaluations (TEEs) are used to assess the amount of learning that occurs from a prescribed training regime and the degree to which the regime results in observable performance changes in the domain environment (that is, transfer of training [ToT]). Transfer effectiveness evaluations are a method for assessing the degree to which a system facilitates training on targeted objectives (see Cohn, Stanney, Milham, Carroll, Jones, Sullivan, and Darken, Volume 3, Section 2, Chapter 17, for a comprehensive review of training effectiveness evaluations). A TEE is critical to perform as it is the primary method for truly understanding the efficacy of a training system or program. Particularly, a ToT evaluation can assist in (1) making decisions regarding the adoption of new training regimes or systems by providing quantitative data to compare their relative value, (2) determining the level of training offered by different training platforms (for example, classroom, simulated environments, live training, and so forth), and (3) determining the correct mix of each of those platforms. Considering the importance of evaluating training effectiveness, it is surprising that there are reports of a general lack of robust evaluation practices in industry (Carnevale & Shultz, 1990; Eseryel, 2002). Similarly, the American Society for Training and Development (ASTD) has reported on the limited assessment of training effectiveness (compare Bassi & van Buren, 1999; Thompson, Koon, Woodwell, & Beauvais, 2002). The application of training effectiveness evaluations has been problematic (for example, Baldwin & Ford, 1988; Saks & Belcourt, 2006) and has resulted in a lack of rigor and a limited application of the trained constructs. Furthermore, as pointed out by Cohn et al. (Volume 3, Section 2, Chapter 17) and others (compare Flexman, Roscoe, Williams, & Williges, 1972), TEEs, in particular transfer of training evaluations, are a resource-intensive process, which is often infeasible to apply in real world settings (this is discussed in more detail later in this chapter). This has made it difficult for practitioners to conduct TEEs in the field.

The application of TEEs in the field is presented with three broad challenges as illustrated by Cohn et al. (Volume 3, Section 2, Chapter 17): (1) the difficulty in developing meaningful measures of skill transfer and the associated large and time consuming data-collection efforts required for their implementation, (2) the need for an operational system or prototype to support the empirical evaluation, and (3) the multiple logistical constraints that surround TEEs (accessibility to trainees, scheduling, and so forth). The authors have tackled many TEE challenges and, in this chapter, they address two of the issues presented above that are encountered while applying training evaluation methods: (1) the logistical constraints that lead to the utilization of untrained undergraduates in TEEs to assess learning and transfer and (2) the resource-intensive nature of traditional transfer of training evaluation methods requiring more efficient approaches. The chapter discusses these challenges and provides a case study in which alternative methodologies were applied in a military operations on urban terrain (MOUT) domain (the sorts of tasks the infantry trains to perform, such as room cleaning). These approaches are believed to render TEEs more feasible and cost-effective to conduct.

ROADBLOCKS TO CONDUCTING TRAINING EFFECTIVENESS EVALUATIONS

Training effectiveness evaluation participants representative of the target domain are required in order to ensure validity of a training regime. Yet, it can be difficult to obtain access to individuals from the target population. This chapter addresses one way to resolve this problem Also, determining an appropriate transfer evaluation methodology (that is, a methodology for calculating the amount of training transfer to operational performance that a training system or program can achieve) and the amount of training a participant should receive in order to evaluate transfer is a challenge, which if incorrectly prescribed, can limit the ability to draw conclusions about transfer effectiveness. This chapter proposes an approach developed to address this issue.

Test Population

In applied fields, it is common to face limitations with respect to obtaining representative domain samples to conduct TEEs early enough in the training life-cycle to influence the training system design. Representative samples are often limited by availability, restricted access, willingness to participate, and resource constraints (for example, cost), and so forth. In these instances, evaluators often recruit from substitute populations to compare design alternatives or to conduct other forms of empirical validation (Ward, 1993). It is imperative that the chosen substitute population sample possesses sufficient knowledge, skills, and attitudes (KSAs) regarding the target domain to ensure validity of results. For example, a common practice is to recruit undergraduate students as experimental participants due to their availability. In these cases, unless these students obtain sufficient knowledge, skills, and attitude levels that bring them closer to those of the target population to be representative enough, they may not provide an appropriate sample from which to generalize to the target population (Wintre, North, & Sugar, 2001). There are other potential differences between a student population and a target population that may limit generalizability, some of which include a lack of context (for example, task meaningfulness) and artificiality of experimental setting (Gordon, Schmitt, & Schneider, 1984). Lack of appropriate KSAs, however, is particularly important given the need for representative data for making inferences regarding the target population. Hence, novel ways of increasing the validity of utilizing undergraduates with limited basic competencies as transfer effectiveness evaluation participants were sought.

The approach adopted and presented herein was to bring a sample of undergraduate students "up to speed" by putting them through a "bootcamp" (Champney, Milham, Bell Carroll, Stanney, Jones, et al., 2006). Specifically, the objective of the bootcamp was to bring undergraduate student participants closer to their target population counterparts in terms of knowledge, skills, attitudes, and an understanding of "what is at stake." In order to accomplish this, the bootcamp consisted of a lecture, practice, rehearsal, feedback, and contextual references (for example, videos and pictures). The bootcamp methodology presented herein is discussed within the context of a MOUT case study reported later in the chapter.

It is important to note that such methodology is not intended to replace a real sample of the target population and should be used only in the absence of a domain sample to bring a substitute sample closer to a domain sample in terms of KSAs and context to ensure validity of conclusions.

Transfer Evaluation Methodologies

Transfer of training refers to the extent with which learning a task is facilitated or hindered by the prior learning of a task (Roscoe & Williges, 1980). In other words, transfer of training can be viewed as the degree with which a trainee's abilities in the real world have been improved (or made worse) by prior training. TEEs utilize this degree of impact on a task to gauge the effectiveness of a training program or system (for example, percent transfer). Nonetheless, there are multiple options for how this impact can be computed, and it is the selection of the more robust measures that proves resource intensive. Of the multiple options for measuring transfer, those approaches that not only consider the impact produced by the training system or program but also its efficiency are the most robust. This is because two distinct training regimes or systems may have the same impact on training transfer, but have completely different efficiencies (for example, one could require more time than the other to produce the same effects). Roscoe (1971) sought to address this issue by using the transfer effectiveness ratio (TER), which takes into account the amount of prior training (that is, in the training system or program under evaluation) by specifying the savings in time (or trials) in the live environment to reach a criterion. This time savings is expressed as a function of a predetermined single amount of time (or trials) in the alternate trainer (for example, X time saved in an airplane for Y time spent in a simulator). One limitation to this approach is that the transfer effectiveness ratio does not consider the incremental gains from variations in the amount of training time/trials in an alternate training platform; it considers only a single instance of training time and its effect on transfer to a live environment (for example, could these differences among 10, 20, or 30 training trials depend on how quickly trainees reach a criterion in the live environment?). Given that there is no guidance regarding at what point the evaluation should occur (for example, how much time or many training trials), this presents the risk of drawing conclusions about training transfer at points along the learning curve that have not stabilized yet, and any inferences made at such points may be of limited validity and utility. To address this issue, Roscoe (1972) adopted the incremental transfer effectiveness ratio (ITER) approach, which takes into consideration the effectiveness of successive increments of training in each training platform by comparing multiple training time regimes in an alternate platform and their associated transfer to a live environment; however, it requires a considerable time (that is, the evaluation must be done for 1 trial, 2 trials, 3 trials, and so forth until a point of diminishing returns is identified) and a large number of participants. As such, it is not always possible to conduct ITER studies given the copious resources required of this approach (compare Roscoe & Williges, 1980).

When faced with a situation with limited resources, an alternative approach may be to use the transfer effectiveness ratio approach and couple it with a technique used to systematically specify the point at which transfer should be evaluated, using the diminishing learning rates in the training system or program under evaluation as a guide. Thus, rather than evaluating increments of training and relative transfer effectiveness (that is, as in the ITER approach), an effort is made to identify the single point at which transfer effectiveness of a system should be evaluated. To address this, a learning curve methodology was developed in which continuous monitoring of performance across trials is used to identify a "plateau" in learning improvements (Champney, Milham, Bell Carroll, Stanney, & Cohn, 2006). The next section presents the learning curve methodology within the context of a TEE for a MOUT trainer, where the goal was to compare the ToT between a low and a high fidelity training solution.

METHODOLOGY AND CASE STUDY

This case study addresses both methodologies discussed above, including (1) a bootcamp to bring the test population up to speed and (2) a feasible transfer evaluation method using the TER where the evaluation point is prescribed using learning curves.

Test Participants: Bootcamp, Bringing Nondomain Participants up to Speed

The use of an undergraduate bootcamp for the TEE of the MOUT trainer included two primary objectives: (1) increase target KSAs and (2) increase contextual understanding of the target domain. In order to develop these into a bootcamp, two activities were conducted: (1) domain understanding through task analysis and training objective identification and (2) use of this knowledge to create an instructional course (training plan, procedure, and material).

Task Analysis and Training Objective Identification

The identification of relevant KSAs was performed using a task analysis of the MOUT domain (that is, specifically the room-clearing task; Milham, Gledhill-Holmes, Jones, Hale, & Stanney, 2004). The task analysis included interviews and collaborations with subject matter experts, observation of task demonstrations, and reviews of military doctrine. The task analysis resulted in a breakdown of the room-clearing task into subtasks, identification of the KSAs necessary to complete these tasks/subtasks, and creation of metrics and performance standards. These data were then used to identify training objectives, which served as a blueprint for developing training materials that matched associated performance metrics and standards (see Table 24.1).

Curriculum

After identifying the training objectives, a training curriculum for the bootcamp was developed with the participation of a subject matter expert. The curriculum focused on the constructs to be learned (for example, knowledge, skills, and attitudes) and domain context to make tasks meaningful. In order to instill the desired KSAs and context, the bootcamp curriculum was built around multiple components, each of which was designed to address a particular component of the training experience, including the following:

1. *Initial classroom instruction:* Used to introduce participants to the domain and teach desired constructs. Beyond conventional lecture conveyance of the requisite KSAs, videos and images were also used to immerse participants into the target domain, thereby supporting contextual training.

Table 24.1.   Sample of Training Objectives and Performance Metrics

Training Objective	Performance Metric
Engagement/acknowledgment	Enemies neutralized Noncombatants acknowledged Missed shots
Room clearing	Percentage of room scanned Time to clear room
Survivability	Shots taken
Exposure	To danger areas (doorways, windows, and entryways) Enemy line of sight

2. *Practical instruction and evaluation:* For physical tasks (for example, maneuvering through corridors while manipulating a rifle), practice opportunities were provided with the subject matter expert providing instruction, evaluation, and direct feedback.

3. *Rehearsal:* Participants were given a rehearsal worksheet, which consisted of a mnemonic that provided an organizational framework for the learned content. Participants were given an opportunity for rehearsal of the mnemonic during the period of time between pre-training (bootcamp) and training with the target training system (that is, days to a few weeks).

4. *Review:* To mitigate any memory decay from delays between when a bootcamp was to take place and when the training system was to be used, a domain refresher review was instantiated in a short video, which highlighted the requisite KSAs utilizing the mnemonic. This was in addition to a familiarization practice, where participants were allowed to familiarize themselves with the training system.

5. *Scenario based feedback:* Subject matter expert feedback was instantiated in different forms in the curriculum, first, following the practical instruction of the physical skills, and later during training (while using the actual training system in evaluation). After the physical skills training, the subject matter expert feedback consisted of a verbal instruction on the aspects of the task execution that were correct or incorrect. During training with the system, participants were periodically evaluated and given feedback using an assessment instrument designed around the mnemonic. In an operational context, this feedback represents the after action review that is provided during field training operations.

The bootcamp was designed to provide quality pre-experimental instruction to trainees, resulting in an experimental group that had the basic KSAs for interacting in the targeted MOUT domain.

Transfer Evaluation Methodology: TER at Point Informed by Learning Curve Analysis

Learning has been shown to follow a universal power law of practice, generally adhering to a pattern of rapid improvement followed by ever-diminishing further improvements with practice (Ritter & Schooler, 2001). This implies that the rate of return for additional practice reaches a point where additional training is no longer cost-effective. Unfortunately, in practice the observation of an asymptote is usually visible only after extensive trials (for example, in some cases over 1,000 trials; Newell & Rosenbloom, 1981); thus an operational definition of plateau in terms of operational metrics (for example, cost) is required.

Plateau analysis of learning curves may be accomplished through several methods, such as by fitting curves and finding asymptotes, or by utilizing parametric or nonparametric statistical tests, such as analysis of variance (Grantcharov et al., 2004) or Friedman tests (for example, Grantcharov, Bardram, Funch-Jensen, & Rosenberg, 2003). Another approach that may be suitable for limited samples involves evaluating the plateau by identifying a period along the curve where improvement variability has slowed to a predetermined level.

Plateau analysis was applied in a MOUT training system ToT evaluation. A plateau analysis involves more than simple visual inspection. While at first glance it may seem tempting to visually determine the location of plateaus, the benefits of such an approach are dismissed once one observes the effect of graphing a scale on one's selection of such plateaus (that is, slopes may appear smaller in larger scales and vice versa). Such an approach lacks objectivity and reliability. This is why it is important to define an objective measure by which to determine where a plateau is present (within the acceptable parameters one defines as a plateau; a description follows).

Parameters

There are three parameters to consider in identifying a plateau within operational constraints. These are (1) percent variability, (2) period of variability stabilization, and (3) general location of plateau, which are illustrated in Figure 24.1 and described below. The process for determining these parameters is explained later in this chapter as applied to the MOUT domain.

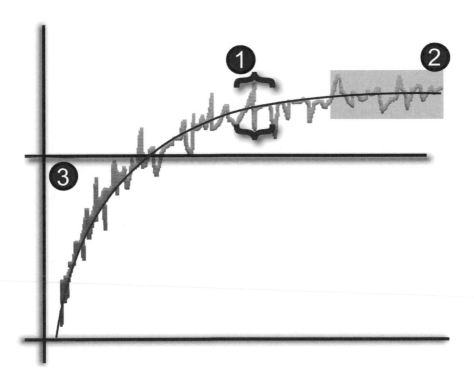

Figure 24.1.   Parameters to consider in identifying a plateau within operational constraints follow: (1) percent variability, (2) period of variability stabilization, and (3) general location of plateau.

1. Percent variability: Variability of a measure's cumulative average between subsequent trials is less than X percent. This parameter is used to define the acceptable gains (or losses) in performance across trials. It is defined as the number of resources needed for one additional trial over the acceptable amount of gains (for example, 10 percent improvement). In cases where limited data are available, the cumulative aggregate data are not as "smooth" so that an acceptable percentage is used as a gauge in variability for either gains or losses in performance. This parameter is of practical importance given that a true plateau may in theory be observed only after very extensive periods (for example, ~1,000 trials; Newell & Rosenbloom, 1981). Values used in the MOUT training system ToT evaluation ranged from 2.5 to 10 percent variability across trials.

2. Period of variability stabilization: In order to determine that a plateau has occurred with an acceptable variability level one must specify a suitable range of continuous performance (for example, performance variability within 10 percent for Y number of trials). Depending on the application, this may range from a few trials to "all remaining trials." Three to five trials have been used with adequate success (for example, Champney, Milham, Bell Carroll, Stanney, & Cohn, 2006).

3. General location of plateau: When using limited datasets it might be possible to observe, while using only parameters (1) and (2), localized plateaus (that is, periods where performance stabilizes and later increases outside the initial parameters). In such cases one might be required to establish a general rule for the location of the plateau. While this might be determined through a visual inspection, specifying a rule ensures consistency across multiple measures. In the MOUT training system ToT evaluation, the latter $1/3$ of the performance range was selected as a general area to qualify as a plateau [that is, the period in the performance scale where performance is hypothesized to stabilize per the universal law of learning; see the shaded area (2) in Figure 24.1]. Given the shape of a typical power curve, one is generally guaranteed that the true plateau would be at the tail of the curve.

Learning Curve Methodology

The proposed learning curve methodology involves a series of phases to arrive at an objective recommendation for the point at which to evaluate transfer under the premise that training on a simulator is to be optimized by maximizing improvement gains before testing for any transfer performance. These steps as applied to the MOUT training system evaluation are discussed below.

1. Understanding the data: Before parameters can be established, the nature and behavior of the selected metrics, which are aligned with specific tasks and training objectives, should be analyzed using data from a pilot study. For the MOUT TEE, this was done by constructing cumulative average plots and tables with the performance data collected (see Table 24.2). Important data characteristics to determine are valid ranges, expected variability from trial to trial, and expected data values (for example, continuous or discrete; note that this method is better suited for continuous data).

2. Determining parameter (1): This is performed by identifying acceptable performance gains per trial per metric through a cost-benefit analysis or in the absence of cost criteria, an arbitrary number (for example, 10 percent). In operational terms, this means that when improvement variability across trials is less than this threshold, the cost of continuing to train outweighs the desired benefits (the expected training

Table 24.2.   Example of Performance Data and Cumulative Data

Trial	Performance	Cumulative Performance
1	50%	50.00%
2	60%	55.00%
3	63%	57.67%
4	70%	60.75%
5	75%	63.60%
6	80%	66.33%

gains). For the MOUT TEE, parameter (1) ranged from 2.5 to 10 percent variability across trials depending on the particular measure.

3. Determining parameter (2): This step involves determining an appropriate parameter Y, which is the number of trials to use to determine that a plateau is present. There is no established standard for determining this parameter other than one's desire for rigor (where a higher number leads to added assurance of a true plateau). Five were used in the MOUT evaluation because, as reported by Champney, Milham, Bell Carroll, Stanney, and Cohn (2006), three produced too many relative plateaus and so a more restrictive number of trials was required.

4. Identifying plateaus in data tables: Utilizing cumulative data tables, the plateau is then identified as the first trial at which the established criteria, parameters (1) and (2), are met. For the MOUT TEE, the plateau was found at different trials depending on the measure; trial 25 was selected as the point of plateau as more than ⅔ of all measures had reached a plateau by that trial number.

5. Visual inspection: Utilizing cumulative data plots, the plateau is next verified to be an absolute plateau with respect to the available data and not just a local one. If it is determined visually that there is considerable learning still occurring past the identified trial, parameter (3) should be applied before reapplying parameters (2) and (3). For the MOUT TEE, it was necessary to apply parameter (3), which resulted in the identification of a more conservative plateau (that is, later in the curve, implying more trials) to minimize the probability of having a relative plateau.

The described TEE learning curve methodology can be used to identify the point at which a TER (Roscoe, 1971) evaluation should be performed; then leverage this calculation to determine the number of trials of live training saved as a result of pre-training. Applying this approach should avoid drawing conclusions about training transfer at points along the learning curve that have yet to stabilize, which is a risk when using the TER.

CONCLUSION

With the methodologies provided in this chapter, practitioners can conduct TEEs with considerably less resource involvement, both in terms of participant and evaluation costs. While the methodologies are not intended to replace the value afforded from the use of target domain participants or incremental evaluations, in the absence of available resources, the approaches can provide data to support informed decisions regarding training effectiveness.

REFERENCES

Baldwin, T. T., & Ford, J. K. (1988). Transfer of training: A review and directions for future research. *Personnel Psychology, 41,* 63–105.

Bassi, L. J., & van Buren, M. E. (1999). 1999 ASTD state of the industry report. Alexandria, VA: The American Society for Training and Development.

Carnevale, A. P., & Schulz, E. R. (1990). Economic accountability for training: Demands and responses. *Training and Development Journal Supplement, 44*(7), pp. s2–s4.

Champney, R. K., Milham, L., Bell Carroll, M., Stanney, K. M., & Cohn, J. (2006). A method to determine optimal simulator training time: Examining performance improvement across the learning curve. *Proceedings of the Human Factors and Ergonomics Society 50th Annual Meeting* (pp. 2654–2658). Santa Monica, CA: Human Factors and Ergonomics Society

Champney, R. K., Milham, L. M., Bell Carroll, M., Stanney, K. M., Jones, D., Pfluger, K. C., & Cohn, J. (2006). Undergraduate boot camp: Getting experimental populations up to speed. *Proceedings of the Interservice/Industry Training, Simulation & Education Conference.* (No. 2976). Arlington, VA: National Defense Industrial Association (NDIA).

Eseryel, D. (2002). Approaches to evaluation of training: Theory and practice. *Journal of Educational Technology and Society, Special Issue: Integrating Technology into Learning and Working, 5*(2), 93–98.

Flexman, R. E., Roscoe, S. N., Williams, A. C., Jr., & Williges, B. H. (1972, June). *Studies in pilot training* (Aviation Research Monographs, Vol. 2, No. 1). Savoy: University of Illinois, Institute of Aviation.

Gordon, M. E., Schmitt, N., & Schneider, W. (1984). An evaluation of laboratory research bargaining and negotiations. *Industrial Relations, 23,* 218–233.

Grantcharov, T. P., Bardram, L., Funch-Jensen, P., & Rosenberg, J. (2003). Learning curves and impact of previous operative experience on performance on a virtual reality simulator to test laparoscopic surgical skills. *American Journal of Surgery, 185*(2), 146–149.

Grantcharov, T. P., Kristiansen, V. B., Bendix, J., Bardram, L., Rosenberg, J., & Funch-Jensen, P. (2004). Randomized clinical trail of virtual reality simulation for laparoscopic skills training. *British Journal of Surgery, 91*(2), 146–150.

Milham, L., Gledhill-Holmes, R., Jones, D., Hale, K., & Stanney, K. (2004). *Metric toolkit for MOUT* (VIRTE Program Report, Contract No. N00014-04-C-0024). Arlington, VA: Office of Naval Research.

Newell, A., & Rosenbloom, P. S. (1981). Mechanisms of skill acquisition and the law of practice. In J. R. Anderson (Ed.), *Cognitive skills and their acquisition* (pp. 1–51). Hillsdale, NJ: Lawrence Erlbaum.

Ritter, F. E., & Schooler, L. J. (2001). The learning curve. In *International encyclopedia of the social and behavioral sciences* (pp. 8602–8605). Amsterdam: Pergamon.

Roscoe, S. N. (1971). Incremental transfer effectiveness. *Human Factors, 13*(6), 561–567.

Roscoe, S. N. (1972). A little more on incremental transfer effectiveness. *Human Factors, 14*(4), 363–364.

Roscoe, S. N., & Williges, B. H. (1980). Measurement of transfer of training. In S. N. Roscoe (Ed.), *Aviation technology* (pp 182–193). Ames: Iowa State University Press.

Saks, A. M., & Belcourt, M. (2006). An investigation of training activities and transfer of training in organizations. *Human Resource Management, 45,* 629–648.

Sadri, G., & Snyder, P. F. (1995). Methodological issues in assessing training effectiveness. *Journal of Managerial Psychology, 10*(40), 30–32.

Thompson, C., Koon, E., Woodwell, W. H., & Beauvais, J. (2002). *Training for the next economy: An ASTD state of the industry report* (Rep. No. #790201). Alexandria, VA: The American Society for Training and Development.

Ward, E. A. (1993). Generalizability of psychological research from undergraduates to employed adults. *Journal of Social Psychology, 133*(4), 513–519.

Wintre, M. G., North, C., & Sugar, L. A. (2001). Psychologists' response to criticisms about research based on undergraduate participants: A developmental perspective. *Canadian Psychology, 42,* 216–225.

Chapter 25

THE APPLICATION AND EVALUATION OF MIXED REALITY SIMULATION

Darin Hughes, Christian Jerome,
Charles Hughes, and Eileen Smith

Mixed reality (MR) is a blending of technologies that leverage the advantages and challenges of combining the real world with virtual objects and processes. Like virtual environment (VR) systems, MR can create entirely synthetic environments, objects, characters, and interactions, but unlike VR, MR can merge these virtual components with real world environments, objects, human characters, and human-to-human interactions. The advantage of MR is in its ability to set experiences in everyday environments and build off of the kinds of interactions that occur in real world experiences—all the while leveraging the power of richly layered visuals, sounds, and physical effects that are generated computationally.

With these benefits come the unique challenges of creating MR simulations. Virtual objects and real objects must be able to exist side-by-side and be properly registered in three-dimensional (3-D) space. Real objects must occlude virtual objects and vice versa depending on their relative location to a user. For example, if a user moves a hand in front of his or her face, virtual objects that are intended to be farther away must not "pop" out of space and into the user's hand. Additionally, virtual sounds and physical effects must be developed in a way that real world sounds and physical phenomenon are not precluded or contradicted.

This chapter describes the MR infrastructure, illustrated in Figure 25.1, developed by the Media Convergence Laboratory at the University of Central Florida, an overview of several unique simulations generated using this infrastructure (MR for training, education, entertainment, and rehabilitation), and last, an evaluation of several of these simulations in terms of training effectiveness and outcomes, knowledge, skill, attitude, presence, and simulator sickness.

MR INFRASTRUCTURE

The MR infrastructure described in the following sections has four central components: visual, auditory, haptic and digital multiplex (DMX) effects, and

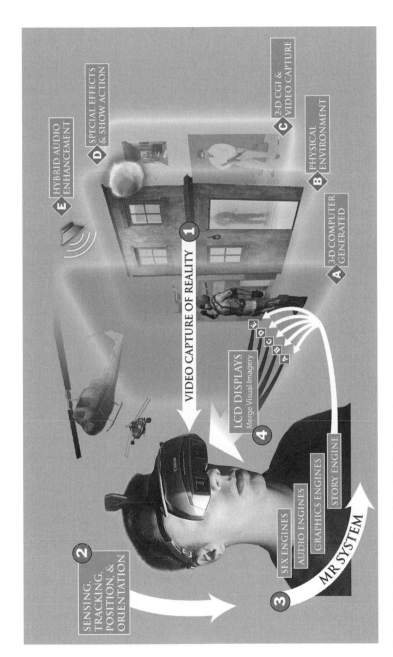

Figure 25.1. A Basic Overview of the MR Infrastructure

scripting. Taken together, these components enable the creation of richly layered and immersive mixed reality simulations.

VISUAL

The visual blending of real and virtual objects requires an analysis and understanding of the real objects so that proper relative placement, interocclusion, illumination, and intershadowing can occur. In the system we describe here, we will assume that, with the exception of other humans whose range of movement is intentionally restricted, the real objects in the environment are known and their positions are static. Other research we are carrying out deals more extensively with dynamic real objects, especially in collaborative augmented virtuality environments. Note, for instance, in Figure 25.2 that two people are sitting across from each other in a virtual setting; each has a personal point of view of a shared virtual environment, and each can see the other. In this case, we are using

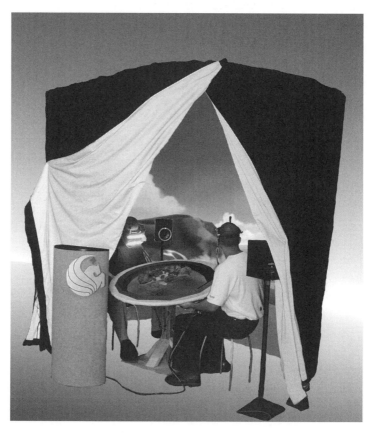

Figure 25.2.   Retroreflective Technology Being Used to Extract Dynamic Silhouettes of Real World Objects

unidirectional retroreflective material so each user can extract a dynamic silhouette of the other (C. E. Hughes, Konttinen, & Pattanaik, 2004). These silhouettes can be used to correctly register players relative to each other and, consequently, relative to virtual assets.

The primary visual issues are (a) lighting of real by virtual and vice versa and (b) shadowing of virtual on real and vice versa. We and our colleagues developed the real time algorithms (see, for example, Nijasure, Pattanaik, & Goel, 2003; C. E. Hughes, Konttinen, et al., 2004), some of which were based on work by Haller, Drab, and Hartmann (2003). Here we just note that each real object that can interact with virtual ones has an associated phantom or occlusion model. These phantoms have two purposes. When used as occlusion models, invisible renderings of phantom objects visually occlude other models that are behind them, providing a simple way to create a multilayered scene; for example, the model of a sea creature is partially or fully hidden from view when it passes behind a display case. When used for lighting and shadows on real objects, these phantom models help us calculate shading changes for their associated pixels. Thus, using them, we can increase or decrease the effects of lights, whether real or virtual on each pixel.

The specific algorithms we have developed can simply and efficiently run on the shaders of modern graphics cards. This graphics processing unit implementation, as well as careful algorithm design, allows us to achieve an interactive frame rate, despite the apparent complexity of the problem (McGuire, Hughes, Egan, Kilgard, & Everitt, 2003).

The virtual objects are lit by the flashlight, and the real box is both lit by the flashlight and darkened by the shadows cast from the virtual teapot and ball. For this simple demonstration, we tracked the box, the "hot spot" on the table and the cylinder using ARToolkit, an image based tracking. In general, though, our preferred tracking method is acoustical, with physical trackers attached to movable objects.

Viewing these scenes can be done with a video see-through head-mounted display (HMD), a mixed reality window (a tracked flat screen monitor that can be reoriented to provide differing points of view), or a mixed reality dome. While the HMD is more flexible, allowing the user to walk around an MR setting, even to stare virtual three-dimensional (3-D) characters in the eye, it is more costly and creates far more problems (for example, hygiene, breakage, and physical discomfort) than the MR window or dome. Both the MR window and the dome require an added navigation interface (for example, control buttons and/or a mouse), since neither is moveable, unlike the HMD whose user can walk around, somewhat freely. The window is more flexible than the dome in that it can be physically reoriented, but it lacks the convenient audience view and the sense of immersion (both visual and auditory) of the dome.

AUDITORY

In the film industry, there is an expression that "the audio is half the experience." This expression is validated by the careful attention to sound and music in film. However, audio production in the simulation community is often given little attention, if any at all. While simulation companies or research institutes may own expensive audio equipment (such as tracked 3-D headphones, and so forth), they often may not allocate resources toward production techniques or sound designers, depending instead upon nonaudio specialists to insert generic sound effects from purchased libraries. The end result of this process is a shallow, unrealistic, nonimmersive auditory environment.

The irony in this arrangement is that audio is at least half of the human experience. Auditory cues are perceived in 360° and on all three axes. Sound can travel through walls and around corners, providing information that is well out of the line of sight. Additionally, audio plays a crucial role in environmental recognition, immersion, and presence and is essential in most forms of communication.

The desire to create an easily configurable, powerful audio engine and a high level interface came about over the course of designing audio for interactive experiences during the last several years. These experiences include exhibits at SIGGRAPH (Special Interest Group on Graphics and Interactive Techniques) 2003, ISMAR (International Symposium on Mixed and Augmented Reality) 2003/2004, I/ITSEC (Interservice/Industry Training, Simulation, and Education Conference) 2002/2003, IAAPA (International Association of Amusement Parks and Attractions) 2002/2003, and the Orlando Science Center, as well as long-term installations at the U.S. Army's Simulation Technology Training Center, Orlando, Florida. Such standard media production tools as sonar, Pro Tools, Cubase, and so forth, while very useful for synthesizing, mixing, and mastering, cannot provide the kind of dynamic control necessary for interactive simulation. In addition to lacking any support for real time spatialization, they do not have features to compensate for suboptimal speaker placement or expanded, multi-tiered surround systems. Both the previously mentioned features are essential since many interactive experiences must occur in environments where optimal speaker placement is not possible and where sounds along the vertical plane are essential both for immersion and for accuracy of training.

The shortcomings of using one of these media production tools in simulations are well documented in the Institute for Creative Technologies' 2001 audio technical report (Sadek, 2001). Their system was based around Pro Tools. Sounds were generated dynamically by sending MIDI (musical-instrument digital interface) triggers to the applications. While this arrangement had some success with "tightly scripted and choreographed scenarios," it was entirely incapable of creating any dynamically created panned sounds. Additionally, their system was further hampered by an inability to trigger more than three pre-mixed sound sets simultaneously.

We also purchased and investigated the AuSIM 3D Goldminer system, which is an integrated hardware and software solution for audio simulation. While producing fairly realistic surround impressions, its main drawback for mixed reality is its use of headphones as its delivery method. It is very important to be able to hear real world sounds in an MR experience. This is especially true in a military training scenario where it is not only essential to hear the footsteps and movements of virtual characters, but also of other human interactors. More information can be found about AuSIM 3D through the Web site http://ausim3d.com.

For the specific demands of a highly dynamic and immersive MR simulation, it became clear that a system must be built either on top of an existing application programming interface (API) or from scratch. Some of our early attempts involved the use of the Java Media Framework that, while providing dynamic cueing, did not support multichannel output, spatialization, and channel control, among many other things (C. E. Hughes, Stapleton, et al., 2004).

An extensive review of the available technology was conducted, including both proprietary and open source hardware and software solutions. Due to the specific demands of the MR audio, EAX (environmental audio extension) 2.0 was selected as the appropriate environment for building interactive audio experiences. However, this technology was also limited in its ability to address specific channels and provide control of multiple hardware arrangement through a single application. In addition, EAX does not provide the kind of low level support for the creation of digital signal processing (DSP) effects—rather a static set of effects are provided. The frustration of working with these technologies due to the particular demands of interactive, immersive simulation led to the design and creation of a custom-built audio engine and high level interface called SoundDesigner.

SoundDesigner was conceived as an intuitive application that allows the user to create or modify entire soundscapes, control all output channels of connected sound cards, designate sound states, and support a variety of delivery systems and classifications. The user of SoundDesigner can individually address audio channels and assign some to a surround system while leaving others open for use with such devices as point source speakers. To achieve this, SoundDesigner needed to be built on top of an API that allows low level control over audio hardware.

Most computer-generated simulations lack the ability to be reconfigured easily and quickly. SoundDesigner allows nonprogrammers and audio novices to assert a high level of control over the soundscape and auditory structure of a scenario. This is particularly useful in simulations where various factors, such as ambient noise, cueing, expectation, and other important variables, can be modified for purposes of evaluation. This software also allows for easy configuration of new audio scenarios or the alteration of previous simulations without the need for reprogramming.

The implementation of a high level interface to an advanced audio architecture such as SoundDesigner requires the definition of new abstractions to represent system components in an intuitive way. The SoundDesigner interface represents individual sound clips in terms of *buffers,* which represent the discrete samples of a sound and *sources,* which represent instances of a buffer currently being played. These are common audio concepts in other libraries, such as OpenAL, but SoundDesigner is unique in providing an explicit representation of individual *speakers,* which it groups and addresses through the interface of *channels.* Each sound source is played on a specific channel, which is to say that the samples generated by that source are mixed, filtered, and output to the speakers bound to that channel. The two fundamental channel types are *point-source* channels (which simply mix and copy channels to all speakers) and *spatialized* channels (which use information about the position of sounds and speakers to perform a per-speaker attenuation on samples in order to associate each source with a specific spatial direction). An overview of SoundDesigner features follows: *support for 3-D sound, assignable channels (3-D and point source), multitiered speaker configurations, configurable speaker placement, real time spatialization, user placement compensation, timeline triggers prescripted paths with waypoints (linear and curved), real time capture and playback of sound (with full SoundDesinger support), basic DSP (echo and reverb), savable configuration files, such standard features as looping, volume control, and envelopes, and the ability to address multiple sound cards.*

Once sounds have been arranged inside a SoundDesigner configuration file or set aside as mono, real time sounds, a "naming and associations" document is passed along to the MR StoryEngine programmer for the final phase of integration into a scenario. This document contains all of the user IDs, their story associations (for example,"trainee_fire" is called when the gun is triggered), an indication of whether the sound is prescripted or real time, and the appropriate file directory to find the configuration file and real time sounds. For full implementation details, see D. E. Hughes (2005).

HAPTICS AND PHYSICAL EFFECTS

The MR infrastructure includes a special-effects engine that employs traditional scenography and show control technology from theme parks to integrate the physical realities of haptic and olfactory devices. For lights, smoke machines, olfactory devices, and so forth, a standard protocol (DMX) is utilized. This allows for dynamic, real time, and variable control over these physical effects. Haptic vests are controlled through a series of actuators that allow for different levels of intensity.

In addition to typical haptic devices, such as vests, haptic audio devices are employed as well. Haptic audio refers to sounds that are felt more than heard. Such effects can be achieved using subwoofers and mounted "bass shakers" that physically vibrate floors, walls, and other surfaces. They can be used to increase the sense of realism and impact, but can also be used to provide informational cues.

In the case of haptic vests, pressure points can be used to alert participants to the direction of targets or potential threats. These feedback mechanisms can be used in coordination with detection devices. As a personalized, tracked audio display within a haptic vest, these devices provide directional cues without cluttering up the already intense acoustic audioscape. With the use of speakers that vibrate more for feeling than for hearing, an intimate communication of stimulating points in the body provides the approximate orientation of potential threats that may not be heard or seen. Thus, a threat may be identified by a vibration or combination of vibrations. This information can give an immediate sense of the direction of a threat and its proximity (for example, by making the vibration's intensity vary with the distance to the threat). It works in essence like a tap on the shoulder to tell the user of a direction without adding to or distracting from the visual or acoustic noise levels. This message is transferred to an alternative sense and thus allows for this critical datum to cut through the clutter of the audiovisual simulation. With targets outside of the line of sight, this approach can significantly reduce a user's response time.

SCRIPTING

The current incarnation of our framework utilizes an XML (Extensible Markup Language) scripting language, based on the concepts of interacting agents, behaviors, guards, and state information. This separates the scriptwriters from the internals of the engine, while providing them a meaningful and effective context in which to encode simple, direct behaviors (O'Connor & Hughes, 2005). The reorganization of the system prompted the development of other supporting engines, dubbed auxiliary physics engines (APEs). These engines are responsible for tasks such as pathfinding and ray casting, since our revised architecture attempts to make distinct and clear the tasks of each engine.

The philosophy of a distributed system was key to the construction of this framework. The StoryEngine is the hub, providing scriptwriters access to any presentation requirements they need. For complex cases, our XML based script language allows one to escape into a special sublanguage, dubbed the advanced scripting language, or ASL. The ASL provides the ability to code behaviors using C-style programming constructs. Such constructs include loops, conditionals, basic arithmetic, and assignment operations.

The script defines a set of *agents,* each of which generally embodies some character the user may interact with (directly or indirectly). Agents are defined in terms of behaviors, which include actions, triggers, and reflexes and a set of state variables that define an agent's current state. Each behavior can perform several tasks when called, such as state modification and the transmission of commands to the presentation engines. Thus, agents are the fundamental building blocks of the system. The ability for agents to communicate with each other allows for a "world-direct" representation to be built: developers define a set of agents in terms of how they want them to act around each other, rather than such actions being a side effect of a more program-like structure.

The graphics and audio engines understand the same basic set of commands. This allows the scriptwriter to easily generate worlds that offer visual and audio stimulations. Each engine also has a set of commands unique to its particular functionality (for example, audio clips can be looped, and visual models can have associated animations).

The SFX engine utilizes the DMX protocol, but control over it originates from the StoryEngine through a series of commands, most of which are defined by loadable "DMX scripts." These scripts are direct control specifications that offer a set of basic functions (typically setting a device to a value between 0 and 255, meaning off to fully on and anything in between). These primitives are hooked together to form complex DMX events.

In our older versions of the system, agent information, such as position and orientation, was managed by the graphics engine. This required the StoryEngine to request regular updates, thus causing network congestion when the number of agents was high. The current incarnation of the system does away with this, and now all physics simulation is performed by the StoryEngine. The data are transmitted as a binary stream, encapsulated in a cross-platform and cross-language format. The data stream is denoted the "control stream," as it controls the position, orientation, velocities, and accelerations of agents. A given control stream is broken up into numbered channels, one channel for each agent (channel numbers are automatically assigned to agents and are accessible through the reserved state variable name *channel*). This enables us to transmit only a subset of the data, usually only that which has changed since the last transmission. The system scales remarkably well.

Many of the distributed capabilities not only involve the major engines, but also a set of utility servers. One type of utility server, the auxiliary physics engine, was referenced earlier. Two APEs were developed for our projects: one to control pathfinding on a walk mesh and another to manage ray casting in a complex 3-D universe. These engines plug in at run time and simply serve the requests of agents.

Another utility server is the sensor server. This basically abstracts data from position and orientation sensors into data streams, which are then transmitted across a network to interested clients. This allows any number of agents to utilize the data. The data stream is transmitted via transmission control protocol/Internet protocol for reliability purposes. The data format follows that of the StoryEngine's control stream data. Thus, to a graphics or audio engine, it is immaterial where control data come from; a given agent's control may be governed by the user's own movements or that of a simulated entity. The sensor server also enables us to record user movement, a vital piece of information for after action review (a military training term, but equally important for such rehabilitation applications as MR Kitchen) and cognitive experimentation.

The ability to define a set of behaviors to be reused in several scripts came to life in the "script component" architecture. This architecture allows "component" files to be written by the scriptwriter and then be included in any number of scripts. Behaviors or entire agents can be scripted and, consequently, included into the main script. This also means that difficult-to-code behaviors and algorithms can be written once and used repeatedly, without having to perform copy-and-paste operations or rename a vast number of states and agents. The StoryEngine allows object-oriented capabilities, such as prototype based inheritance and delegation, to make coding agents reasonably straightforward and simple.

A final and rather recent innovation to the architecture was that of a remote system interface. Originally designed as an interface to allow remote (over-the-network) access to agent state information for display on a graphical user interface, the remote graphical user interface (GUI) protocol also provides a way to transmit information back to the StoryEngine. It is, in effect, a back door into the virtual world controlled by a given script, whereby agent command and control can be affected by a purely remote, alien program. We recently took advantage of this capability to link our system to DISAF (Dismounted Infantry Semi-Automated Forces), an artificial intelligence system that provides behaviors used in many distributed interactive systems applications.

Used for its original purpose, the remote GUI protocol and program architecture allow graphical interfaces to be defined by a simple XML file. The file specifies graphical components to be used, as well as options for each component that link it to agents and states in the script. An example of this would be to display the number of times a particular agent was encountered by the user. A simple state variable in the agent itself would keep track, and changes to that information would be retrieved and displayed by the remote GUI. This approach is amenable to a drag-and-drop approach for creating such GUIs, something we will do in the next release of the software.

All of the previously described technologies have been configured or hardwired into a number of various testbeds for different applications. The following section will describe notable configurations that have shown promise as useful training testbeds.

MR FOR TRAINING—MR MOUT

The MR MOUT (military operations in urban terrain) testbed is a training simulation that re-creates urban façades to represent a 360° mini MOUT site. Tracking employs the InterSense IS-900 acoustical/inertial hybrid system. The tracked area contains virtual people (friends, foes, and neutrals), real props (crates, doors, and a swinging gate), a realistic tracked rifle, real lights, and building façades. Standing inside the mini MOUT creates the sense of the reality faced by a dismounted soldier who is open to attack on all sides and from high up.

Using a combination of blue screen technology and occlusion models, the real and virtual elements are layered and blended into a rich visual environment. The trainee has the ability to move around the courtyard and hide behind objects with real and virtual players popping out from portals to engage in close-combat battle. The most effective and powerful result of this mixed reality training is the fact that the virtual characters can occupy the same complex terrain as the trainees. The trainees can literally play hide-and-seek with virtual foes, thereby leveraging the compelling nature of passive haptics.

Figure 25.3 shows the mini MOUT from the observer station. In the middle, to the right of the observer, you can see the participant with HMD and a rifle. That person's view is shown on the screen mostly blocked by the observer; the other

Figure 25.3. MR MOUT: Demonstrates Observer Views (Mixed and Virtual) with a Direct View of the Real World Containing an MR Participant

three views are from an observer camera (middle right) and two virtual characters (lower right and top center).

Notice the crates in the view on the middle right side. The models that match these physical assets are rendered invisibly, providing appropriate occlusion (they clip the rendered images of characters that they would partially or totally occlude in the real world). Special effects complete the creation of a realistic combat scenario where the real world around the trainee feels physically responsive. This is done using the SFX engine to control lights, smoke from explosions, and other types of on/off or modulated actions. The system can react to the trainee based on position, orientation, or actions performed with a gun that is tracked and whose trigger and reload mechanism are sensed. For example, the lights on the buildings can be shot out (we use a simple ray casting auxiliary physics engine that returns a list sorted by distance of all intersected objects), resulting in audio feedback (the gunshot and shattered glass sounds) and physical world visual changes (the real lights go out).

With all the compelling visual and haptic effects, users' hearing and training can provide a competitive edge, due to a heightened acoustical situational awareness (D. E. Hughes, Thropp, Holmquist, & Moshell, 2004). They cannot see or feel through walls, around corners, or behind their heads. However, their ears can perceive activity where they cannot see it. In urban combat where a response to a threat is measured in seconds, realistic audio representation is vital to creating a combat simulation and to training soldiers in basic tactics. Standard 3-D audio with earphones shuts out critical real world sounds, such as a companion's voice or a radio call. The typical surround audio is still two dimensional (x and z axis) with audio assets designed for a desktop video game that tend to flatten the acoustical capture. Our system allows audio to be synchronized temporally and spatially, leading to an immersive experience.

MR FOR EDUCATION—MR SEA CREATURES

The experience begins with the reality of the Orlando Science Center's Dino-Digs exhibition hall—beautiful fossils of marine reptiles and fish in an elegant, uncluttered environment. As visitors approach the MR dome, a virtual guide walks onto the screen and welcomes them to take part in an amazing journey. While the guide is speaking, water begins to fill the "hall" inside the dome. As it fills, the fossils come to life and begin to swim around the pillars of the exhibit hall. The dome fills with water and visitors experience the virtual Cretaceous environment. The visitors are able to navigate a rover through the ocean environment to explore the reptiles and fish. The viewing window of the rover is shown in the heads-up display of the MR dome. (See Figure 25.4.)

As the experience winds down, the water begins to recede within the dome, and the unaugmented science center hall begins to emerge again. At about the point where the water is head high, a pterodactyl flies overhead, only to be snagged by a tylosaur leaping out of the water. Holding the pterodactyl in its mouth, the tylosaur settles back down to the ocean floor. When all the water

Figure 25.4.   MR Sea Creatures: Free Choice Learning Using Mixed Reality at Orlando Science Center

drains, the reptiles and fish return to their fossilized reality at their actual locations within the hall. A walk into the exhibit space reveals that the tylosaur was trapped in time with the pterodactyl in its mouth. This connection of the MR experience back to the pure real experience is intended to permanently bond the experiences together in the visitor's mind.

The purpose of an informal education experience is to inspire curiosity, create a positive attitude toward the topic, and engage the visitor in a memorable experience that inspires discussion long after the visit. One of our research initiatives is in creating experiential learning landscapes, where the currently harsh boundaries between learning in the classroom, learning at a museum, and learning at home become blurred. MR Sea Creatures is our first MR museum installation intended for this purpose. We have, in fact, already experimented with a non-MR installation that supported extended experiences to the home and school

(C. E. Hughes, Burnett, Moshell, Stapleton, & Mauer, 2002). Its success, though on a small scale, has helped to strengthen our convictions.

MR FOR ENTERTAINMENT—MR TIME PORTAL

MR Time Portal, publicly shown at SIGGRAPH 2003, was the first widely seen experience we developed that involved complex 3-D models with rich animations and a nontrivial storyline. Its goal was to immerse participants within a story, with some people at the center of the action and others at the periphery. Figure 25.5 (left) is a scene from an animatic[1] we produced that helped us to test story elements while still in VR mode. Figure 25.5 (right) shows the full MR with one person on the right wearing an HMD in order to be embedded in the experience and two people at a vision dome on the left observing the experience from the perspective of an unseen second participant. In essence, this is an MR version of a theme park experience employing those venues' notion of divers (the ones who get in the action), swimmers (those who influence the action), and waders (those who observe from afar) (Stapleton & Hughes, 2003, 2005).

In 2003, our StoryEngine was based on the concept of Java objects holding the states and primitive behaviors of actors, each having an associated finite state machine (Coppin, 2004) that controlled the manner in which these behaviors were invoked based on stimuli such as timed events, GUI inputs, and interactions with other actors. Most actors reflected the virtual and active real objects of the MR world, but some existed to play the roles of story directors, encouraging players in directions deemed most supportive of the underlying story.

For instance, the MR time portal contained actors associated with a back-story movie, the portal through which threats to our world arose, various pieces of background scenery, each robotic threat, each friendly portal guard, a futuristic physical weapon, a ray-tracing beam to make it easier to aim the gun, a number of virtual explosions, the lighting rig above the exhibit area, several abstract objects operating as story directors, and, of course, the real persons who were experiencing this world. Each of these actors had optional peers in our graphics engine, audio engine(s), and special effects engine. The reason these are optional is that, at one extreme, abstract actors have no sensory peers; at the other extreme, robotic threats have visual representations, audio presentations that are synchronized in time and place with the visuals, and special effects when the robots hit the ground (a bass shaker vibrates the floor under the shooter); in between are such things as the lighting rig that has only a special effects peer.

An actor component, when added to the authoring system, had a set of core behaviors based on its class. An actor class sat at the top of this hierarchy providing the most common default behaviors and abstract methods for required behaviors for which no defaults exist. A finite state machine, consisting of states and transitions, was the primary means of expressing an actor's behavior. Each transition emanated from a state, had a set of trigger mechanisms (events) that enabled the transition, a set of actions that were started when the transition was selected, and a new state that was entered as a consequence of carrying out the

Figure 25.5.   MR Time Portal: Experiential Movie Trailer: (Left) Animatic and (Right) Mixed

transition. A state could have many transitions, some of which had overlapping conditions. If multiple transitions were simultaneously enabled, one was selected randomly. The probability of selecting a particular transition could be increased by repeating a single transition many times (the cost was just one object handle per additional copy). States and transitions could have associated listeners causing transitions to be enabled or conditions to be set for other actors.

MR FOR REHABILITATION—MR KITCHEN

The goal of the MR Kitchen was to demonstrate the use of MR in simulating a cognitively impaired person's home environment for purposes of helping that individual regain some portion of independence. More broadly, the goal was to experiment with the use of MR as a human experience modeler—an environment that can capture and replicate human experiences in some context. Here the experience was making breakfast, and the context was the individual's home kitchen (Fidopiastis et al., 2005).

Experience capture starts by recording the spatial, audio, and visual aspects of an environment. This is done at the actual site being modeled (or a close approximation) so we can accurately reproduce the space and its multisensory signature. To accomplish this we employ a 3-D laser scanner (Riegl LMS-Z420i), a light capture device (Point Grey Ladybug camera), and various means of acoustical capture (Holophone H2-PRO, stereo microphones on grids, transducers to pick up vibrations and sounds in microenvironments, and even hydrophones for underwater soundscapes). Once captured, models of this real environment can be used to augment a real setting or to serve as a virtual setting to be augmented by real objects. This MR setting immerses a user within a multimodal hybrid of real and virtual that is dynamically controlled and augmented with spatially registered visual, auditory, and haptic cues.

For our MR Kitchen experiment, we went to the home of a person who had recently suffered traumatic brain injury due to an aneurism. Spending about two hours there, we "captured" his kitchen (see the bottom right monitor of Figure 25.6 for an image of him in his home kitchen). This capture included a point cloud, textures, and the lighting signature of the kitchen and its surrounds (audio was not used for the experiment). We then built parts of the real kitchen out of plywood to match the same dimensions and location of critical objects (pantry, silverware drawers, and so forth). We purchased a refrigerator, cupboard doors, a coffee maker, and toaster oven and borrowed common items (cups, utensils, and favorite cereal). Figure 25.6 shows two participants in this kitchen. The screen on the left shows the view from the man on the right. Notice that the real objects are present; however, the textures of the counter and doors are the same as in the subject's home.

All aspects of the subject's movement and his interaction with objects and the human therapist are captured by our system as seen in the center monitor of Figure 25.6. This capture includes a detailed map of the subject's movement and head orientation, allowing for analysis and replay. Additionally, cameras can be

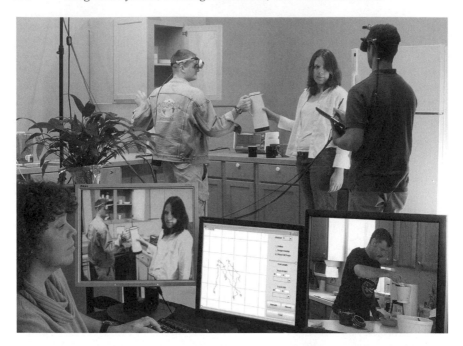

Figure 25.6.   MR Kitchen: Cognitive Rehabilitation—Demonstrates Real and Mixed Views, Captured Movement Data for One Participant, and Subject's Home Kitchen

positioned to capture any number of observer viewpoints of his activities and those of his therapist. Replaying the experience allows viewing events from multiple perspectives and with appropriate augmentation (for example, data on the user's response times and measured stress levels).

EFFECTIVENESS

The lessons learned throughout the MR evolution have led to the design and production of a robust training environment that is capable of simulating real world tasks in a safe environment. Field testing across various applications, including installations for entertainment, free-choice learning, training, and cognitive rehabilitation, provided much feedback to the iterative design lifecycle, as well as revealing new training insights. The success of our MR environment as a training tool has been assessed in a number of ways during a number of research efforts. The following section will summarize how effective this environment can be as part of a training program.

TRAINING OBJECTIVES AND/OR OUTCOMES

Training effectiveness can be operationally defined a number of different ways summarizing various facets of the participants' personal change or improvement over their pre-training state. The overall objectives or outcomes of interest of the training system must be decided upon in order to determine how well the system meets these training goals. Effectiveness is typically quantified by measures of how much information is obtained during and retained after the training tasks, by various measures of skill improvement via task performance, and by changes in motivation or attitude.

Knowledge

One way to assess training effectiveness is to observe the amount of information a participant is able to recall after training has occurred in the MR environment. MR has high face validity as a learning tool since it presents information from multiple sensory modalities, which improves recall and recognition. Visual images are especially helpful; much research has shown that people seem to be better at remembering images than remembering words (Bower & Winzenz, 1970); however, this imagery benefit can be exploited to improve memory for words as well (Sweeney & Bellezza, 1982). This can be seen nonexperimentally with MR Sea Creatures and MR Kitchen experiences.

MR Sea Creatures presents fossils of marine fish and reptiles from the Cretaceous period in an uncluttered learning environment (C. E. Hughes, Stapleton, Hughes, & Smith, 2005). The goal was to present an environment to the users in such a way that would be highly memorable, as well as fun, in order to inspire curiosity and create a positive attitude toward the topic. Subjective questionnaires revealed that 98 percent of the users felt that the environment encouraged longer time spent exploring the sea creatures and that more than 80 percent felt that the system encouraged them to return to the exhibit again. Although this reveals more of the user's attitude than knowledge gained, more time in the system would likely lead to more information retention; 83 percent of users even reported that they felt as though they had learned more about the sea creatures of the Cretaceous period.

Skill

Skill or task performance can also be used to assess training effectiveness. If training has been successful, an improvement in task performance can be seen over the performance when the training system is not used. Recent research by Jerome, Witmer, and Mouloua (2005) and Jerome (2006) have used the MR MOUT environment to investigate human perception, attention, and performance. The first goal was to determine if the user can successfully locate where stimuli are being displayed visually, auditorily, or haptically using the MR system. Further, this research explored whether people can focus attention on specific spatial locations of the visual scene when cued from either visual, auditory, or tactile cues; if there are any differences when the user is cued using similar cues, but with no spatial information; if the user can be cued to focus attention of differing breadths; and how workload may interact with these modalities. In the first experiment, the effectiveness of the cues was assessed, that is, whether spatial information can be determined from the computer-generated stimuli. Spatial cues were three different types: visual cues, auditory cues, and a combination of visual and audio cues (unfortunately, the tactile vest was not ready in time for the first study). While visual cues led to more targets being accurately acquired than audio cues (the mean difference was −0.537, standard error (SE) of 0.044, and 95 percent confidence interval (CI) −0.628 to −0.447, which does not include zero, indicating that the null hypothesis of zero difference can be rejected), the combination of audio and visual cues together produced even better performance than the visual cues alone (mean difference was 0.100, SE of 0.044, and 95 percent CI 0.009 to 0.191). The presence of visual cues in the visual and audiovisual conditions allowed participants to pinpoint the targets more accurately. The primary value of including audio cues is to direct participants to targets that are not within the immediate line of sight. Audio cues and visual cues did not differ significantly in affecting the speed of acquiring targets. In contrast, the audiovisual combination produced significantly faster acquisition times than either cue modality used alone.

In the second experiment, these spatial cues (including tactile cues) were incorporated into an interactive simulated scenario and were also varied in size (small, medium, large, none) to determine the performance effects of different levels of spatial information. During the experimental task, 64 participants searched for enemies (while cued from visual, auditory, tactile, combinations of two, or all three modality cues) and tried to shoot them while avoiding shooting the civilians (fratricide) for two 2 minute low workload scenarios and two 2 minute high workload scenarios.

The results showed significant benefits of attentional cuing on visual search task performance as revealed by benefits in accuracy from the presence of the tactile cues ($M = 0.44$, $SD = 0.13$) was significantly better than the presence of visual cues ($M = 0.31$, $SD = 0.10$), $t(42) = -2.57$, $p < 0.01$; and the presence of audio cues ($M = 0.36$, $SD = 0.07$) was significantly better than no cues ($M = 0.26$, $SD = 0.07$), $t(42) = 1.99$, $p < 0.05$) when displayed alone; and the combination of the visual and tactile cues together ($M = 0.43$, $SD = 0.09$) was significantly better than the combination of audio cues and visual cues together ($M = 0.32$, $SD = 0.13$), $t(42) = 2.31$, $p < 0.05$. Fratricide occurrence was shown to be amplified by the presence of the audio cues, that is, significantly higher for the audio cues ($M = 0.047$, $SD = 0.035$) than for the control group ($M = 0.016$, $SD = 0.022$), $t(42) = 2.63$, $p < 0.01$. The two levels of workload produced differences within individual's task performance for accuracy, $F(1, 56) = 6.439$, $p < 0.05$, and reaction time $F(1, 56) = 11.426$, $p < 0.001$. Accuracy and reaction time were significantly better with the medium-sized cues than the small cues [accuracy = $F(1, 56) = 13.44$, $p < 0.01$, reaction time = $F(1, 56) = 4.31$, $p < 0.05$], large cues [accuracy = $F(1, 56) = 17.37$, $p < 0.01$, reaction time = $F(1, 56) = 8.56$, $p < 0.01$], and the control condition (cues with no spatial information) [accuracy = $F(1, 56) = 63.62$, reaction time = $F(1, 56) = 17.67$, $p < 0.01$] during low workload and marginally better during high workload. Generally, cue specificity resulted in better accuracy and reaction time with the medium cues.

Attitude

Attitude and motivation can also be used to assess the effectiveness of a training system. Although knowledge gained and skill improvement are more objective measures that more clearly represent training effectiveness, training could be undermined without motivation and a positive attitude toward the system. Subjective data that represents the user's feeling toward the system was also collected with the Sea Creatures research described above. Results revealed that 88 percent of users felt that the experience was entertaining, and 84 percent would be motivated to visit this exhibit or similar exhibits in the future.

PRESENCE

Similar to attitude and motivation, presence does not usually represent training effectiveness; however, research shows that without a feeling of presence, training effectiveness can be undermined, and with presence, training effectiveness can be enhanced. To become immersed, that is, to have the feeling of "being there" in the artificial environment and to become removed from real world stimuli is referred to as presence and enhances the individual's experience in a VE (Witmer & Singer, 1998). Individual differences can moderate the effects of a particular immersive VE on an individual's feeling of presence. Individual differences in immersive tendency, aspects of the technology affecting the sense of presence, and negative side effects of the VE causing sickness symptoms may mediate VE task performance and training effectiveness using these systems. Measuring these mediating effects is of great importance toward understanding the relationships among them and, of course, maximizing the effectiveness of the training simulation.

In a study by Jerome and Witmer (2004), data from 203 Orlando, Florida, college students from five separate studies using various VEs and tasks were analyzed using the structural equation/path analysis module in STATISTICA. To judge the fit of the hypothesized model to the data, the goodness-of-fit index (GFI) is generally used and should be above 0.90 to show good fit. The GFI for the hypothesized model is 0.94, showing good fit of the data. The results suggest that a sense of presence in VEs may have a direct causal relationship upon VE performance, and immersive tendency and simulator sickness may have an indirect relationship with VE performance, both fully mediated through presence. The findings imply that improving the virtual experience to enhance the feeling of presence may improve human performance on virtual tasks. Also, since immersive tendency is a characteristic of the individual, it may not (easily and/ or quickly) be manipulated. Therefore, VE task performance may not be improved by immersive tendency. However, it may be used as a post hoc explanation of low/high task performance. VE designers may benefit from such results such that they could tweak an environment to capitalize on presence-enhancing features. Consequently, the VE may be a more usable, entertaining, and trainable tool.

SIMULATOR SICKNESS

Sickness symptoms caused by simulator exposure has long been thought to undermine training effectiveness (Kennedy, Fowlkes, & Lilienthal, 1993). Recent research has suggested that the relationship might be indirect, that is, that simulator sickness does not directly cause poor task performance and reduced learning, but first causes a reduction in the feeling of presence, and the reduced presence is directly related to the reduction in training effectiveness. Jerome and Witmer (2004) showed these results in their structural equation modeling analysis described above. The analysis showed a negative zero order correlation between simulator sickness and performance; however, when put into the structural equation modeling model considering the effects of presence and immersive tendency, all of the variance explaining the relationship between sickness and performance is rerouted through the presence construct. These results suggest that reducing the occurrence and severity of simulator sickness may improve task performance indirectly by increasing the sense of presence felt. The MR environments used in the Media Convergence Laboratory have shown very few simulator sickness symptoms and at the same time have created a high level of presence as subjectively reported from the users. Therefore, these MR environments will not have significant reductions in training effectiveness due to these issues.

CONCLUSION

This chapter describes the evolution of one specific system for authoring and delivering MR experiences. We make no specific claims about its comparative benefits over other systems, such as AMIRE (authoring mixed reality) (Traskback, 2004), MX Toolkit (Dias, Monteiro, Santos, Silvestre, & Bastos, 2003), Tinmith-evo5 (Piekarski & Thomas, 2003), and DELTA3D (http://www.delta3 d.org). Rather, our goal is to note the challenges we faced creating complex MR experiences and, within this context, to describe our means of addressing these issues.

As in any project that is coping with an evolving technology, we must sometimes provide solutions using existing and new technologies (for example, solving clipping problems with blue screens and then employing unidirectional retroreflective material in contexts that require the dramatic effects of changing real light). Other times we need to develop new scientific results, especially in the algorithmic area as in addressing realistic illumination and associated shading and shadowing properties in interactive time (Konttinen, Hughes, & Pattanaik, 2005). Yet other times we must create new artistic conventions to deal with issues not easily solved by technology or science (for example, taking advantage of people's expectations in audio landscapes) (D. E. Hughes et al., 2004).

We believe that the most important properties of the framework we evolved are its use of open software, its protocols for delivering a scalable distributed solution, and its flexible plug-in architecture. In general, flexibility in all aspects of the system has been the key to our success and is helping us to move forward with new capabilities, such as a bidding system for story based rendering.

In its present form, our framework still requires scripts to be written or at least reused to create a new experience. Our goal (dream) is to be able to use our experience to capture capabilities to evolve the behaviors of virtual characters in accordance with actions performed by human participants, as well as those of other successful virtual characters. For instance, in a training environment, the actions of an expert at room clearing could be used to train virtual SWAT (special weapons and tactics) team members by example. In a rehabilitation setting, the actions of a patient could be used as a model for those of a virtual patient that is, in turn, used to train a student therapist in the same context. Of course, this is a rather lofty goal, and just making authoring more intuitive, even with drag-and-drop, would help.

The MR framework described here is a system that is intended to generate, deploy, capture, analyze, and synthesize an interactive story. Whether these stories are designed to train, teach, sell, or entertain is immaterial. The point is that we drive an MR experience by generating a world within, on top, beneath, and around the real world and real senses in which we live. Our goals for this framework and for mixed reality in general are bounded only by our temporal imagination. Tomorrow, we will conceive of new applications of MR, leading to new requirements that continue to guide the evolution of our system and place new demands on our creativity.

NOTE

1. An animatic is a simple visual rendering of the story from a single point of view. Its purpose is to communicate the vision of the creative team. This allows the art director, the audio producer, and the lead programmer to effectively exchange ideas and determine each team's focus.

REFERENCES

Bower, G. H., & Winzenz, D. (1970). Comparison of associative learning strategies. *Psychonomic Science, 20,* 119–120.

Coppin, B. (2004). *Artificial intelligence illuminated.* Sudbury, MA: Jones and Bartlett Publishers.

Dias, J. M. S., Monteiro, L., Santos, P., Silvestre, R., & Bastos, R. (2003). Developing and authoring mixed reality with MX toolkit. In *IEEE International Augmented Reality Toolkit Workshop* (pp. 18–26). Tokyo, Japan.

Fidopiastis, C. M., Stapleton, C. B., Whiteside, J. D., Hughes, C. E., Fiore, S. M., Martin, G. A., Rolland, J. P., & Smith, E. M. (2005, September). *Human experience modeler: Context driven cognitive retraining and narrative threads.* Paper presented at the 4th International Workshop on Virtual Rehabilitation (IWVR2005), Catalina Island, CA.

Haller, M., Drab, S., & Hartmann, W. (2003). A real-time shadow approach for an augmented reality application using shadow volumes. *Proceedings of ACM Symposium on Virtual Reality Software and Technology—VRST'03* (pp. 56–65). Osaka, Japan.

Hughes, C. E., Burnett, J., Moshell, J. M., Stapleton, C. B., & Mauer, B. (2002). Space-based middleware for loosely-coupled distributed systems. *Proceedings of SPIE, 4862,* 70–79.

Hughes, C. E., Konttinen, J., & Pattanaik, S. N. (2004). The future of mixed reality: Issues in illumination and shadows. *Proceedings of the 2005 Interservice/Industry Training, Simulation & Education Conference.* Arlington, VA: National Training Systems Association.

Hughes, C. E., Stapleton, C. B., Hughes, D. E., & Smith, E. (2005). Mixed reality in education, entertainment and training: An interdisciplinary approach. *IEEE Computer Graphics and Applications, 26*(6), 24–30.

Hughes, C. E., Stapleton, C. B., Micikevicius, P., Hughes, D. E., Malo, S., & O'Connor, M. (2004). Mixed fantasy: An integrated system for delivering MR experiences [CD-ROM]. *Proceedings of the VR Usability Workshop: Designing and Evaluating VR Systems.*

Hughes, D. E. (2005, July). *Defining an audio pipeline for mixed reality.* Paper presented at the Human Computer Interaction International 2005 (HCII2005), Las Vegas, NV.

Hughes, D. E., Thropp, J., Holmquist J., & Moshell, J. M. (2004, November 29–December 2). *Spatial perception and expectation: Factors in acoustical awareness for MOUT training.* Paper presented at the 24th Army Science Conference (ASC 2004), Orlando, FL.

Jerome, C. J. (2006). Orienting of visual-spatial attention with augmented reality: Effects of spatial and non-spatial multi-modal cues. *Dissertation Abstracts International, 67* (11), 6759. (UMI No. 3242442)

Jerome, C. J., & Witmer, B. (2004, October). *Human performance in virtual environments: Effects of presence, immersive tendency, and simulator sickness.* Poster presented at the Human Factors & Ergonomics Society's Annual Conference, New Orleans, LA.

Jerome, C. J., Witmer, B., & Mouloua, M. (2005, July). *Spatial orienting attention using augmented reality.* Paper presented at the Augmented Cognition Conference, Las Vegas, NV.

Kennedy, R. S., Fowlkes, J. E., & Lilienthal, M. G. (1993). Postural and performance changes following exposures to flight simulators. *Aviation, Space, and Environmental Medicine, 6*(10), 912–920.

Konttinen, J., Hughes, C. E., & Pattanaik, S. N. (2005). The future of mixed reality: Issues in illumination and shadows. *Journal of Defense Modeling and Simulation, 2*(1), 51–59.

McGuire, M., Hughes, J. F., Egan, K. T., Kilgard, M. J., & Everitt, C. (2003). *Fast, practical and robust shadows* (Brown University Computer Science Tech. Rep. No. CS-03-19). Retrieved September 27, 2004, from http://www.cs.brown.edu/publications/techreports/reports/CS-03-19.html

Nijasure, M., Pattanaik, S. N., & Goel, V. (2003). Interactive global illumination in dynamic environments using commodity graphics hardware. *Proceedings of Pacific Graphics 2003* (pp. 450–454). Canmore, Alberta, Canada.

O'Connor, M., & Hughes, C. E. (2005). Authoring and delivering mixed reality experiences. *Proceedings of 2005 International Conference on Human-Computer Interface Advances in Modeling and Simulation—SIMCHI'05* (pp. 33–39). Las Vegas, Nevada.

Piekarski, W., & Thomas, B. H. (2003). An object-oriented software architecture for 3D mixed reality applications. *Proceedings of the IEEE and ACM International Symposium on Mixed and Augmented Reality—ISMAR 2003* (pp. 247–256). Tokyo, Japan.

Sadek, R. (2001). 3D sound design and technology for the sensory environments evaluations project: Phase 1 [Online]. http://www.ict.usc.edu/publications/ICT-TR-01-2001.pdf

Stapleton, C. B., & Hughes, C. E. (2003). Interactive imagination: Tapping the emotions through interactive story for compelling simulations. *IEEE Computer Graphics and Applications, 24*(5), 11–15.

Stapleton, C. B., & Hughes, C. E. (2005). Mixed reality and experiential movie trailers: Combining emotions and immersion to innovate entertainment marketing. *Proceedings of 2005 International Conference on Human-Computer Interface Advances in Modeling and Simulation—SIMCHI'05* (pp. 40–48). Las Vegas, Nevada.

Sweeney, C. A., & Bellezza, F. S. (1982). Use of the keyword mnemonic in learning English vocabulary words. *Human Learning, 1,* 155–163.

Traskback, M. (2004). Toward a usable mixed reality authoring tool. In the *2004 IEEE Symposium on Visual Languages and Human Centric Computing* (pp. 160–162). Rome, Italy.

Witmer, B. G. & Singer, M. J. (1998). Measuring presence in virtual environments: A presence questionnaire. *Presence: Teleoperators and Virtual Environments, 7*(3), 225–240.

Chapter 26

TRENDS AND PERSPECTIVES IN AUGMENTED REALITY

Brian Goldiez and Fotis Liarokapis

Training in the real environment is not easy, mainly due to sociotechnological barriers. This chapter explores the potential effectiveness of augmented reality (AR) applied to training. We discuss previous applications of AR in live training and findings arising from formative evaluations of these systems. Various approaches for applying AR to training are discussed. Overviews of the most characteristic evaluation methods, as well as suggestions on assessing the performance of AR in training, are provided.

INTRODUCTION

AR describes a technology where the real world is the baseline, and additional information from a computer-generated sensory display is added. AR is contrasted with virtual reality (VR) where the baseline is a synthetic (artificial) environment and the desired state is complete immersion of the human sensory system within a computer-created environment. As one adds more computer augmentation to a real world, the demarcation between virtual and augmented becomes blurred. Rapid advances in technology have contributed to blurring. Milgram (2006) has characterized a variety of continuums between the real and virtual worlds that reflect different ways one can view the interaction and use of the technology. The confluence of views of realities provides opportunities for adapting technologies from one domain to another and the opportunity to adapt human performance studies across domains.

There are three major characteristics of AR systems described by Azuma (1997). First, AR systems must seamlessly combine the real world with virtual information. This combination is typically considered in the visual domain, but is not exclusively restricted to it. Second, an AR system must operate in real time. That is, an AR system must provide responses commensurate with the system using the AR, in this case a human. Third, an AR system must spatially register the display with the real world in three-dimensional (3-D) space. Currently there are two broad application areas for AR: decision-making tasks, where mobility is not critical, and tasks where mobility is of primary importance (for example, navigation).

Technical Requirements for AR Systems

Research in tracking, display, and interaction technologies contributes to the immersiveness of AR systems, and new mathematical algorithms improve the effectiveness of the software system and realism of the visualization output. To correctly register the computer-generated objects and the real world, accurate tracking of the coordinates of the participant's point of view is required. When training requires a stationary user position, the registration process is much easier than if the user is moving (Azuma, 1997). In both systems, accurate alignments between the virtual and real objects are required to avoid the appearance of floating objects. The common tracking techniques are vision or sensor based. Sensor based systems typically use magnetic or sonic devices to detect the user's position, while vision based systems use cameras and visually distinctive markers to locate the user's position in an environment. These markers, called fiducials, are used in indoor systems with good results (Kato, Billinghurst, Poupyrev, Imamoto, and Tachibana, 2000). Other indoor systems employ hybrid-tracking technologies, such as magnetic and video sensors, to achieve good registration results. In outdoor and mobile systems, other sensor devices are used, such as the global positioning system (GPS) coupled with orientation sensors (for example, digital compasses).

AR systems typically use special displays to immerse participants in augmented environments. Head-mounted displays (HMD) for AR include optical or video see-through, which project merged computer-generated and real images onto the user's eyes. However, there are other ways to immerse the user, such as large-area displays or stereoscopic glasses. Regarding interaction technologies, many commercially available hardware devices can be used to increase the level of interaction between participants and computing devices. Thus, a robust AR system must integrate an ergonomic software and hardware framework and address the following issues: (1) *calibration and accurate user-viewing position,* (2) *natural interaction,* and (3) *realistic rendering of virtual information.*

REVIEW OF STATE OF THE ART IN AR

The literature is organized by successively considering AR applications, prototypes, components, and concepts. AR applications exist (principally in laboratories or as prototypes) and have been subjected to some type of evaluation by humans. Prototypes are essentially AR systems that have been created, but generally have not been evaluated by users. Components are subsystems of AR systems. Concepts are ideas or concerns that have not been reduced to practice.

A further categorization of AR literature supports work by Goldiez (2004) that pointed to technological hurdles in AR in the areas of tracking and visualization.

Tracking can currently be accomplished at a precise level in small spaces and in gross terms in larger areas. Visualization that makes added content indistinguishable from the real world requires graphics processing and display technology that does not currently exist in a mobile computing environment and minimally exists in a fixed setting. Many technical problems are mitigated when mobility is restricted, but with large impacts on cost and/or flexibility in AR usage. As AR systems are deployed for experimentation and demonstration, new issues arise, principally in human factors and ergonomics, because the focus of AR expands from technical only to encompass usage. A more complete review of the literature can be found in Goldiez, Sottilare, Yen, and Whitmire (2006).

It is worth mentioning that there is no perfect AR technology, and all existing ones have some advantages as well as limitations. To overcome the limitations of each technology hybrid, AR systems can be employed to meet requirements that do not fall strictly into one category noted above. These hybrid AR systems combine different vision techniques and hardware devices to achieve results that better meet a user's requirements. Obviously, hybrid systems can further immerse participants, but they will also generally increase the overall cost of the AR system because they might stretch the limits of the technology and have special integration and operational needs.

AR Application Domains

AR systems have been developed to facilitate improved human performance in such areas as entertainment, medicine, communications, navigation/decision making, and military-oriented operations.

Entertainment

AR is being used in several areas in the entertainment industry. As examples, Liarokapis (2006b) describes how to transform a traditional arcade game into 3-D and then into an AR interactive game. Initial studies found that users preferred the AR experience in terms of enjoyment. Cavazza, Martin, Charles, Marichal, and Mead (2003) created an interactive system that immerses the storyteller into the background environment, while Gandy et al. (2005) integrates users into a scenario based on the *Wizard of Oz.* A simple tennis game has been developed using commercially available Bluetooth cellular technology (Henrysson, Billinghurst, & Ollilia, 2005).

Medicine

The medical field currently benefits from AR systems. For example, a virtual retinal display is being used for patients who suffer from poor vision and as a surgical display (Viirre, Pryor, Nagata, & Furness, 1998). Scheuering, Rezk-Salama, Barfufl, Schneider, and Greiner (2002) report on using a video see-through HMD to overlay imagery during surgical procedures. Also, Vogt, Khamene, Sauer, Keil, and Niemann (2003) developed a system to visualize X rays, CT scans, and so forth, onto a person or mannequin by utilizing a retroreflective marker tracking system.

Communication

Several AR systems have been developed to facilitate communication and collaboration. Regenbrecht et al. (2003) describe an AR conferencing system allowing users to meet without leaving their desks. Billinghurst, Belcher, Gupta, and Kiyokawa (2003) describe two experiments investigating face-to-face collaboration using a multiuser AR interface. These results, however, found no advantage in using AR due to limitations from restricted peripheral vision.

Navigation

AR has been used to facilitate navigation and wayfinding. As part of the LOCUS project, Liarokapis (2006c) developed a system that uses AR and VR techniques to enhance mobile navigation by guiding pedestrians between locations in urban environments. Two prototypes were developed for outdoor navigation, one based on manually placed fiducials and another based on natural feature selection. The first prototype has robust tracking, but limited range, while the opposite is true for the second prototype. A hybrid approach using natural features and GPS is being researched that should provide better tracking efficiency. Goldiez (2004) utilized the Battlefield Augmented Reality System (BARS) to study the benefits of using AR in search and rescue navigation by exploring using different map displays to facilitate navigation through a maze. Results determined that BARS does improve user performance in specific situations.

Spatial Relations Using AR

Bennet and Stevens (2004) describe a projection augmented, multimodal system to explore how interaction with spatially coincident devices affects perception of object size. Results showed that performance in combined (visual/ haptic) conditions was more accurate in distance estimation, verifying the theory that a person's perception of size is magnified by using more than one sense. Grasset, Lamb, and Billinghurst (2005) investigated how a pair of users, one user utilizing AR (an exocentric view of the maze) and one utilizing VR (an egocentric view of the maze) can accomplish a collaborative task. Results concluded mixed space AR collaboration does not disrupt task efficiency.

Military-Oriented AR Systems

BARS is an important military based AR application that was developed by the Naval Research Laboratory for use in urban settings. BARS has served as a de facto integration platform for a number of technological and human-performance research efforts. For example, it has been used in several experiments investigating the impact of various technological innovations on human performance (for example, Goldiez, 2004; Livingston, Brown, Julier, & Schmidt,

2006). Livingston et al. developed innovative algorithms to facilitate pointing accuracy and the sharing of information among BARS users. Additionally, Franklin (2006) discussed experiments using a system similar to BARS, but developed by QinetiQ to assess the maturity of AR to supplement live training. In the QinetiQ-developed system a virtual aircraft was inserted into live ground assets, which could see and interact with live participants, but live participants had no knowledge of the virtual world. The results suggested that a more robust interface to the live environment was necessary and the bulkiness of the AR equipment was an impediment to performance. To overcome limitations in the field of view, users suggested the use of small visual icons on the display periphery to cue the user to the aircraft position. Discrepancies between the real and synthetic worlds with respect to environmental effects were problematic to training.

AR Components

At a top level, AR components include visual software and hardware, spatial tracking devices, other sensory devices, computing, and consideration of ergonomics. Integrating components creates an AR system.

Visual Components

Visual software and hardware are key factors distinguishing AR from VR. Superimposing virtual images onto a real background is challenging and relies on efficient processing to create realistic scenes, compensation for motion, and tracking tools for placing images in the correct position. Several factors contribute to the VR-AR distinction, including the need in AR to accommodate differences in dynamic changes in brightness and contrast between the real and virtual parts of the scene, latency in overlaying the virtual image onto the real world, image fidelity differences, helmet-mounted display weight, and so forth.

A variety of visualization research has been conducted to enhance AR. A novel approach was taken by Fischer, Bartz, and StraBer (2005), who reduced the visual realism of the real environment to better match the computer-generated object(s) being superimposed onto the real world. An alternative approach for interacting with smaller 3-D objects in AR is suggested by Lee and Park (2005), who use blue augmented foam as a marker. Mohring, Lessig, and Bimber (2005) describe the technology of video see-through AR and its development on a consumer cell phone achieving 16 frames per second. Ehnes, Hirota, and Hirose (2005) have developed an alternative to the HMD based on a computer-controlled video projection system that displays information in the correct place for a user.

Tracking Components

Tracking in AR is the operation of measuring the position of real 3-D objects (or humans) that move in a defined space. Six degrees of freedom (6 DOF) tracking is referred to as the simultaneous measurement of position and orientation in some fixed coordinate system, such as the earth. It is normally required that the location of the tracking device (for example, a camera) and the item being tracked (for example, a trainee) be simultaneously and continuously known in 6 DOF. The most significant technologies available for tracking in AR environments can be subdivided into six broad categories: mechanical, electromagnetic, optical, acoustic, inertia, and GPS. As with visual systems, tracking systems drive AR implementations into fixed or limited motion situations to allow for display rendering and for precisely tracking human appendages or important components. Wider-range motion AR systems are less precise and therefore limit the degree the virtual image aligns with the real world.

Computer vision tracking is also a major area of research for AR. Vision based tracking (Neumann & You, 1999) enables the potential recognition of an object in a natural environment that serves as a fiducial. Software algorithms have been developed by Behringer, Park, and Sundareswaran (2002) to use vision tracking to recognize buildings and/or structures. Naimark and Foxlin (2005) describe the development of a hybrid vision-inertial self-tracker that utilizes light emitting diodes (LEDs). Tenmoku, Kanbara, and Yokoya (2003) describe an alternative to vision based tracking that integrates magnetic and GPS sensors for indoor and outdoor environments. The user's location is tracked utilizing a combination of radio frequency identification (RFID) tag(s) deployed in the environment, GPS (outdoors), and magnetic (indoors) sensors.

Human Factors/Mobility

Even a system with flawless tracking and visual augmentation would be worthless if the user were unable to perform the desired tasks comfortably and effectively; thus, ergonomics cannot be overlooked in AR development. Weight, location of controls, and mobility all influence user performance. Liarokapis (2006a) presents an overview of a multimodal AR interface that can be decomposed into offline, commercially produced components. A variety of interaction paradigms, such as the use of fiducial based icons, support physical manipulation of an object. Vogelmeier, Neujahr, and Sandl (2006) from the European Aeronautic Defence and Space Company discuss the need for similarity in various sensory interactions when wearing AR and/or VR equipment as compared to the real world. An attractive feature of AR is mobility and with it possible extensions in the variety and range of human interactions. Tappert et al. (2001) and Espenant (2006) discuss the possibilities of using AR based wearable devices as visual memory prosthetics or for training. Mobility in AR will also require considering user location. For example, Butz (2004) discusses approaches that consider using radio links and infrared or third-generation cellular technology to support mobility, enabling the acquisition of the user's location for subsequent processing of relevant data.

ADVANCED CONCEPTS IMPACTING AR

The markets will determine when several technologies important to AR emerge, as it appears that several needed technical innovations are dependent upon developments in the commercial sector. These interrelated areas include advances in power management, computer packaging, and communications. Power management (power sources and power-consuming devices) is important to sustained mobility and operations in AR. Computer packaging is another area where the commercial market will determine what products become available. The literature alludes to the need for devices that consume less power and are more compactly packaged.

Handheld and mobile computing may become an advantageous platform for hosting AR applications. Emerging mobile technology employs on-board computing and graphics rendering resources that are useful for AR applications. Researchers (for example, Liarokapis, 2006c) are exploiting this technology, but are not creating the hardware or software operating systems. They are dependent upon the mobile industry to create products that are useful to AR while also serving the wider cellular marketplace. This type of leveraging is advantageous as development costs and economies of scale are borne by someone other than the AR community. However, the AR community must stand by the sidelines and wait for developments that may or may not occur.

A review of the literature suggests that when real and virtual environments are mixed, handling interruptions is a major unresolved issue. Unanticipated items (for example, people) crossing the field of view could result in unacceptable anomalies in the AR visualization. The work of Drugge, Nilsson, Liljedahl, Synnes, and Parnes (2004) showed that interruptions in AR occur due to unforeseen events (for example, someone walking across a scene causing visual anomalies), but also are due to the tasks conducted by the user (for example, divided attention tasks). This work could be significant to AR in providing a strategy for handling events that occur in the virtual world when mixed with the real world. Conceptually, one could envision an AR user marking an item of interest and having the AR system report back if the item's situation had changed, thereby possibly mitigating divided attention related issues.

Understanding context is another concept where a better understanding of the impact of mixing environments to create viable AR implementations is needed. Because AR uses the real world, which is naturally multimodal, it is not yet clear what information needs to be captured prior to and during an AR experience to understand human activity that occurs during the AR experience. A wide range of environmental data and externally originated sensory stimuli could be relevant to creating an appropriate and dynamic AR experience.

In conclusion, AR systems-oriented research and development progress in the United States has been principally technological. Formal evaluations of this technology are not yet evident in the training-related areas. Future work currently sponsored by the European Commission will create new VR and AR systems along with formal evaluations for various purposes.

AR UTILITY FOR TRAINING

AR seems ideally suited to support training in navigation, manipulation of items, and decision making. Experimentation has indicated benefits for using AR in training for certain applications. Early work demonstrated its usefulness in manipulation and spatial experiments (Goldiez, 2004). AR's role in supporting decision making requires a longer-term view with enhancements needed in technology before human performance benefits can be realized (Franklin, 2006). AR has shown benefits to enhance human performance in navigating, and near-term benefits in training appear promising. In live (or live/virtual) exercises, AR could serve as an on-board instructor, guiding the trainee should he or she become lost or venture outside the desired training area. This capability could greatly simplify the tracking problems in AR by allowing the use of GPS or RFIDs for gross tracking and a more precise tracking mechanism at critical locations. Thus, for training, the aforementioned tracking problem can be controlled by appropriate scenario design coupled with the use of AR as a surrogate instructor.

AR offers the opportunity to improve various training subsystems. Visual simulation immediately comes to mind because of the potential for video or optical see-through devices to add (or subtract) content from a scene. AR, though, can also augment the instructor by providing in situ tutoring (such as hints when the trainee is lost while learning to navigate) and individualized after action review of trainee activity in live and/or virtual exercises. Mobile AR also offers the potential for personalized training by providing information in a form most suitable for the user's needs.

At a conceptual level AR can also be envisioned as a technology that will facilitate better methods in team training. Because of its ability to provide additional information display, as well as information storage and persistence, AR can facilitate mitigating team situational awareness issues by providing pointers and nonverbal communication into areas for team attention. It is logical to envision this sharing of information and enhanced situational awareness being used as a tool for training.

Dr. Walter Van de Velde, Program Officer for the European Commission's Future and Emerging Technology Initiative, noted the following in a brochure e-mailed to said author on August 11, 2006:

> Current virtual and augmented reality environments try to provide the best display realism, taking for granted that this automatically leads to the best user-experience. Practice shows that this is not true: users do not easily feel fully engaged in high-tech VR worlds. On the other hand they can feel extremely present in simpler environments, like when chatting on line or when reading a book. A better understanding of this [presence] will give rise to new immersive interface technologies that exploit human perceptual, behavioral, cognitive and social specificities for stimulating a believable and engaging user-experience of presence, in spite of using artificial stimuli.
>
> (Van de Velde, 2006)

Investigations into measuring and controlling presence are potentially critical for training using AR because users will be interacting with real and virtual items and could need to distinguish between the two. Properly structured research in this area would thus yield valuable insights into strategies for handling interruptions. After action review systems for live and virtual training have been prototyped; however, AR adds new complexities. An appropriate after action review for AR should include the following: capturing relevant contextual information in the real world, identifying interruptions, and handling or correlating varying spatial positions and poses of the trainee with his or her real and virtual positions.

Moreover, AR has the huge potential for improving training by integrating new and existing skills in training. In some cases, this might be done by providing AR training systems that have unique capabilities for testing and evaluating trainees. From another perspective more research into AR interface issues will likely help answer some key questions, as well as help foster better training solutions and applications. Some additional aspects of the utility of AR for training could include enhanced assessment and diagnostic capabilities in the real time portion of the system allowing trainees the ability to review actions and decisions from different perspectives. Potentially such AR systems could have the capability to visually compare the trainee's paths, actions, decisions, and so forth to those of experienced experts such that trainees could see (and the instructor could discuss) differences between the novice's and the expert's actions.

Several aspects of human-centered design should be studied with respect to making AR better suited to supporting training in various vocations. These include personalizing the software for training to certain classes of individuals and human factors considerations for hardware, noted above. The work of Liarokapis (2006c) using mobile technology adapted for VR and AR shows great promise for training, using virtual scenes at modest prices and good operating performance. Coupling location awareness (through techniques such as RFIDs) with a digital compass provides reasonable information on user location. Rendering time and data transfer rates are currently insufficient for real time operation, but advances are being made by the cellular community. These types of devices represent a viable future delivery mechanism.

CONCLUSIONS

AR is an exciting technological development offering the opportunity to overcome many of the limitations in individualized virtual environment systems. These include performance limitations, such as self-motion, and programmatic limitations, such as high costs and relatively large facility requirements. AR has its own set of issues, as noted in this chapter, that are being addressed by research teams across the globe. Most AR activity has been focused on computer graphics fused to the real world to create an immersive environment. While fully immersive systems are beneficial, there are more immediate and near-term opportunities for less immersive AR systems. A principal benefit in using AR is its apparent ease of deployment. Such deployable systems employing wearable computers provide increased flexibility for AR's use when and where needed. Moreover, coupling the broader view of AR with its classification into three categories and two usage areas encourages experimentation and development along more focused lines of research.

ACKNOWLEDGMENTS

Part of the work presented herein was supported by the U.S. Army Research Institute for Behavioral Sciences. Also, part of the work presented has been conducted within the LOCUS project. The views expressed herein, though, are those of the authors and do not reflect an official position of a government agency.

REFERENCES

Azuma, R. T. (1997). A survey of augmented reality. Presence: *Teleoperators & Virtual Environments, 6*(4), 355–385.

Behringer, R., Park, J., & Sundareswaran, V. (2002). Model-based visual tracking for outdoor augmented reality applications. *International Symposium on Mixed and Augmented Reality, 01,* 277–322.

Bennet, E., & Stevens, B. (2004). The effect that haptically perceiving a projection augmented model has on the perception of size. *Third IEEE and ACM International Symposium on Mixed and Augmented Reality, 03,* 294–295.

Billinghurst, M., Belcher, D., Gupta, A., & Kiyokawa, K. (2003). Communication behaviors in collocated collaborative AR interfaces. *International Journal of Human-Computer Interaction, 16*(3), 395–423.

Butz, A. (2004). Between location awareness and aware locations: Where to put intelligence. *Applied Artificial Intelligence, 18*(6), 501–512.

Cavazza, M., Martin, O., Charles, F., Marichal, X., Mead, S. J. (2003). User interaction in mixed reality interactive storytelling. *The Second IEEE and ACM International Symposium on Mixed and Augmented Reality,* IEEE Computer Society, 304–305.

Drugge, M., Nilsson, M., Liljedahl, U., Synnes, K., & Parnes, P. (2004). Methods for interrupting a wearable computer user. *Proceedings of the Eighth International Symposium on Wearable Computers* (pp. 150–157). Washington, DC: IEEE Computer Society.

Ehnes, J., Hirota, K., Hirose, M., (2005). Projected augmentation-augmented reality using rotatable video projectors. *Third IEEE and ACM International Symposium on Mixed and Augmented Reality, 03,* 26–35.

Espenant, M. (2006). Applying simulation to study human performance impacts of evolutionary and revolutionary changes to armoured vehicle design. In *Virtual Media for Military Applications* (RTO Meeting Proceedings No. RTO-MP-HFM-136, pp. 17-1–17-2). Neuilly-sur-Seine, France: Research and Technology Organisation.

Fischer, J., Bartz, D., & StraBer, W. (2005). Stylized augmented reality for improved immersion. *IEEE Virtual Reality, 01,* 195–202.

Franklin, M. (2006). The lessons learned in the application of augmented reality. *Virtual Media for Military Applications* (RTO Meeting Proceedings No. RTO-MP-HFM-136, pp. 30-1–30-8). Neuilly-sur-Seine, France: Research and Technology Organisation.

Gandy, M., Macintyre, B., Presti, P., Dow, S., Botter, J., Yarbrough, B., Oapos, Initial, & Rear, N. (2005). AR karaoke acting in your favorite scenes. *Fourth IEEE and ACM International Symposium on Mixed and Augmented Reality, 04,* 114–117.

Goldiez, B. F. (2004). Techniques for assessing and improving performance in navigation and wayfinding using mobile augmented reality. *Dissertation Abstracts International, 66*(02), 1206B. (UMI No. 3163584)

Goldiez, B. F., Sottilare, J., Yen, C., & Whitmire, J. (2006, November). *The current state of augmented reality and a research agenda for training.* (Tech. Rep., Contract No. W74V8H-06-C-0009). Orlando, FL: U.S. Army Research Institute for Behavioral Sciences.

Grasset, R., Lamb, P., & Billinghurst, M. (2005). Evaluation of mixed-space collaboration, *Fourth IEEE and ACM International Symposium on Mixed and Augmented Reality, 04,* 90–99.

Henrysson, A., Billinghurst, M., & Ollilia, M. (2005). Face to face collaborative AR on mobile phones, *Fourth IEEE and ACM International Symposium on Mixed and Augmented Reality, 04,* 80–89.

Kato, H., Billinghurst, M., Poupyrev, I., Imamoto, K., & Tachibana, K. (2000). Virtual object manipulation on a table-top AR environment. *Proceedings of the International Symposium on Augmented Reality* (pp. 111–119). Washington, DC: IEEE Computer Society.

Lee, W., & Park, J. (2005). Augmented foam: Tangible augmented reality for product design. *Fourth IEEE and ACM International Symposium on Mixed and Augmented Reality, 04,* 106–109.

Liarokapis, F. (2006a). An augmented reality interface for visualizing and interacting with virtual content. *Virtual Reality. 11(1),* 23–43.

Liarokapis, F. (2006b). An exploration from virtual to augmented reality gaming. *Simulation & Gaming, 37*(4), 507–533.

Liarokapis, F. (2006c) Location based mixed reality for mobile information services. *Advanced Imaging: Solutions for the Electronic Imaging Professional, 21,* 22–25.

Livingston, M. A., Brown, D. G., Julier, S. J., Schmidt, G. S. (2006). Mobile augmented reality: Applications and human factors evaluations. *Advanced Information Technology Code 5580.* Washington, DC: Naval Research Laboratory.

Milgram, P. (2006). Some human factors considerations for designing mixed reality interfaces. *Virtual Media for Military Applications* (RTO Meeting Proceedings No. RTO-MP-HFM-136, pp. KN1-1–KN1-14). Neuilly-sur-Seine, France: Research and Technology Organisation.

Mohring, M., Lessig, C., & Bimber, O. (2005). Video see-through AR on consumer cellphones. *Third IEEE and ACM International Symposium on Mixed and Augmented Reality, 3,* 252–253.

Naimark, L., & Foxlin, E. (2005). Encoded LED system for optical trackers. *Fourth IEEE and ACM International Symposium on Mixed and Augmented Reality, 4,* 150–153.

Neumann, U., & You, S. (1999). Natural feature tracking for augmented reality. *IEEE Transactions on Multimedia, 1,* 53–64.

Regenbrecht, H., Ott, C., Wagner, M., Lum, T., Kohler, P., Wilke, W., & Mueller, E. (2003). An augmented virtuality approach to 3D videoconferencing. *The Second IEEE and AC International Symposium on Mixed and Augmented Reality, 02,* 290–291.

Scheuering, M., Rezk-Salama, C., Barfufl, H., Schneider, A., Greiner, G. (2002). Augmented reality based on fast deformable 2D-3D registration for image guided surgery.

In S. K. Mun (Ed.), *Medical Imaging 2002: Visualization, Image-Guided Procedures, and Display* (pp. 436–445). Bellingham, WA: International Society for Optical Engineering.

Tappert, C. C., Ruocco, A. S., Langdorf, K.A., Mabry, F. J., Heineman, T. A., Brick, D. M., et al. (2001). Military applications of wearable computers and augmented reality. In W. Barfield & C. Thomas (Eds.), *Fundamentals of wearable computers and augmented reality* (pp. 625–647). Mahwah, NJ: Lawrence Erlbaum.

Tenmoku, R., Kanbara, M., & Yokoya, N. (2003). A wearable augmented reality system for navigation using positioning infrastructures and a pedometer. *The Second IEEE and ACM International Symposium on Mixed and Augmented Reality, 2,* 344–345.

Van de Velde, W., (2006). Presence and interaction in mixed-reality environments (FET Proactive Initiative). Unpublished manuscript.

Viirre, E., Pryor, H., Nagata, S., and Furness, T. A. (1998). The virtual retinal display: A new technology for virtual reality and augmented vision in medicine, In D. Stredney & S. J. Weghorst (Ed.), *Proceedings of Medicine Meets Virtual Reality* (pp. 252–257). Amsterdam: IOS Press and Ohmsha.

Vogelmeier, L., Neujahr, H., & Sandl, P. (2006). Interaction methods for virtual reality applications. In *Virtual Media for Military Applications* (RTO Meeting Proceedings No. RTO-MP-HFM-136, pp. 14-4–14-8). Neuilly-sur-Seine, France: Research and Technology Organisation.

Vogt, S., Khamene, A., Sauer, F., Keil, A., Niemann, H. (2003). A high performance AR system for medical applications. *The Second IEEE and ACM International Symposium on Mixed and Augmented Reality.* Los Alamitos, CA: IEEE Computer Society.

Chapter 27

VIRTUAL ENVIRONMENT HELICOPTER TRAINING

Joseph Sullivan, Rudolph Darken, and William Becker

Through the last decade, the virtual environment (VE) community became interested in the use of VEs for training a variety of tasks, particularly spatial tasks. Because VEs, unlike conventional interactive computing environments, are inherently spatial, it is reasonable to assume that spatial tasks might be performed better in VEs and possibly trained better in VEs. We began a line of research focused on spatial navigation in VEs, initially confining ourselves to terrestrial navigation in both urban and natural terrains (Banker, 1997; Darken & Banker, 1998; Goerger et al., 1998; Jones, 1999). We were able to show how a VE could be used to develop spatial knowledge of a real environment via exposure to a VE simulation of that place. While much progress was made, a common criticism in the training domain (as opposed to mission rehearsal) was that it was cheaper to train land navigation by practicing map and compass skills in a physical environment rather than in a VE. The same could not be said, however, when we began to look at helicopter navigation where every hour in the air is extremely expensive and consequently limited. What we did not know is how much of what we had learned about navigation on the ground would translate to the air.

In addition to advancing our basic understanding of human spatial cognition, we needed to significantly improve how aviators are trained to navigate from the air. We started by contrasting the tasks of helicopter navigation and land navigation. In the military setting, navigation is rarely a primary goal; it is a necessary component of a larger mission. Land navigation and helicopter navigation both rely on terrain association skills: the ability to match a two-dimensional map representation of a terrain feature with a feature within the field of view. The primary differences are altitude, speed, maneuvering limitations, and available field of view. At typical helicopter speeds and altitudes, fewer terrain features will be in view for a shorter amount of time. In the time it takes a novice to look down to reference a map, an entirely new set of terrain features may come into view. It is rarely feasible for a helicopter crew to stop to regain orientation. This makes error recovery a notoriously difficult task. Navigation is inherently a crew task. All crew members share the task of avoiding terrain and obstacles and relaying information on navigation cues. Ironically, the aircraft proves to be a difficult platform for training. The instructor is responsible for basic aircraft control and obstacle avoidance. This affords few opportunities to observe a trainee's procedures and provide guidelines. Given these unique challenges, VEs are an appealing training solution. They are inherently spatial and many of the important characteristics of the real world can be faithfully re-created while many of the real world limitations can be removed.

There were several practical issues involved in applying VE technology to helicopter training. As is the case in many application domains, cost was an issue. The form of the solution was also key. Because most of our work was for naval aviators (U.S. Navy and U.S. Marine Corps), any training device had to be small and rugged enough to fit aboard ship. We were also concerned about usage modes. The VE training literature is rife with examples of unsuccessful automated intelligent tutoring systems (ITS) for complex skills. For a good discussion of general issues of ITS, see Psotka, Massey, & Mutter (1988). We did not want to rely on ITS technology, but requiring that an instructor be physically present at all times was not an attractive alternative either. Thus, knowing what modes of usage were appropriate and how they affect training was a necessary component of this line of research.

This introduction frames our research program that has spanned over 10 years of applying emerging VE technology to helicopter training and evaluating the results. We begin with our earliest efforts to apply VE simulation to this unique problem domain and take the story to the present where the characteristics of the problem remain unchanged, but the form of the solution has changed dramatically. See Figure 27.1.

MAP INTERPRETATION AND TERRAIN ASSOCIATION VIRTUAL ENVIRONMENT

Helicopter pilots are currently trained to navigate in a number of ways. The navy has a course of instruction called the map interpretation and terrain association course, or MITAC, that specifically teaches these skills. Conventional classroom instruction is used to teach the basic concepts, such as the use of displays, map coordination, dead reckoning, and compass use. Noninteractive video is sometimes used to practice the task where the view from a flight is shown and the trainee must follow along on a paper map. However, navigation is an inherently interactive task. A video that does not respond to the action of the trainee is of little use in learning the cause and effect relationship of movement to spatial orientation. To a large degree, the shortcomings of this video were a key motivator for the use of simulation.

Instruction then moves to the aircraft where the trainee must perform a complex navigation task in the cockpit under the pressure of all the other things that a pilot has to be aware of in flight. Scaffolding techniques in the aircraft are difficult if not impossible to achieve. As noted by Ward, Williams, and Hancock (2006, p. 252), "intuition and emulation" tend to guide this process more than

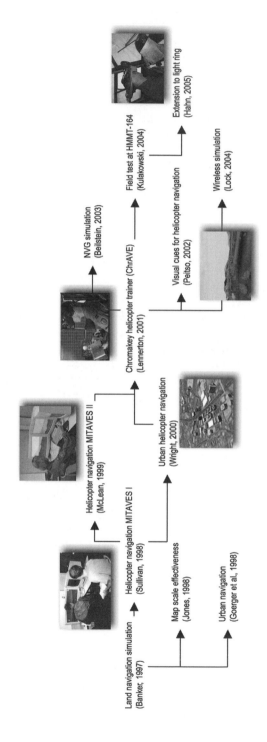

Figure 27.1. A Road Map of Navigation Research for the Office of Naval Research

evidence based practices. There are not many options for the instructor to use in between classroom materials and the real task in which to train. The result is that, upon graduation from flight school, many pilots are far from expert navigators. Can simulation be used to remedy this?

We first prototyped a simple navigation training device that would run on standard hardware and that would be small enough and inexpensive enough to function in the training space of a typical squadron. We developed a three-screen display format on a Silicon Graphics desktop graphics computer that used a simple joystick for control (Sullivan, Darken, & McLean, 1998; Sullivan, 1998). We called the system the map interpretation and terrain association virtual environment system or MITAVES, as these were exactly the skills our simulator was intended to address. See Figure 27.2.

The screen resolution was poor on the initial prototype, and we needed to move to more common personal computer (PC) hardware, so we reimplemented the system two years later using an Intergraph PC with Wildcat graphics cards (McLean, 1999). With much improved resolution and an improved interface, the results were encouraging. See Figure 27.3.

Most importantly, pilot navigation became active as opposed to the passive video training that was used previously. The wide field of view display was a critical element for the helicopter domain because a typical flight profile for a helicopter is slow enough that a feature, such as a hilltop or ridgeline, will remain in view long enough to be useful as a navigation aid. Had we opted for a single-screen narrow field of view display, we would have been training pilots to focus only on features directly in front of them rather than looking side to side, which is not a good practice.

Figure 27.2.   Initial Prototype of the VE Navigation Training Device, MITAVES

Figure 27.3.   The Second Version of MITAVES

Another element of the MITAVES design was the joystick control, which purposely did not simulate flight dynamics. The trainer is for navigation skills, and in real flight the navigating pilot will not be at the controls, so we simplified the control mechanism. The joystick was used to "point" the aircraft in a direction, and it would follow the terrain until the direction or altitude was changed.

The last part of the design central to this discussion is the use of maps, especially "you-are-here" maps, in MITAVES. Being disoriented is a part of navigational training, but being hopelessly lost, which frequently occurs in video training, can be damaging. Providing some sort of you-are-here map was appropriate, but making it available while navigating would prove counterproductive. Pilots stop looking at the "out the window" view and stare at the moving map display, which is completely artificial. The solution was to provide a you-are-here map, but to provide it in a mode that stops active navigation. When the pilot requests the map, flight is stopped while the map is displayed and the pilot reorients on the paper map. Then the map is dismissed and flight continues. We can also count map views as a measure of performance in addition to actual navigation performance. See Figure 27.4.

We tested both versions of MITAVES at Helicopter Antisubmarine Squadron Ten (HS-10) at Naval Air Station North Island in San Diego, California. We discovered several modes of use for MITAVES. The most obvious mode is as a practice tool for asynchronous use by trainees. After completing classroom material, student pilots can use MITAVES on their own to further develop their

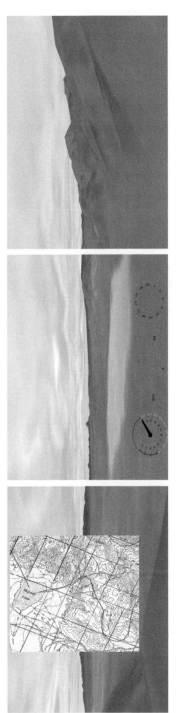

Figure 27.4. The MITAVES Map View

navigation skills. Instructors want to use MITAVES in the classroom when they realize how much more effective it is than the MITAC videotapes in current use. Experienced pilots can use the system for refresher training. Since navigation is a perishable skill, pilots can use MITAVES to retain sharpness in their navigation skills when they are in nonflying status.

Using a subjective evaluation method, HS-10 instructors evaluated student pilots on a standard navigation flight normally scheduled as part of their syllabus. They did not know which students had received MITAVES training. Trainees were evaluated on overall flight performance, ability to recover from errors, and ability to correctly identify features. The number of pilots in the initial studies was low, but data suggest that pilots who received VE training were better prepared for the navigation flight than those who did not.

These early studies revealed that time and distance estimation is a key element to successful air navigation and that the cues provided might be inadequate. The ability to perceive relative motion based on surface detail is an important cue that expert pilots use effectively. VEs often have such poor resolution on the ground that it is difficult if not impossible for the pilot to determine airspeed and distance traveled from anything other than cockpit displays, where in the actual aircraft the pilot develops the ability to estimate relative distance based on optical flow of the surface detail on the ground. That surface detail is often nonexistent in a VE. We conducted a separate study to determine how much detail was needed to sustain reasonable performance on this element of helicopter navigation. Using the amount of surface detail as the independent variable, we were able to show that just a 1 percent density of ground detail will allow a helicopter pilot to maintain a reasonable hover as compared to when only a flat texture is provided (Peitso, 2002). See Figure 27.5.

Although a low cost PC based system showed promise as an effective trainer for helicopter pilots, results were limited exclusively to natural terrains. It could be assumed that the system would be equally effective in urban terrains, but rendering a full fidelity urban terrain in the VE to facilitate positive training remained a challenge.

Using the same implementation as the second iteration of MITAVES, we developed a model of the northern Virginia area around Tysons Corner with the assistance of Marine Helicopter Squadron One (HMX-1, the Presidential Helicopter Squadron). See Figure 27.6.

Using only experienced Marine Corps helicopter pilots (because Marine Corps pilots often fly in urban environments), we studied the effectiveness of this new approach to rapidly building urban terrains for training and mission rehearsal. Two groups of pilots received paper maps and charts with which to prepare for an evaluation flight in the simulation in which they would be asked to identify key buildings and features. One group was also able to practice the flight path using the VE, while the control group was not. We then used a high resolution video produced for us by HMX-1 as the transfer task. Pilots who received the VE training were significantly better at identifying features and checkpoints during flight than the control group.

Figure 27.5.   Surface Detail Used to Hold a Hover

Thus, the nature of the task can be used to identify where to concentrate the effort in building virtual databases of real places. Since Marine Corps pilots were most interested in essential features, such as roads and intersections, rivers, bridges, and the most salient buildings, nonessential features, such as houses, cemeteries, and shopping malls, were not included. Using this approach, we can quickly produce large-scale virtual urban terrains that can be effectively used for mission rehearsal (Wright, 2000).

Although the VE training device built for helicopter pilots operated on low cost hardware, with a relatively small footprint and positive effects on navigation performance in the air, and terrain databases had been built for it, significant shortcomings remained. For example, within MITAVES, the pilot can stop the simulation anytime to take a break, to think about an error, or for any other reason. In the air, stopping is never an option except in an emergency. Successful navigation has to be performed under conditions of extreme stress. In addition, even though MITAVES ran on an off-the-shelf PC using standard displays, a typical deployed squadron would be hard-pressed to find room for it on board ship. Could it be made any smaller?

THE CHROMAKEY AUGMENTED VIRTUAL ENVIRONMENT

We tried to address these issues as well as several others using a new approach to VE simulation for aviation. We called it the Chromakey Augmented Virtual

Figure 27.6. Virtual and Real Tysons Corner, Virginia

Environment or ChrAVE (Darken, Sullivan, & Lennerton, 2003; Lennerton, 2003). The approach was to use chromakey technology to mix the real environment of the actual cockpit with the virtual environment computed separately and delivered to the pilot's eyes via a head-mounted display. The glass of the cockpit canopy is covered in blue material. The head-mounted display (HMD) has a camera and a 6 degrees of freedom tracking device mounted on it. As the pilot looks around, the video feed from the camera is passed to a video mixing unit. The position and orientation of the head is passed to the simulation software, which renders the appropriate image for that frame. The mixing unit replaces anything the camera sees as blue with the virtual environment, resulting in an augmented, composite image where the cockpit interior and the pilot's body are seen, but the glass is replaced with the VE. See Figure 27.7.

In our laboratory, we built a helicopter cockpit mock-up with no glass canopy, so the blue-screen material was placed in frames in front and to the side of the apparatus. We used illumination to create an even distribution of light across the surface of the fabric. During training in the mock-up, the seated pilot wears the HMD and uses a paper map, the cockpit displays, and the VE to navigate the course. Flight commands are given using the same verbal protocol used on an actual flight. In practice, the system would be composed of a shock-mounted rack for the PC and the video mixer with cables going to the HMD. Further reductions in size may be possible, but, most importantly, *the simulation platform becomes the helicopter itself* and does not require a simulation device that would need space on board ship. See Figure 27.8.

The design of the HMD apparatus was nontrivial. We needed a robust mounting of the camera to the display so it would not loosen or break easily. We considered using a small form factor camera mounted inside the housing of the HMD, but decided against that because it would be too fragile and we would lose the ability to alter the focal length of the camera. We decided to mount the camera on top of the HMD with the spatial tracker mounted above it. While this gave

Figure 27.7. The First Version of the Chromakey Augmented Virtual Environment

Figure 27.8. The Camera (Left), VE (Center), and HMD (Right) Views

us a rugged apparatus, it effectively moved the placement of the pilot's eyes to a different location. This altered perception considerably and therefore caused us to investigate if this had adverse effects on performance overall. See Figure 27.9.

We used a ball-tossing exercise as a measure of hand-to-eye coordination. We measured the participants' abilities to catch a small ball tossed to them from a few feet away and compared their performance unhooded to hooded with the HMD initially, hooded with the HMD post-exposure, and finally unhooded again post-exposure. As expected, eye displacement minimally impairs hand-to-eye coordination, but performance returns to baseline levels very quickly after exposure.

The study measured the performance of 15 experienced pilots as they prepared for and then executed an approximately hour-long flight over virtual Southern California. Several subjective evaluations were used to compare performance in the simulator to actual performance. Visual scan patterns in the ChrAVE were similar to actual cockpit scan patterns, as was the response to added stress in the task. For example, a secondary task required pilots to listen to simulated radio calls for an individually assigned call sign and report whether it had been called. The simulated environment of the task became so difficult at times that participants lost track of the radio calls altogether, and performance on this secondary task dropped significantly. This indicated that task complexity was mirroring reality and that performance was becoming comparable as well.

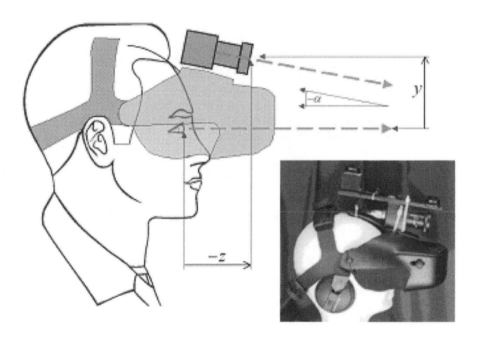

Figure 27.9.   The ChrAVE Head-Mounted Display Showing Visual Offset

One of the things we gave up when we moved to the HMD in the ChrAVE was wide field of view. While wide field of view is critical for helicopter pilot navigation, narrow field of view in an HMD is less critical because (a) the field of view can be artificially extended by back and forth head movement and (b) this is exactly how night vision goggle (NVG) usage works. In the military context, a large percentage of flights are made under NVG conditions, so having NVG simulation capabilities is critical to the adoption of the system. In the next phase we focused on how to address NVG flight using the ChrAVE (Beilstein, 2003).

We replicated the initial ChrAVE study using the same apparatus, but instead of a daylight simulation for the VE, we replaced it with an NVG simulator. The NVG simulation image generator used had two modes: physically based and nonphysically based. When using the physically based mode, the material properties of the environment are used to calculate the illumination of each pixel in the display. When not physically based, everything is approximated using simple heuristics to create a believable display, but it is not capable of changing based on the type of materials or the moon state.

Using the same criteria as the initial study, we determined that performance in the NVG simulator was a close approximation to real flight and that again, the simulated stress of the primary task was valid. Interestingly, even the most experienced subject pilots were not affected by the physics of the NVG simulation. They performed equally well with or without physical realism. Given the heavy pre-simulation and run-time costs of computing NVG imagery to be physically authentic, questions remain as to which sorts of tasks require physical reality and which do not. We believe the default should be to approximate NVG imagery given our results.

In addition to knowing how it would perform, before taking the ChrAVE to the fleet it was necessary to determine if the fleet saw it as a plausible way to train and rehearse missions if adopted for large-scale usage. At this time, the Office of Naval Research program adopted the ChrAVE as the third part of its three-platform research, development, and concept demonstration for the U.S. Navy and the U.S. Marine Corps. The ChrAVE became VEHELO (Virtual Environment Helicopter) and was soon to perform in a networked simulated training environment along with VELCAC (Virtual Environment Landing Craft, Air Cushion) and VEAAAV (Virtual Environment Advanced Amphibious Assault Vehicle) in the Virtual Environments and Technologies program.

Transitioning from laboratory to field experimentation involved changes in the physical equipment, the software, and the experimental design. Up to this point the design goal had been to re-create the environment, task, and stress in a virtual setting. Success was measured by comparing virtual performance and behavior to real world performance and behavior in order to determine if the virtual experience could serve as a surrogate for real world experience. The next step was to quantify the value of a virtual experience to determine if training with VEHELO could improve real world performance.

Changes to the physical equipment for the first two experiments involved mounting the PC and electronic equipment in a portable rack-mount container and simplifying the mock cockpit to make it more portable. The portable mock cockpit included only the seat, flight controls, and a liquid crystal display (LCD) representing the instrument panel mounted in front of the user. It did not include the cockpit frame structure. In the laboratory setting, the pilots had a restricted field of view based on the physical cockpit. In the first two field experiments pilots had a less restricted field of view. The physical setup for the first two studies is shown in Figure 27.10.

In designing the first transfer study we attempted to leverage and extend previous work on helicopter overland navigation. The goal changed from similar performance and behavior to improved performance. The key questions centered on ideal treatment for experiment and control groups, as well as performance measurement. Boldovici, Bessmer, and Bolton (2002) provide an excellent summary of the typical difficulties encountered in conducting and drawing meaningful conclusions from such field studies. To eliminate the potential bias of comparing a group that received training to a control group that received no additional training, an alternative treatment was devised consisting of a detailed review of the techniques of overland navigation applied to the specific navigation route that students would fly. The experimental group flew the route using VEHELO, while the control group reviewed the route using "best practices." The experimental and control groups received equal time in their respective training treatment. This training followed normal classroom training leading up to syllabus flight events, which were used as the real world transfer task.

It is difficult to determine reliable performance measurements for overland navigation. Experience from previous studies highlighted the fact that simple navigation plots are not always indicative of a trainee's ability. A navigator who is accidentally on the route would be incorrectly rated as better than a pilot who is aware of his or her position and surroundings, but intentionally deviates from the route. Similarly, navigation plots from training sessions are not necessarily a reliable indicator of the potential value of a training session. A trainee who is consciously trying to practice a variety of navigation techniques may appear less proficient than a pilot who is inappropriately relying on a single

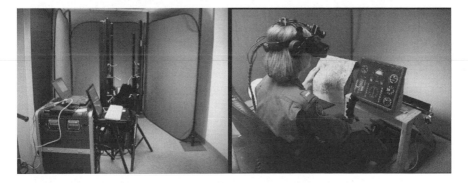

Figure 27.10. VEHELO Portable Configuration

method, such as dead reckoning. This is closely related to the concept of deliberate practice outlined by Ericsson, Krampe, and Tesch-Romer (1993). As noted by Ericsson et al., it is difficult to evaluate and measure via observation when deliberate practice occurs.

To augment the navigation data, subjects completed questionnaires that included self-efficacy and workload ratings. Subjects also completed static terrain association tests designed to measure a subject's ability to match a plotted position on a map with the corresponding out-the-window view. We also extended the normal grading criteria associated with the syllabus flight events. Instructor pilots who flew with test subjects completed a more detailed grade card with more specific assessment of navigation performance and workload. Members of the experimental design also flew as aircrew on syllabus flights to provide subjective assessment. While the small sample size made it difficult to provide statistically significant differences between the control and experimental groups, several conclusions were clear. Pilots who flew the route virtually were subjectively assessed by their instructors as having better overall navigation performance in terms of ability to location position, identify key navigation features, and manage the cockpit workload. Pilots who flew in VEHELO were also better at maintaining track.

Based on this initial field study, two additional studies involving several major changes were conducted at the Marine Corps' H-46 Fleet Replacement Squadron (FRS)—HMM(T)-164 in Camp Pendleton, California (Kulakowski, 2004; Hahn, 2005). The experimental design was changed significantly. In the initial experiment we wanted to avoid any confounds associated with individual instruction. Thus, the protocol limited the feedback that the individual running VEHELO could provide to warnings at prescribed distances from intended track. Given the positive indications of potential training value from the first study, we removed this constraint for the second study. In the second study, the evaluator [a former FRS instructor at HMM(T)-164] was allowed to provide whatever instruction he deemed appropriate. The task was extended from terrain association and navigation to explicitly include crew resource management. In practice, navigation is a collective responsibility in which every member has a role. In the H-46 the aerial observer and the crew chief provide information on salient visual cues within their field of view. The nonflying pilot is responsible for coordinating the overall effort. To provide trainees experience coordinating these efforts, the evaluator filled the role of other crewmen by simulating typical intercockpit communications system (ICS) calls for various crew positions. To attempt to increase sample size and minimize the impact on the syllabus, the terrain association test and student questionnaires were omitted. In the second study, the difference in ability to maintain track was significantly better for subjects who trained using VEHELO. Additionally, their instructor pilots rated them better at navigation performance, as well as crew resource management.

While these results were promising, there were still issues with the VEHELO configuration. The equipment was cumbersome, difficult to set up and adjust, and did not provide the immersive environment originally intended.

Coincidentally a system that would address these issues was being developed by the entertainment industry, which also needed highly portable and easy-to-set-up chromakey. The solution was a ring of light emitting diodes (LEDs) that could be mounted around a camera lens with a special retroreflective material that worked with the wavelength of the LEDs and only reflected light directly back at the source. This configuration was used on one of HMM(T)-146's helicopters dedicated for maintenance training. The windscreen area was covered with a sheet of retroreflective material; a ring of LEDs was added to the camera/HMD, and an LCD panel was used to render the instrument panel. With the new configuration, the setup time changed from hours to minutes. Subjects now had many of the constraints of the operational environment: the same field of view and obstructions, communication via the aircraft's ICS system, limited space to manage maps, checklists, and route kneeboard cards. We were also able to upgrade the HMD from a 640 × 480 display to a 1,280 × 1,024 display. See Figure 27.11.

The experimental design for the third round of transfer studies did not change. Results from the third round of studies echoed previous work. Subjects who were exposed to virtual environment training maintained track better. They were also subjectively evaluated as superior at crew resource management, navigation, and management of cockpit workload.

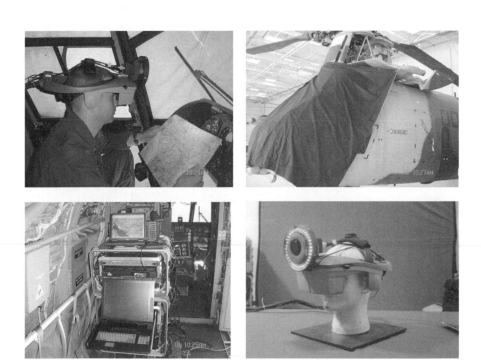

Figure 27.11.   The Field Tested HMM(T) apparatus for VEHELO

CONCLUSIONS

Our years of study in the use of VEs for training navigation skills in helicopter pilots have proven fruitful. We have learned how to build low cost VEs for training that are valuable tools for spatial skill and knowledge development. We have refined our techniques to include night vision capabilities and urban terrains. We have extended the concepts toward embedded training that will be far more suitable for many training domains.

In all cases, a thorough understanding of the training domain is key to properly constraining the design so that it fits the training requirements, the trainees' needs, and the environment and situations for use. These are all critical for military training. Most importantly, this work revealed that conventional training transfer experiments are simply not practical in the military setting most of the time. The work represents a creative approach that controls exposure, treatment, and trainees in such a way that researchers can learn what works without sacrificing effective training for experimental subjects. The results of this work contribute to real training systems for military helicopter pilots and crew.

REFERENCES

Banker, W. P. (1997). *Virtual environments and wayfinding in the natural environment.* Unpublished master's thesis, Naval Postgraduate School, Monterey, CA.

Beilstein, D. L. (2003). *Visual simulation of night vision goggle imagery in a chromakeyed, augmented, virtual environment.* Unpublished master's thesis, Naval Postgraduate School, Monterey, CA.

Boldovici, J., Bessmer, D., & Bolton, A. (2002). *The elements of training evaluation.* Alexandria, VA: U.S. Army Research Institute for Behavioral and Social Sciences.

Darken, R., Sullivan, J., & Lennerton, M. (2003). A chromakey augmented virtual environment for deployable training. *Proceedings of I/ITSEC* [CD ROM]. Arlington, VA: National Training and Simulation Association.

Darken, R. P., & Banker, W. P. (1998). Navigating in natural environments: A virtual environment training transfer study. *Proceedings of Virtual Reality Annual International Symposium* (pp. 12–19). Washington, DC: IEEE Computer Society.

Ericsson, K., Krampe, R., & Tesch-Romer, C. (1993). The role of deliberate practice in the acquisition of expert performance. *Psychological Review, 100*(3), 363–406.

Goerger, S., Darken, R., Boyd, M., Gagnon, T., Liles, S., Sullivan, J., et al. (1998, April). *Spatial knowledge acquisition from maps and virtual environments in complex architectural spaces.* Paper presented at the 16th Applied Behavioral Sciences Symposium, U.S. Air Force Academy, Colorado Springs, CO.

Hahn, M. E. (2005). *Implementation and analysis of the Chromakey Augmented Virtual Environment (ChrAVE) version 3.0 and Virtual Environment Helicopter (VEHELO) version 2.0 in simulated helicopter training.* Unpublished master's thesis, Naval Postgraduate School, Monterey, CA.

Jones, Q. (1999). *The transfer of spatial knowledge from virtual to natural environments as a factor of map representation and exposure duration.* Unpublished master's thesis, Naval Postgraduate School, Monterey, CA.

Kulakowski, W. W. (2004). *Exploring the feasibility of the Virtual Environment Helicopter system (VEHELO) for use as an instructional tool for military helicopter pilots.* Unpublished master's thesis, Naval Postgraduate School, Monterey, CA.

Lennerton, M. (2003). *Exploring a chromakeyed augmented virtual environment as an embedded training system for military helicopters.* Unpublished master's thesis, Naval Postgraduate School, Monterey, CA.

McLean, T. (1999). *An interactive virtual environment for training map-reading skill in helicopter pilots.* Unpublished master's thesis, Naval Postgraduate School, Monterey, CA.

Peitso, L. (2002). *Visual field requirements for precision Nap-of-the-Earth helicopter flight.* Unpublished master's thesis, Naval Postgraduate School, Monterey, CA.

Psotka, J., Massey, L. D., & Mutter, S. A. (Eds.). (1988). *Intelligent tutoring systems: Lessons learned.* Hillsdale, NJ: Lawrence Erlbaum.

Sullivan, J., Darken, R., & McLean, T. (1998, June 2–3). *Terrain navigation training for helicopter pilots using a virtual environment.* Paper presented at the 3rd Annual Symposium on Situational Awareness in the Tactical Air Environment, Piney Point, MD.

Sullivan, J. A. (1998). *Helicopter terrain navigation training using a wide field of view desktop virtual environment.* Unpublished master's thesis, Naval Postgraduate School, Monterey, CA.

Ward, P., Williams, A., & Hancock, P. (2006). Simulation for performance and training. In K. Ericsson, N. Charness, P. Feltovich, & R. Hoffman, (Eds.), *The Cambridge handbook of expertise and expert performance* (pp. 243–262). New York: Cambridge University Press.

Wright, G. T. (2000). *Helicopter urban navigation training using virtual environments.* Unpublished master's thesis, Naval Postgraduate School, Monterey, CA.

Chapter 28

TRAINING EFFECTIVENESS EXPERIMENTATION WITH THE USMC DEPLOYABLE VIRTUAL TRAINING ENVIRONMENT— COMBINED ARMS NETWORK

William Becker, C. Shawn Burke, Lee Sciarini,
Laura Milham, Meredith Bell Carroll,
Richard Schaffer, and Deborah Wilbert

Military teams are increasingly confronted with a need to be adaptive as they operate against asymmetric threats across a wide variety of environments and mission types ranging from combat to stability, support, transition, and reconstruction operations (U.S. Department of Defense, 2006). In efforts to achieve this goal there has been a movement toward training troops to operate independently in smaller, "scalable" units. It is argued that these units, acting in concert with commander's intent, will be able to exercise initiative to locate, close with, and destroy enemies (*Fleet Marine Force Field Manual 6-5,* 1991). While the move toward smaller, scalable units may make military forces more agile, it does not guarantee that the skills needed to foster adaptation of strategy, structure, and process will be present. In order to promote such skills, the military relies on a wide variety of training methods, not the least of which is the use of experiential based simulation. Due to the current operational tempo where resources and time are at a premium, it is essential that these simulations be based on the science of learning paired with the appropriate use of technology such that maximum efficiency and learning occurs (Salas & Burke, 2002).

One of the many units seeking to utilize such training is Fire Support Teams (FiSTs). A FiST is a small group of marines within the Marine Air-Ground Task Force charged with the tactical coordination of air and indirect fire assets. Key FiST members include the FiST leader, forward observers (FOs) (for artillery and mortar crews), and a forward air controller (FAC). Together the FiST creates and communicates an attack plan to supporting artillery, mortar, and aircraft units to achieve mission success.

Currently, the majority of FiST training is conducted through the use of classroom instruction in which declarative and procedural knowledge is received. Trainees then practice the application of such knowledge through the use of mental simulation that incorporates a sand table or battle board. Finally, environment and organizational resources permitting, trainees engage in a live-fire exercise. This type of training is commonly referred to as the "crawl, walk, run" method. Unfortunately, the majority of the Marine Corps' current training options are limited in the ability to provide accurate visualizations or temporal accuracy until live-fire training.

In an effort to address these gaps an Office of Naval Research program known as the Multiplatform Operational Team Training Immersive Virtual Environment was field-tested in the fall of 2006. This program, sponsored under the Virtual Technologies and Environments (VIRTE) program office, later became a program of record for the Marine Corps as the Deployable Virtual Training Environment (DVTE), a subpart of a larger system known as the Deployable Virtual Training Environment—Combined Arms Network (DVTE-CAN). DVTE-CAN is a network of simulation and support systems, which incorporate advances in simulation technology along with knowledge about individual and team learning to provide a virtual training environment for FiST members, as well as air and ground support personnel. The system provides an environment in which the forward air controller and forward observers can train to provide coordinated close air support, mortar, and artillery fires. For a more detailed description, the reader is referred to Volume 3, Section 2, Section Perspective.

The program objective for DVTE-CAN was to provide a simulation based experiential learning environment that focuses on the tasks requiring mastery for scalable marine infantry units, such as FiST teams. The requirement placed upon the experimental team was to determine the usability, utility, and effectiveness of DVTE-CAN as a supplemental training tool during each phase of the FiST training process. This chapter documents a portion of this process and lessons learned.

METHODS

One of the primary goals of this effort was to determine the usability and utility of the DVTE-CAN system across a variety of marine populations. To this end, experimentation was conducted that examined a variety of experience levels with units at several locations, including students at the Infantry Officers Course (IOC) (introductory training), Expeditionary Warfare School (EWS) (advanced refresher training), members of the active reserve (basic refresher training), and marines preparing for deployment (advanced training). Other United States Marine Corps (USMC) entities involved in field-testing included Expeditionary Warfighter Training Group Atlantic and Expeditionary Warfighter Training Group Pacific.

Introductory Training

The research team had three separate opportunities to insert DVTE-CAN into the curriculum at IOC. IOC prepares officers to be platoon leaders and is the second school a new Marine Corps officer attends. Included in this training are several days dedicated to call for fire (CFF), close air support (CAS), and FiST procedures and tactics. Participating in data collection efforts at IOC were 154 marines representing 2nd lieutenants with an average age of 23, with a majority having less than one year in service. In line with their time in service, this population of marines had minimal exposure or experience with CFF or CAS.

Basic Refresher Training

Data collection efforts with the 1st Battalion, 23rd Marines (1/23) afforded the team an opportunity to examine DVTE-CAN within an activated reservist unit conducting field training. The reservists reported having no experience as actual FiST members and minimal training in supporting arms or CAS. Twelve marines representing a range of ranks (that is, lance corporal to sergeant) and an average of five years in service used the system to train with regard to the FO role within a FIST environment.

Advanced Refresher Training

Data collection at EWS afforded DVTE-CAN to be implemented and feedback collected from experienced, active-duty marines (1st lieutenants to lieutenant colonels with an average age of 32). DVTE-CAN was utilized as a refamiliarization tool and training device with 12 squads approximately six months into their yearlong training cycle at EWS. Over half of the marines reported that they had prior training experience as a FiST member, and 80 percent had less than 500 hours of training in supporting arms or close air support. As compared to many of the other implementations of DVTE-CAN, the marines at EWS reported a wider variety of military occupational specialties (MOSs), including infantry, artillery, logistics, administration, police, aircrew, armor, medical, and combat engineer. Within the MOSs reported, 55 percent were classified as ground, 25 percent air, and 18 percent supporting agencies.

Advanced Training

An intact FiST team organic to 3rd Battalion, 4th Marines (3/4) used DVTE-CAN as an advanced team training device while training at the USMC Air-Ground Combat Center; 3/4 was about halfway through its 180 day training cycle prior to deployment. FiST members ranged in age from 21–31 and time in service ranged from 2–12 years. DVTE-CAN was used to supplement the pre-deployment FiST training package with a focus on ensuring the team was capable of correctly executing 3/4's FiST battle drill in a variety of scenarios.

EXPERIMENTAL DESIGN AND PROCEDURE

The design of choice was a between-subjects design with two levels of the independent variable, practical application method. Specifically, marines within the control condition used mental simulation or sandbox-type exercises for the basis of their practical application, while those in the experimental condition used DVTE-CAN to develop and execute their battle plans. Marines in both conditions were given the same commander's intent (for example, rules of engagement, available assets, and deadlines) and operating picture. In developing their battle plans, marines in the control condition used standard planning tools (for example, battle board, compass, and map), while those in the experimental condition used the planning tools embedded within the DVTE-CAN system augmented with the standard tools.

Upon arrival marines completed an informed consent document, demographics, and self-efficacy questionnaires. Marines in the experimental condition then participated in a brief system orientation session, facilitated by their instructor or resident contractor. At the conclusion of this session, commander's intent was delivered to the FiST teams for the first scenario. FiST members then either used DVTE-CAN or standard practical application tools (as described above) to develop their battle plans. Instructors and squad members then subjected FiSTs within both conditions to a critique of their battle plans. At the conclusion of the critique the plan was executed. Within the control condition execution occurred via mental simulation, while within the experimental condition DVTE-CAN allowed marines to see execution take place in real time and space. At the conclusion of the practical application, FiST teams again completed a series of questionnaires. Upon completion of the in-class execution phase, squads participated in live-fire training, when available.

While the preferred experimental design was as detailed above, resource constraints often dictated that modifications were made to the initial design (see Table 28.1). For example, on several occasions, training schedules did not permit the creation of true control and experimental conditions; consequently, experimental design across the populations reflected a combination of true control groups and experimental pre-/post-test-only groups.

CORRESPONDING INSTRUMENTS

Depending on the exact purpose of each data collection opportunity the exact questionnaires varied, but potential questionnaires included pre-/post-self-efficacy, reaction/utility, usability, and ability of the system to be used to achieve specified learning goals.

Self-Efficacy

A 33-item questionnaire assesses the confidence in a FiST member's own ability, as well as the team's ability to accomplish key FiST tasks ($\alpha = .93$).

Reaction/Utility

An 18-item questionnaire assesses reactions. System utility in terms of promoting FiST-related skills, the degree of confidence marines had that DVTE-CAN

Table 28.1. DVTE Application Variants

Training Unit	DVTE Configuration	Training Environment	Training Details
IOC	Joint Semi-Automated Forces (JSAF), forward observer artillery (FO ARTY), forward observer mortars (FO MORT), and FAC	Schoolhouse	• DVTE replaced the traditional sand table indoctrination training in the experimental group IOC classes. • Three training days: two schoolhouse; one live-fire day. • Two classes received live-fire training. One class used DVTE in place of live fire due to the lack of available assets. • Four member teams: (1) FiST leader, (2) FOs, (1) FAC. • FOs & FACs used DVTE; FiST leader used battle board.
1/23	JSAF, FO ARTY, and FO MORT	Field tent	• DVTE replaced the traditional sand table and mental simulation. • System powered by deployable generator. • Small teams worked together to build battle plans for multiple scenarios. • Participants trained in the role of the FO providing the instructor solutions to the scenarios. • Individuals practiced simulated FO communications with the instructor while the plan was executed in the environment and projected for observation.

| EWS | JSAF, FO ARTY, FO MORT, FAC, Combined Arms Planning Tool (CAPT), and AH1 Simulation | Schoolhouse | • DVTE replaced the traditional sand table C4 (command, control, communications, and computers) training
• FiSTs practiced developing, critiquing, and executing battle plans
• JSAF operator also played fixed-wing role
• One trainee (a pilot) operated the AH1 Cobra simulation.
• Acting commanding officers (COs) received scenarios from the instructor, developed intent, and passed it to the FiST.
• FiST members created plan and communicated it to supporting agencies
• Plan was collaboratively critiqued by the instructor, acting CO, and the FiST.
• Finalized plan was executed; corresponding effects and coordination requirements were observed. |
| 3/4 | JSAF, FO ARTY, FO MORT, and FAC | Schoolhouse | • DVTE replaced the traditional sand table training and spanned two days.
• Day One: CO (also the instructor) received personal JSAF training while trainees received group instruction on the other DVTE components.
• Marines used free play to reinforce guided instruction.
• Marines received system setup instruction.
• System training culminated with execution of two basic scenarios.
• Day Two: Trainees set up DVTE system in <30 minutes.
• CO operated JSAF and conducted training the way he would have in the traditional setting. |

prepared them for live fire and to lead a FiST during live fire was assessed. Items also assessed if marines would recommend that DVTE-CAN be given to other trainees or used while on deployment. Finally, several open-ended items encouraged the marines to identify system areas especially useful, as well as areas in need of improvement.

Usability

A 25-item questionnaire reflected screen aspects, terminology/system information, learning, interactions, and system capabilities.

Learning Goals

Learning goals assess the degree to which DVTE-CAN could be used to train key FiST learning objectives and tasks. Tasks and corresponding learning objectives were identified based on a task analysis (Bell et al., 2006) and examination of Marine Corps documents.

Instructor and observer ratings of FiST performance were also collected during training and live fire when possible, but these results will not be reported here. Only a brief subset of those that most closely relate to spiral development and user acceptance will be discussed in detail within the current chapter. Results that relate to utility, learning, and transfer are to be documented in future publications.

RESULTS

Infantry Officer Course

The collection of data in this applied setting posed unique challenges with respect to logistics, training tempo, and availability of assets. As a result there are varying levels of data associated with each event. In order to keep aligned with the other events presented in this chapter, results concerning the degree that the system met the training goals at IOC and system usability for novice trainees will be presented.

Overall, DVTE-CAN was able to meet IOC's specific training goals. Trainees reported that the DVTE-CAN system had utility over the conventional training in terms of promoting individual, team, and other communication skills related to key FiST tasks. Trainees also reported that DVTE-CAN provided them with a more complete understanding of the FiST domain and enhanced classroom instruction.

Efficacy data revealed that trainees were confident that DVTE-CAN prepared them and their team for live fire, as well as to serve in any position within the FiST. Perhaps most importantly, efficacy data revealed that they were highly confident to serve as a FiST leader during live fire. Finally, trainees showed increases in self-efficacy in all areas of mission performance including planning, coordination, terminal control of aircraft, communication, and equipment use, with significant increases over those who had not received training on DVTE-CAN for both terminal control of aircraft and equipment use.

At each event, the marines who used DVTE-CAN rated their overall experience with the system and its tools favorably. The use of the tools provided and the visualizations were consistently ranked high across each event. All three events presented a variety of usability areas for improvement. After analysis, these areas were prototyped and then incorporated in improved versions of DVTE-CAN for testing at subsequent events in efforts to improve usability and training effectiveness. Regardless of the training event, one of the most frequently recorded usability issues was the students' initial confusion with the operating the system. This highlights the need for a comprehensive and standardized training package that allows enough time for familiarization on the system prior to classroom instruction and scenario execution. Overall, free response comments from the trainees were encouraging. Many of the students shared the opinion that experiencing the terrain, targets, and impact of ammunition from the first-person point of view was beneficial and that it provided a level of realism not available on the table-mounted terrain replica. Additionally, the marines indicated that the ability to experience the temporally valid scenarios provided by DVTE-CAN increased their understanding of the timing, coordination, and communications required for successful FiST operations.

1st Battalion, 23rd Marines

After using the DVTE-CAN system, marines reported that it had utility in terms of promoting individual, team, and communication skills related to key FiST tasks. Overall, participants felt that the practical application provided an effective training tool for live-fire exercises, that it enhanced classroom instruction, and that it was moderately challenging. However, perhaps most insightful were the free-form comments provided by the reservists with regard to how the system was used for their particular purpose. Specifically, the marines of 1/23 reported that system components dealing with hands-on practice, call for fire, the map, system tools, and feedback were particularly useful and noted as key system components. For example, with regard to hands-on practice, marines reported the realistic simulation and hands-on experience of what occurs within a FiST to be especially useful. Marines also reported that the system had strengths with regard to call for fire, specifically helped with the work-up of CFF, assisted in learning CFF in the classroom environment, provided instruction of verbal commands, and covered step-by-step instruction given the mission. Marines also highlighted features of the map, such as quick access to the map and protractor, map reading techniques, and help with plotting targets on the map. Similarly, in terms of system tools, marines reported that the tools used for calculations were especially useful. Finally, marines reported liking aspects that could be categorized as related to feedback, including the ability for the program to tell users to "say again" if incorrect, the process of getting used to making adjustments, and the point system in showing faults.

While several system components were specifically mentioned as standing out, a few areas were also noted for improvement. Specifically, it was noted that certain aspects of the map (for example, thick lines on markers) made it difficult to read; a need to practice application with the field radio was noted, and it was suggested that not all the tools presented within the system are available to all marines (so there was some unfamiliarity, for example, the Viper). Finally, a note was made of a need for more training on the system.

Expeditionary Warfare School

After using the system, marines rated the usefulness of the system in its ability to support several critical knowledge, skills, and abilities related to fire support, confidence in its ability to prepare them for live fire, its accurate simulation of FiST operations, and added value over traditional practical applications. Overall the system received very favorable results in all rated areas. Similar to the quantitative findings, overall qualitative data were also favorable. Those aspects of the simulation that received the most consistent praise can be categorized into three areas: visualization, practicing communication/coordination skills, and active participation. While many comments were made that reflected the importance of the visualization capabilities of the system for these marines, a few illustrative examples, including the following program visuals, were useful and assisted in (a) timing, deconfliction, and finding the enemy, (b) graphic depiction of feedback, (c) ability to see aircraft and marks as specified by the mission timeline, and (d) the depiction of planning/execution shortfalls. Overall comments reflected the utility of the system allowing visualization of how things come together and play out in "real" time. The ability to actually practice communication and coordination in real time was also noted as a valued system component. For example, marines made such comments as the system (a) enabled participants to work together as a team and communicate together in real time, as they would in a real situation and (b) it provided the capability to practice communication and coordination and realize errors in a safe environment. Finally, with regard to active participation, marines valued the fact that it "puts students in the position to do versus merely watching." A few other notable comments indicating system benefits as seen by the marines at EWS include (a) it was extremely helpful in pulling all aspects of running a FiST together, (b) capability to apply skills learned in class, (c) timeline integration, and (d) real time effects on target.

While the predominant number of comments regarding the use of DVTE-CAN as configured at EWS were positive, a few areas of improvement were suggested. While there were several suggested areas of improvement, many dealt with the need for participants to have a better orientation to the tool and more time to run through the simulated missions. In many instances observers noted that in light of time constraints, instructors would "simulate" a key aspect of execution, thereby constraining some of the benefits of using simulation. A few notable suggested system improvements included the inability to plot units on the map, difficulty in plotting and drawing the fixed-wing initial point on the map, and a general suggestion for more map interactivity. Also noted were a lack of fixed- and rotary-wing attacks on the battle board timeline within CAPT and a lack of a FiST specific common operational picture map.

3rd Battalion, 4th Marines

Overall, DVTE-CAN was able to meet 3/4's specific training goals, and, similar to 1/23's results, the marines reported that the DVTE-CAN system had utility in terms of promoting individual, team, and communication skills related to key FiST tasks. Additionally, the members of 3/4's FiST reported that DVTE-CAN was exceptionally useful for plotting asset and target locations, planning suppression of enemy air defense missions, as well as observing the impact of rounds for effectiveness, adjustment, and/or determining battle damage assessment. On more than one occasion, the researchers observed individual marines recognizing procedural slips through the visualization provided by DVTE-CAN; self-correcting; and then sharing those slips with the team, thus creating a team learning experience. While team learning behavior was not being directly investigated, the potential of DVTE-CAN to strengthen shared mental models and to provide an improved team training experience cannot be overstated. Overall, the marines of the 3/4 FiST felt that the practical application provided an effective training tool and that they were, in fact, better prepared for a live-fire training exercise.

The marines that used the forward observer PC simulator rated their overall experience with the program and its tools favorably. It is important to emphasize that several aspects and tools were specifically identified as being exceptional. There were a few areas in which the marines made suggestions for improvement. These suggestions were directly related to their prior experience in live-fire and/ or combat situations. Specifically, it was noted that map icons should be in operational terms in order to match what the users would experience in the actual domain. Additionally, in reference to the compass tool, the marines suggested that the virtual M2 and lensatic compasses needed minor modifications to match their live counterparts. One of the most positive ratings came from a free-form comment stating that even in high complexity missions, the system allowed the user to easily perform his operations.

JSAF received an overall high user experience rating from the instructor's point of view. Some of the more notable items were the usefulness and accessibility of the system tools, the display of virtual units and other elements, the correct and consistent use of domain relevant terminology, and the ease of interaction. These high rankings are encouraging since the operator was also instructing and evaluating the FiST marines while he was playing the role of the supporting agencies. There were areas in which JSAF received average ratings. One of these areas was a navigational issue resulting in difficulties setting a time on target for all participating entities. Additionally, the lack of availability of different information formats presented difficulties for the user. These items were directly related to the execution of the scenarios and may have been the result of the lack of comprehensive JSAF functionality training.

OVERALL LESSONS LEARNED

The uniqueness of each data collection effort provided a robust opportunity to assess the perceived utility and flexibility of DVTE-CAN, as well as the effectiveness of utilizing a spiral development process. Based on usability data collected throughout IOC events and other opportunities, usability issues were identified and problem solution tables were provided to system designers who implemented system improvements. Comparing version 1 to version 4 of DVTE-CAN, usability gains resulted in over a 40 percent decrease in heuristic violations.

While each marine population indicated that DVTE-CAN had value over traditional classroom training, each group valued different system components based on experience and stated training goals (see Table 28.2). For example, insertions at IOC served to illustrate how the system could be used during indoctrination training to FiST tasks. At IOC, DVTE-CAN was seen as being able to provide marines with a realistic understanding of FiST communications and timelines that are often not available through traditional classroom methods. Marines also reported that DVTE-CAN provided a visual aspect not available on wall-mounted maps, promoting a better understanding of synchronization.

Similar to IOC, Marines at EWS saw the system as having utility, with nearly all officers recommending it for training while on deployment and indicating that it added value over traditional classroom training. Implementation at EWS illustrated DVTE-CANs use add value to an experienced set of marines who have been deployed, but have varying levels of familiarity with FiST operations and a diverse set of occupational specialties. Instructors also noted its usefulness by expressing interest and querying as to its ability to train teams of teams.

A marked contrast to the previous environments was DVTE-CAN's use by 1st Battalion, 23rd Marines. This case demonstrated DVTE-CAN's portability, adaptation for field use, and versatility to meet a unit's training goals. With 1/23, DVTE-CAN was primarily used to augment the standard lecture format by allowing the marines to practice the actions required of a FO within a FiST. Finally, it illustrated the perceived utility of DVTE-CAN with a set of reservists with minimal knowledge of FiST-related tasks.

Moving to the other end of the spectrum, 3rd Battalion, 4th Marines served to illustrate how the system could be used and valued with an experienced FiST whose members had deployment and FiST-related combat experience. With this population several firsts were noted. Specifically, the marines heavily weighted the value of the planning component as compared to execution, deployed the system and brought the system online quickly, and demonstrated the ability of a designated marine to operate JSAF and execute training scenarios. Ultimately the marines saw the system as having utility and tended to rate it highly; the officers and enlisted marines recommended DVTE-CAN for training in garrison, in the field, and while on deployment.

While there are lessons learned with each marine population (see Table 28.1), there are also lessons learned at the program level that facilitate others

Table 28.2. Combined Lessons Learned

Training Group	System Use	Unit Lessons
Infantry Officer Course		
• Active • Infantry MOS • Primarily LTs • No prior experience in FiST roles	• Indoctrination training • Schoolhouse • Four-person ad hoc FiST • Instructor led	• System seen as offering utility over traditional classroom training by both instructors and students. • System ability to be used for both planning and execution portions. • Visualization aspect highly valued; of particular value are identification of targets on deck in terrain, watching timelines unfold. • Instructor value indicated by use of system beyond original implementation population when live-fire canceled.
EWS		
• Active • Wide variety of MOSs • Captain to Lt. Col. • Some prior experience in FiST roles	• Refresher training • Schoolhouse • Four-person ad hoc • FiST + CO	• Perceived utility for refresher training on FiST operations within marines who had deployed across a wide variety of MOSs. • Instructors expressed interest in DVTE-CAN's ability to train teams of teams. • Additional training on system was requested. • Suggested a need for more time on system to gain full benefits (average time on system @ 40 minutes).

1st Battalion, 23rd Marines		
• Reservists • MOS • Rank	• Refresher training • Field environment • Instructor led	• Portability and successful use of DVTE-CAN for field; even with generators as power source. • Incorporated into lectures and simulation training. • Hands-on experience, calculation tools, practicing adjustments, plotting targets, working up transmissions and calls for fire, and hands-on practice of what goes on in FiST were valued system components. • Used primarily as a FO trainer and how this role relates to larger FiST.
3rd Battalion, 4th Marines		
• Active • Intact FiST, FAC role simulated • Operational experience in FiST roles	• Sustainment training • Classroom • CO-led training • JSAF operated by CO	• Ability of a designated marine to operate JSAF and execute training scenarios. • Learning curve on system reduced. • Ability of marines to easily deploy and bring system online quickly. • System was perceived to meet primary learning goals within an experienced intact FiST team.

conducting field research with marines. A representative sample of these appear below.

Lesson 1. Do not forget the importance of prepackaged tutorials to coincide with implementation of new systems.

At each data collection event, a day was dedicated to train-the-trainer sessions. However, as the trainers proceeded to later introduce the system to FiST members, it was done in many different ways due to differences in instructor style, resource constraints, and training schedules. Repeatedly the experimental team witnessed comments that suggested the need for additional time for user familiarization. While constraints on training time are unavoidable, there needs to be a prepackaged tutorial that goes along with newly implemented systems. A short tutorial can be designed to not only deliver systematic training, but to ensure that there is an appropriate baseline understanding of the instructional equipment such as not to hinder the learning of FiST procedures.

Lesson 2. A spiral development process incorporating a close partnership between learning specialists, programmers, and user groups is essential to move past the prototype stage.

Overall DVTE-CAN has received very positive feedback from the marines. The key to moving from a prototype system to transitional product has been the close partnership between subject matter experts, users, learning specialists, and programmers. The subject matter experts understand where the trainees need to be in terms of proficiency, the learning specialists assist in determining what features need to exist from a human learning standpoint, and the programmers are the ones who must develop the capabilities to meet these needs. Based on system or resource constraints, alternative designs are proposed, which in turn create new discussions on a way forward. Finally, if flexibility in terms of user population is warranted, then users of various experience levels should be interviewed.

Lesson 3. Prepare a secondary (and third) experimental plan that anticipates changes dictated once in the field environment.

When working within field environments, adaptability is key. Out of the six data collections that are represented within the chapter, not one went according to the original plan. Most often, due to weather conditions or training time constraints, experimental opportunities changed once the reality of Marine Corps training unfolded. As a result, the experimental team typically prepared a three-tiered experimental plan. Having a three-tiered plan that specified the ideal design, an alternative moderate design, and a minimum design allowed flexibility once on site as it had been determined a priori what sacrifices in terms of design would be acceptable, based on the questions being asked.

Lesson 4. Having a champion within the organization is essential to creating the relationships needed for access and success.

Within a relatively short amount of time DVTE-CAN was partially implemented or demonstrated to a wide marine population. In all, data were collected at six locations, with some data collection being done more informally than others. The key to gaining access within each population was having a champion within the Marine Corps who could assist in explaining the purpose, benefit to the marines, and promote the system through word of mouth. This, in turn, allowed the experimental and programming teams to ultimately provide a flexible system based on feedback gathered across a variety of marine populations. Access was further promoted through instructors recognizing that their feedback was valued and when possible was implemented into system.

Lesson 5. Unobtrusiveness is crucial when working with operational marines.

Within most field environments the researcher collecting data is often in the middle of actual training or work in progress. The importance of being unobtrusive in such environments cannot be emphasized enough. This is important from both an experimental and a practical standpoint. The more obtrusive methods and data collection instruments are, the more the researcher may be seen as a burden and a drain on already limited training time. In collecting team data, it is difficult to be unobtrusive, but several mechanisms can be used to mitigate the perception of obtrusiveness. First, coordinate with instructors and/or the point of contact(s) well in advance of arriving so they are aware of the data collection strategy and can provide feedback. Gaining instructor feedback early can also serve to ensure instruments will make sense to the population of interest. Second, collect only the information that is truly needed; the "nice to have" data may have to wait. Third, attempt to keep questionnaires to a minimum in terms of number and length. Finally, part of being unobtrusive is blending in with the environment to the best of the researcher's ability, so dress accordingly.

CONCLUDING COMMENTS

DVTE-CAN was developed through a close partnership among learning specialists, software engineers, subject matter experts, and the operational customer. As such, the resulting product served to score high in its perceived utility and generally engendered high levels of positive affect by the marines. Due to the logistical constraints often encountered in field settings, in many cases, true experimental and control groups were not able to be created. This, in turn, places some constraints on the ability to make a definitive statement regarding DVTE-CAN's impact as compared to traditional methods across the variety of marine populations that used the tool. However, the use of experimental and control groups at the basic training level (IOC) and the use of pre-/post-test quasi-experimental groups at many of the other locations gives us a fair amount of confidence in the findings presented with regard to utility and the perceived ability of the system to meet the training goals for a diverse set of marines.

By presenting a glimpse of the evaluation process for a system that has been successfully received by the user, it is our hope that the methods used and corresponding lessons learned can assist others in transversing the challenging environment of field research. Specifically, we hope that this chapter will assist those charged with the development and evaluation of training systems to develop systems that can easily transition into operational products that are not only scientifically based, but used by the operational community.

REFERENCES

Bell, M., Jones, D., Chang, D., Milham, L., Becker, W., Sadagic. A., & Vice, J. (2006). *Fire support team (FiST) task analysis surrounding eight friction points* (VIRTE Program Report, Contract No. N00014-04-C-0024). Arlington, VA: Office of Naval Research.

Fleet Marine Force field manual 6-5: Marine rifle squad. (1991). Retrieved April 23, 2008, from http://www.lejeune.usmc.mil/2dfssg/med/files/FMFM%206-5.pdf

Salas, E., & Burke, C. S. (2002). Simulation is effective for training when . . . . *Quality and Safety in Health Care, 11,* 119–120.

U.S. Department of Defense. (2006). Quadrennial Report. Washington, DC: Department of Defense.

Chapter 29

ASSESSING COLLECTIVE TRAINING

Thomas Mastaglio and Phillip Jones

Assess: to estimate or determine the significance, importance, or value of; evaluate.
— *Webster's New World College Dictionary, 4th Edition*

we can know more than we can tell . . .
—Michael Polanyi, *The Tacit Dimension*

THE UNIQUE CHALLENGES TO ASSESSING COLLECTIVE TRAINING SYSTEMS

Virtual simulations supporting collective training—the training of teams and teams of teams—present a unique challenge when it comes to determining their effectiveness and value to the intended user community. Virtual simulations designed to support collective training, at some level, further complicate that challenge, but because their usage is more pervasive and homogeneous at the event or trainee level, and because they offer controlled access to the training audience, it is possible to collect data to support subjective assessments. The methodology described in this section uses inherently subjective data, but plans for and conducts data collection followed by a structured analysis of that data in an objective manner to ensure the process focuses on the goals that the assessment is designed to accomplish.

The Challenges of Empirically Based Evaluations or Testing

A complete training effectiveness analysis of any system requires controlled-use scenarios, access to and control of subject units throughout their training life-cycle, significant data collection, and analysis (Boldivici, Bessemer, & Bolton, 2002). Some argue that it must include a comparison to alternative methods (for example, a baseline training approach, such as a field training exercise) to truly evaluate whether the system is worth the investment in development and commitment of time required from the training audience who will serve as subjects. Such an effort would be cost-prohibitive and time consuming (Burnside, 1991).

The significant issues for virtual simulations are what value its users perceive they attain: whether users are satisfied with the technology and implementation approach and whether they use it to their advantage to improve performance. This type of assessment can be more properly termed a training utilization assessment or study. In an empirical sense, a system cannot be training effective unless it is used properly; training utilization is a necessary condition for training effectiveness. Therefore, studying the training utility of a virtual simulation provides valuable insight, from a customer perspective, into the technology and the context in which it is being used. The results can also help determine where technology enhancements warrant investments in preplanned product improvements and how to improve training strategies or site operational policies.

Another challenge arises from virtual simulation's role as an enabler for operational performance. Although virtual simulations are independent systems, they are in reality a substitute, either for operational performance or for some other simulation system—live, virtual, or constructive. Thus, while improvement in individual or team performance from using a virtual simulation is a valid measure, the more useful measure is performance improvement within an overall training program and relative to other intervention methods. Relative improvement should be measured in terms of total improvement, speed of improvement, and resources expended to achieve that improvement.

Controlling Dependent versus Independent Variables in Collective Training Assessments

Measuring the resources invested to achieve virtual simulation results is another challenge. It is difficult to identify and measure the total simulation investment, which includes research, development, production, and maintenance, plus individual simulation-supported event management and creation costs. Finally, it should also include the post-event costs of transforming virtual simulation results into individual and organizational skills, knowledge, and competencies.

Impact of Assessments on the Training Events Supported

It is a challenge to assess virtual simulations while minimizing disruption to those using the simulations. Users turn to virtual simulations in order to obtain efficiencies; more desirable methods are too resource intensive to use effectively or safely. It is difficult to justify to users the disruptive costs of an extensive system assessment.

An Alternative Approach to Assessment Is Required

This chapter describes an alternate method for assessing virtual simulations, one that provides an analysis with depth, breadth, granularity, and rigor to support evaluation. The methodology leverages the human inclination to assess, our environment and experiences innately and continuously. Those responsible for virtual simulations programs should plan to take advantage of an abundance of organic assessment that is already taking place by the users who innately assess their training tools and events in terms of their utility and value. The challenge is recognizing and harvesting the results of the organic assessments users are already doing.

LEVERAGING CUSTOMER RELATIONSHIP MANAGEMENT PRINCIPLES

Previous interest in and efforts to assess the effectiveness of training systems have focused on technology per se. We suggest that effective use of a virtual simulation, as a system, is akin to a customer relationship management (CRM) challenge (Hurwitz Group, 2002). CRM, as the term is used in industry, focuses on meeting the needs of a particular customer or class of customers. Effective training utilization can be viewed as a CRM challenge for users of virtual environment technology. Regardless of the quality of the technology or investment of development funds, the system will be of value to its ultimate customers only if they are able and motivated to use it effectively.

Identifying Appropriate Stakeholders—Customers

A critical step in the assessment process is to identify the user level stakeholders. Many complex virtual simulations will have multiple stakeholders, from those who direct or approve investment in them to the contractors who implement the technology. All have a role and are important to a successful program, but the focus of assessment has to be on the direct users of the simulation. From a CRM perspective, these users are the customers who must embrace both the technology per se and the capabilities it provides.

Diagnosing Customer Commitment to Fielded Technology

Cost savings are often cited as valid proof that the use of war games or constructive simulations to support collective training is a wise investment of resources. It is appropriate to determine the value that those who train with virtual simulation have realized from the investment in them. We are interested in how committed those customers are to the product—the virtual simulation—being evaluated. The assessment will likely focus on the following high level issues:

- Is the device being used as it was designed?
- Do its users perceive the device as having value to them?
- Are the results of training events that use the device integrated with and used to plan for other training events using the same simulation, another simulation, or alternative training modality (for example, a live exercise)?

Importance of Customer Input to Programs

Assessment based on the approach described here is highly dependent on customer input, which is the rationale for clearly identifying the end-user customer (Goodwin & Mastaglio, 1994). It follows that feedback on both the technology design solution and the training program within which that simulation is imbedded is needed to deliver a successful and meaningful assessment.

Technology Design Solutions

Virtual simulations, like many of our information technology based products, are too often developed based on what the engineer believes is the requirement (Mastaglio, 1991). The decision to conduct an assessment is an opportunity to collect end-user feedback on the implemented design. During the development of the assessment process, it is important to identify the key technical features, those that impact user acceptance and facilitate achieving the training goals. Such features as operational-environment detail and realism, scene resolution, simulator controls, and fidelity to actual operational systems are examples of such technology design features.

A METHODOLOGY FOR PLANNING AND EXECUTING ASSESSMENTS

The first step in assessment is to determine its purpose. The study manager should advise the study sponsor as to what can be reasonably achieved. The tendency, once an organization becomes serious about conducting an assessment, is to expand its scope. The study manager must help the organization focus on its key goals.

Assessment should focus on the level of performance achievable within the virtual simulations, the efficiency or speed those levels are achieved, the duration of results, the resources saved by the virtual simulation including time and money, the opportunity costs of using the virtual simulation compared to another methodology, and the negative impact of using the virtual simulation. This last focus item is often overlooked. However, because virtual simulations cannot equal live execution, and because virtual simulation users are usually success oriented, there is a tendency to perform when using the virtual simulation in a manner that optimizes performance within the virtual environment in lieu of the target environment for the training.

We have developed and recommend the use of an approach called the study of organizational opinion (SO3). It is comprised of eight steps.

Map the Organization

The sponsoring organization is the source for data supporting the assessment and also it is the consumer of the assessment results. Understanding the sponsoring organization is essential to organizing the assessment, gathering the required data, and providing actionable results or findings. Mapping the organization is an important step. The assessment team should start with the organization's objectives. The team may have to assist the organization in identifying its goals. Organizational aspects to consider during mapping may include the following:

- What is the organizational structure? Who makes decisions? What are those decisions? What information do the decision makers need?
- What are the existing assessment/decision processes? How important is continuity versus evolution?
- What management processes must the information support?
- What are the various organizational agendas? How do you work around them?

Continuity with previous assessment efforts will be important to the organization, and there will be resistance to new assessment methodologies. It should be emphasized that continuity lies in the information and insight that comes from the assessment, not in the particular assessment methodology.

Map the Respondent

Concurrent with mapping the organization, the assessment team must examine the respondent population that possesses the required knowledge and will be the focus of the collection effort. The team works with the organization to identify those demographic factors—age, gender, rank, specialty, and so forth—that are important to understanding population responses. The team should divide the respondents into groups based on significant or critical demographic factors.

Disaggregate Goals into Questions

Organizational assessment goals are usually broad, high level queries. The assessment team must disaggregate these broad goals into deliverable questions that respondents could be expected to understand and answer. The process disaggregates goals through three levels: issues, subissues, and questions—all in the form of a query. Each level supports the one above, so the questions support subissues, and the subissues support the issues.

Questions should be phrased for best understanding by each respondent group, keeping the information requirement constant. As a simple example, if the information requirement is to determine the operational environment-specific realism of a training program, the question should be written in future tense for those who have not yet experienced the operational environment and present tense for those who have.

Question form should follow from the required information. Question form refers to the type of question asked—yes/no, true/false, multiple choice, rank order, fill in the blank, essay, and so forth (Schuman & Presser, 1996; DeVellis,

2003). Effective assessments are open to all forms of questions and may include open and closed questions, as well as qualifiable and quantifiable questions. Open, qualifiable questions add to the analysis burden of the assessment team, as well as to the burden of the respondent, because they are more difficult to answer.

In studies of complex organizations, with different echelons of command or management and with various channels of authority, the disaggregation should be tied to those echelons, such that issues are executive level information requirements, subissues are management level requirements, and questions are targeted toward the assessment population.

Link Questions to Respondents

The final step in preparation is to link questions to respondents to generate questionnaires. More questions will be prepared than are appropriate for delivery to the entire population, so this step requires deciding which questions NOT to deliver to a respondent group. While every situation is different, our experience shows that a single questionnaire should be limited to approximately 50 questions.

In addition to questions prepared through the disaggregation process, a questionnaire includes "filtering" questions. These are questions designed to identify biases or other critical aspects of an individual respondent. Filtering questions must be subtle so respondents do not perceive that they are being screened.

Question sequencing on questionnaires is important (Schuman & Presser, 1996). The most critical questions should be front loaded within the questionnaire as there is a natural drop-off after answering approximately 20 questions. This drop-off can be physical—respondents departing the questionnaire—or cognitive—respondents not putting sincere effort into their answers.

Questionnaires must be rehearsed by having sample respondents execute the entire questionnaire as delivered in order to obtain internal feedback on the ease of the completion and understandability. In addition, each questionnaire should be delivered to an exemplar respondent via an interview regardless of the intended deployment means. Our experience is that the best way to vet the questionnaire is by verbalizing questions to a sample respondent.

Deploy, Track, and Receive Questionnaires

Questionnaires are deployed via interviews, focus groups, written surveys, or the Web. The objective of deployment, tracking, and reception is to maximize questionnaire response and throughput. "Response" is the number of respondents who initiate the questionnaire. "Throughput" is the number of respondents who finish the questionnaire and provide valid data. The most important metric is response quality, which we define as a combination of total response, response by respondent group, quality of responses, and quantity of data.

Store Data Points/Answers

As data, in terms of respondent answers, are received, the assessment team must ensure it is properly stored in an appropriate "facility" with adequate backup. The design of the storage facility, most often a database, is critical. During this step the results are loaded into that data storage so that they can be readily retrieved during the analysis process.

Analyze Data Points/Answers

Analysis within SO3 should be considered a discovery process as the analyst develops insight from the data. Analysis is done in three sweeps.

1. The first sweep occurs during data collection using results of the closed, quantifiable questions, including demographic questions. This sweep provides an initial assessment of the data and informs later analysis. It also allows the assessment team to provide immediate feedback to the sponsoring organization.

2. The second sweep is the primary sweep. It mirrors the disaggregation conducted in step 3, aggregating the data into consensus answers for each question, subissue findings, and issue findings. For closed-end answers, this is a straightforward aggregation of responses. For open-end answers, the analyst must identify main points in the respondent's answer. A respondent could provide multiple recommendations or bits of knowledge. These points are combined across the respondent group to provide weighted answers, showing both the emphasis and totality of the respondent group's opinion or knowledge. Respondent group consensus answers are aggregated into by-respondent group subissue findings. These are then aggregated into consensus subissue findings, which are aggregated into issue findings. Multiple analysts can be used to support each aggregation and to obtain differing perspectives, but a single analyst should be responsible for the aggregation chain in order to maintain consistency and maximize insight.

3. The third sweep is a final inspection of the data and consists of unplanned or informal data analysis. SO3 analyses frequently yield some unexpected results, and the third sweep accommodates that; it can include more sophisticated methods, such as cluster analysis.

Report Results

The value of the assessment is only as good as the organization's ability to understand and act on it. To formulate decisions and drive change, results must be reported in a fashion that supports use by the sponsoring organization. Reports must be presented with clarity while still allowing organizational users to drill down into the details of the data.

Relationship of SO3 to Other Methodologies and Approaches

The SO3 methodology is a superb tool to integrate into existing assessment processes. The thoroughness and flexibility of this methodology will provide both the data and the analysis to support Kirkpatrick's four level assessment

(Kirkpatrick, 1994), the formative and summative assessments part of the Dick and Carey systems approach model (Dick, Carey, & Carey 2005), or other methods. The SO3 methodology provides a means to extract, organize, and comprehend the information required to support these others processes. However, SO3 also has the ability to go beyond these processes to capture or access tacit knowledge.

Tacit knowledge is first-person knowledge—it is knowledge of the world as the world impacts the individual. By extracting and analyzing tacit knowledge, the SO3 methodology leverages the organic expertise encapsulated in respondents' tacit understanding. Thus, the SO3 methodology serves to assess not only the targeted virtual simulation training, but also of the training embedded within it. This includes such aspects as the overall training program, the relationship to other training and training systems, and to the conditions of the user. Other processes, which approach assessment from a third-person perspective, may not yield this level of detail or insight.

METHODOLOGY APPLIED

The authors have applied the SO3 approach in a series of training effectiveness analyses of networked virtual simulation technologies for the U.S. Army. These include the close combat tactical trainer (CCTT; Callahan & Mastaglio, 1995) and simulation network (SIMNET) (Alluisi, 1991). CCTT is the virtual simulation of the U.S. Army's primary heavy fighting vehicles, the M1A1 Abrams main battle tank and the M2/3 Bradley infantry fighting vehicle. SIMNET is the technological predecessor to CCTT. To further the reader's understanding of the process, the following section describes one of those studies.

Overview of CCTT and Challenges of Assessing Its Effectiveness

In January 2004, the Army Research Institute and the TRADOC (U.S. Army Training and Doctrine Command) Program Integration Office–Virtual contracted a study to assess the effectiveness of the CCTT (Jones & Mastaglio, 2006). This study assessed the general effectiveness of the CCTT (Goldberg, Johnson, & Mastaglio, 1994) via a process of interviewing and surveying users, then consolidating their opinions to develop general findings. The process provided sufficient insight and validated the approach; therefore, a second study was contracted to evaluate the contribution of virtual simulations to combat effectiveness. We will discuss this second study. For more details on the first study and the fundamental research to develop the methodology, refer to Mastaglio, Peterson, and Williams (2004) or Mastaglio, Goldberg, and McCluskey (2003).

Virtual Simulations in Preparing Army Units for Combat Operations in Iraq

This project involved two closely related, but separate studies; one was how active U.S. Army units used the CCTT during preparation for deployment to an anticipated high intensity close combat environment, and the other was an assessment of National Guard use of available virtual simulations to prepare their units to deploy to that environment. The initial goals were as follows:

- To determine if virtual training impacts combat effectiveness and
- To evaluate if changes should be made to the CCTT simulation or site operations to better meet pre-deployment training needs.

The study methodology consisted of collecting and consolidating user opinions. The respondents were from eight units across the United States that had returned from Operation Iraqi Freedom (OIF).

The study scope was extended to include assessing the use of mobile virtual simulations in the U.S. Army National Guard (ARNG). The ARNG uses a mix of virtual simulation systems to train close combat tasks. These included mobile and fixed CCTT SIMNET systems. These systems are referred to collectively as virtual maneuver trainers (VMT).

Methodology

Both studies were each conducted in three phases.

Preparatory Phase

For each study a formal research plan that incorporated the eight SO3 steps was prepared for U.S. Army review. Study goals were disaggregated through two levels. From the study goals, the team developed several issues, and each was further separated into subissues for which questions were developed. This preparation process and decomposition is shown in Figure 29.1.

Simultaneously, the team mapped the organization, identifying a desired sample of type units and respondents to whom the questions would be delivered. For each study, a list of respondents was created. These were personnel who would likely have the knowledge being sought, both officer and noncommissioned officer small unit leaders, from platoon to battalion level.

For the OIF study, respondent lists were matched to the same list of respondents as in the earlier CCTT study: battalion commanders, battalion command sergeants major, battalion executive officers, battalion operations officers, company commanders, platoon leaders, and platoon sergeants. For the ARNG study, it was decided not to interview executive officers and to collect input from battalion master gunners rather than command sergeants major. The team also collected input from full-time support staff in the National Guard's distributed battle simulation program and active U.S. Army advisors to the ARNG, as well as site staff supporting training.

In the OIF study, there was one identified respondent demographic factor: duty position. The ARNG study included the following seven demographic factors to support a more detailed analysis:

- Duty position,
- Months in position,

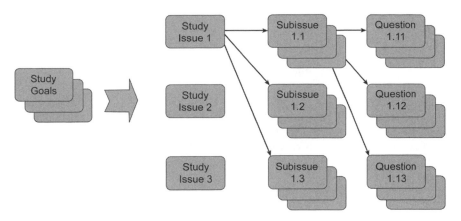

Figure 29.1. Study Goal Disaggregating

- Primary VMT,
- Level of VMT experience,
- Home state,
- OIF veteran, and
- Under orders for OIF.

Questions were cross-referenced to the class of respondents. Physical questionnaires were prepared for each respondent within each sample unit.

MySQL databases were developed as a repository for the results of all questionnaires. A set of data analysis interface tools was developed to support analysis of the large amount of data anticipated.

Data Collection Phase

Teams visited each unit within the unit samples providing first an in-briefing to battalion leadership, then conducting interviews. The interviews consisted of the prepared questionnaires, but included follow-up questions and time for discussion. Frequently the follow-up provided insights that were of use during the analysis phase.

Both studies also used the Internet for delivering questionnaires via a Web site. Respondents were solicited on the Army Knowledge Online home page and via e-mails to visit a Web site to complete questionnaires. Respondents on the Internet went through an informal, multilayer screening process. The initial announcement served as the first screening layer. The next screenings were self-screening. Respondents first had to decide if they fit the survey criteria. Once on the site, they had to complete a demographic questionnaire targeted either to their duty position or to "other respondent." The final screening was performed after-the-fact by a subject matter expert. He reviewed each answer to determine its validity. Answers that did not follow from the question were not included.

Data Analysis Phase

The final phase consisted of aggregating the data in reverse order of the earlier disaggregating process. Individual answers were combined into consensus answers for each question, by position. These by-position consensus answers were then used to determine, by position, subissue findings. By-position subissue findings were combined into general subissue findings. Finally, general subissue findings were combined into issue findings.

The final phase was aggregating the data. Individual answers were combined into consensus answers for each question, by position or by job title. These by-position consensus answers were then used to determine by-position subissue findings. By-position subissue findings were combined into general subissue findings. Finally, these were combined into issue findings.

The final step was reporting results via briefings and in writing. Use of the SO3 methodology permitted the identification and reporting of insights not related exclusively to the use of CCTT and SIMNET, but gleaned from the respondents' perspectives of virtual simulations and their opinions on needed changes to the training environment based on the operational environment experienced in fighting the Global War on Terror. Some examples are as follows:

- The change from using virtual simulations to live training because of the greater emphasis on individual tasks,

- A need for instruction on how to integrate virtual simulations into an overarching training program, and

- The need for virtual simulation systems to present more degraded mode conditions, that is, model unexpected equipment failures.

PERSISTENT TRAINING ASSESSMENT—THE NEXT EVOLUTION AND RECOMMENDATIONS

Virtual simulations will improve and be more readily accessible while becoming more common, more realistic, more distributed, and more integrated. As the types and capabilities of virtual simulations expand, assessing the training for which they are used will become more critical. Execution of assessment, however, may become more difficult. Heretofore virtual simulation systems have been centrally located and controlled and their users relatively easy to identify. As virtual simulations become embedded systems or distributed via Web-like solutions, identifying the value and impact of their usage will become more difficult.

We make several recommendations to conducting assessments in the future, primarily efforts to expand and integrate assessment.

- Integrate assessment into the system or program design—make it organic. Generally, assessment is external to system design and development, and assessing a simulation requires extra effort. Software and hardware should include the ability to automatically capture, store, and report data to support assessment.

- Maximize the collection of data by capturing it whenever the user operates the virtual simulation. Major marketers now use "point of sales" collection. Instead of extracting a large amount of data from relatively few customers, marketers now extract a small amount of data from many customers. Each time a product passes a scanner, a customer accesses a Web page or calls the company, or other contact between the customer and the company occurs, that information is recorded. Often, the customer is not aware that he or she is supporting an assessment. The increase in the amount of data available over time can aid in providing a much more complete assessment.

- View virtual simulations as a system within a system of systems and expand the assessment across the breadth of that system of systems. Virtual simulations are normally a component within a training process that uses a variety of live, virtual, and constructive simulation tools. The performance of a single system can best be viewed in the context of the entire environment; assessment should be done across that environment.

CONCLUSION

Assessment is not a simple process and will not—particularly for the virtual simulations supporting team and collective training—yield to traditional comparative studies of normative use. Assessments will be inherently based on an analysis of qualitative data. However, the collection of that data can and should be planned and organized to focus on accumulating information that will best support an analytic effort to address the sponsoring organization goals or specific needs. Assessments performed using SO3 can provide the insight needed to help designers and developers of virtual simulations and training programs to deliver more effective training.

REFERENCES

Alluisi, E. A. (1991). The development of technology for collective training: SIMNET, A case history. *Human Factors, 33*(3), 343–362.

Boldovici, J. A., Bessemer, D. W., & Bolton, A. E. (2002). *The elements of training effectiveness.* Alexandria, VA: U.S. Army Research Institute for Behavioral and Social Sciences.

Burnside, B. L. (1991). *Assessing the capabilities of training simulations: A method and Simulation Networking (SIMNET) application* (Tech. Rep. No. 1565). Alexandria, VA: U.S. Army Research Institute for Behavioral and Social Sciences.

Callahan, R., & Mastaglio, T. (1995). A large-scale complex virtual environment for team training. *IEEE Computer, 28*(7), 49–55.

DeVellis, R. F. (2003). *Scale development: Theory and applications* (2nd ed.). Thousand Oaks, CA: Sage Publications.

Dick, W., Carey, L., & Carey, J. (2005) *The systematic design of instruction* (6th ed.). New York: Allyn & Bacon.

Goldberg, S., Johnson, W., & Mastaglio, T. (1994). Training in the Close Combat Tactical Trainer. In R. Seidel (Ed.), *Learning without boundaries.* New York: Plenum Press.

Goodwin, E., & Mastaglio, T. (1994, December). *Integrating users into systems development: User evaluations in CCTT.* Paper presented at the 16th Interservice/Industry Training Systems Conference, Alexandria, VA.

Hurwitz Group. (2002). *Customer life cycle management* [White paper written for J. D. Edwards]. Framingham, MA: Hurwitz Group, Inc.

Jones, P., & Mastaglio, T. (2006). *Evaluating the contributions of virtual simulations to combat effectiveness* (Study Rep. No. 2006-04). Alexandria, VA: U.S. Army Research Institute for the Behavioral and Social Sciences.

Kirkpatrick, D. (1994). *Evaluating training programs: The four levels.* San Francisco: Berrett-Koehler Publishers.

Mastaglio, T. (1991, April). *Designing simulation systems to support collective training: Lessons learned developing military training systems.* Paper presented at the ACM Conference on Human Factors in Computing Systems, CHI 91 Workshop on Advances in Computer-Human Interaction in Complex Systems, New Orleans, LA.

Mastaglio, T., Goldberg, S., & McCluskey, M. (2003, December). *Assessing the effectiveness of a networked virtual training simulation: Evaluation of the Close Combat Tactical Trainer.* Paper presented at the 2003 Industry/Interservice Training Systems and Education Conference, Orlando, FL.

Mastaglio, T., Peterson, P., & Williams, S. (2004). *Assessing the effectiveness of the Close Combat Tactical Trainer* (Research Rep. No. 1920). Alexandria, VA: U.S. Army Research Institute for the Behavioral and Social Sciences.

Schuman, H., & Presser, S. (1996). *Questions & answers in attitude surveys.* Thousand Oaks, CA: Sage Publications.

SECTION 3
FUTURE DIRECTIONS

SECTION PERSPECTIVE
Rudolph Darken and Dylan Schmorrow

Did anyone ever think that virtual environments would not be useful for training? Certainly they were crude and extremely limited at the start, but this seems like an obvious conclusion to draw from the set of technologies we call "virtual environments." How long after the Wright brothers took their first flight at Kitty Hawk did someone think that someday that invention could be used for transportation? The airplane went from flights of 100 feet to 24 miles in less than 18 months. Virtual environments (VEs) did not mature quite so fast, but even the wireframe images the NASA (National Aeronautics and Space Administration) Ames laboratory was producing early on had obvious practical uses—training being one of these.

As we read "In the Uncanny Valley" by Judith and Alexander Singer, we see a vision of a symbiotic relationship between the trainee and his VE. In fact, Craig gets confused and cannot tell what is the VE and what is real. At one point he says, "When is this happening? Or is this some kind of replay . . . exercise?" His sensory system was enclosed in the VE in a closed loop. There was no clear separation between Craig's VE and the real world. How similar is this to Orson Scott Card's *Ender's Game* (1991), which is the focus of Jack Thorpe's "Trends in Modeling, Simulation, Gaming, and Everything Else"? Ender's life is consumed by "training" for a mission he thinks he is preparing for in the future. Only later does he come to realize that he was *doing* the mission when he thought he was only *training*. There is no separation for him. Thorpe takes this further by suggesting that a host of activities might someday converge through what we call a virtual environment. Personnel selection, recruitment, socialization and initial training, specialization training, crew and unit training (the military would call this "collective training"), planning, rehearsal, reconstruction, and review all meld into one activity within the VE. Think this is purely a military phenomenon? Think again. What NFL (National Football League) team would not pay dearly for such a system that helps it field a superior team? What airline would not pay for a system to assure the highest performing and safest aircrew? This idea is pervasive and universal.

But this is a philosophical argument. Is it what we really want, and, if so, what will it take to get there? Randall Shumaker fuels the discussion in "Technological Prospects for a Personal Virtual Environment" with a thorough discussion of where we are today and where we might be in 2020, given a reasonable set of assumptions. What might we expect in terms of visual and aural displays, cognitive computation, and human behavior modeling? Is the "personal VE" that Shumaker describes capable of being the symbiotic environment that Singer and Singer as well as Thorpe describe? Furthermore, how well does this vision match what potential users of such a system think they will want?

Alfred Harms Jr. describes in "The Future of Navy Training" where the U.S. Navy is headed, with a strong belief that VEs will play a critical role in training. He believes VEs will supply a unique learning experience that allows for "discriminating realism, contextual fidelity dynamic and interactive participation, confirming repetition and controlled assessment of cognitive, affective and psychomotor skills." This is desperately needed as he sees the missions of the U.S. Navy becoming less predictable and being activated on short notice.

William Yates, Gerald Mersten, and James McDonough carry this theme in "The Future of Marine Corps Training" into the Marine Corps, which is arguably becoming less predictable than the U.S. Navy. They describe a scenario where simulation based training using a VE is utilized to achieve a desired end. They make the case for immersive mixed reality that is neither purely real nor purely virtual. But they also call for increased sensory realism as it fits the training requirements of a mission. Are you ready for snakes falling from trees?!

The U.S. Army has long taken a leadership role in the development of VE technologies for training. In fact, a large portion of the innovations that we regard as virtual environment technologies were initially funded from the Department of Defense, some specific to certain branches, but many were shared across the services. Roger Smith tells us in "The Future of Virtual Environment Training in the Army" that this has been changing. We are past the tipping point where the government is no longer the prime catalyst for technology development; it is industry (often in service to the military) that is innovating. Most importantly, the video game industry has been a key motivator for the improvement of computing hardware, especially graphics computing and display. Smith takes us into U.S. Army missions and how VE applications might serve them in the future.

The U.S. Air Force recognizes the need for improved performance assessment and continuous learning. Will there be a day when each airman has his or her own "learning VE" for all training and education needs? In "Future Air Force Training," Daniel Walker and Kevin Geiss relate these ideas to existing trends in live-virtual-constructive simulation technology, which really starts to sound a lot like *Ender's Game* if you take it to its logical conclusion.

Admittedly, the mission profiles of all the services are extremely large—so large that the generalization we are about to make can never be absolutely accurate, but in general it holds. The natural interface between soldiers or marines and their working environment is their own bodies. There is no abstraction beyond their physical bodies. This is not generally true for the U.S. Navy or the U.S.

Air Force where a ship, an aircraft, or a display of some sort is the abstraction between the operator and his or her world. This is important because if we think Ender's game or the uncanny valley might never happen, it will happen there first because it is easier to mimic an existing abstraction than to mimic the real world. Simply, the breadth of stimuli is far too large if we have to simulate physics than if we can abstract it to a small subset.

Medical training is another area where VEs will play a large role. The extraordinary precision needed for many medical training tasks has somewhat delayed the development of medical VE trainers, but recently this is beginning to change. James Dunne and Claudia McDonald discuss how this might happen in "Factors Driving Three-Dimensional Virtual Medical Education." Their "perfect storm" refers to the enormous shortfall in medical capabilities combined with the increased performance and VE capabilities offered by the technology. Put together, we are going to see progress in the medical field happening at a much faster rate than we have seen previously.

Last, in industry, we see an increased interest in the use of VEs for design and production. The automotive industry has long been a participant in advancing VE technology, but what about training? Dirk Reiners discusses how industry views the use of VEs for training in "Virtual Training for Industrial Applications." The business case for the use of VEs is still problematic. He refers to this as the cost of entry. If you cannot afford to invest enough to do something useful, you cannot do anything. Software issues are as much a problem for industry as they have been for the military. Yet decreased costs, increased capabilities, and the need for a competitive edge is driving industry to seek new ways to exploit VEs for training.

Robert Gehorsam gets more specific in terms of training applications based on game technologies that focus on the identity, social, and interpersonal aspects of the VE in "Corporate Training in Virtual Environments." He explains how avatar based applications are better able to deal with human-to-human interactions so important to many corporate training problems. Projecting this into the future, will there be "always-on" training environments that we all will use on an "on demand" basis to meet our training needs?

Do the military, medical, and industrial user communities agree with the vision of convergence described in *Ender's Game* and "In the Uncanny Valley"? Yes and no. There still seems to be a focus on increased fidelity. We need more visual fidelity, realism, haptic feedback, or as Yates et al. suggest, a "smack in the face" where appropriate. Imagine that we could construct VEs that were indistinguishable with the real world. Would that solve our training problems? In fact, we can do this now. What is the difference to a radar operator deep inside a ship's combat information center between a real mission where he sees blips on a screen that are stimulated by sensors and a simulation that virtually stimulates his display in exactly the same way? What he sees and hears is exactly the same. That radar operator already knows something about Ender's game. If we had a perfect place to practice, what we would have is a perfect place to practice, not necessarily perfect performers. In other words, even if we could build a virtual experience that was indistinguishable with the real experience, we would still need more. What would we need?

Several of the chapters in this section touch on this. We would need a model of learning. How does human and/or team performance improve? What techniques accelerate learning or performance improvement? Anyone who has played golf knows what a performance plateau is. You improve to a point and then you do not get better no matter what you do. Play more golf? No improvement. Take more lessons? No improvement. Then, for what appears to be no reason whatsoever, you get better. There are lots of theories as to why this occurs, but until we understand it well enough to build mechanisms in our VEs that help us overcome performance barriers, VEs will just be another place to practice. Why replicate the real world when the real world is so limited? Cannot our VEs for training do better? They should. Roger Smith has it right when he says, "VEs are not a training technology." They are not in and of themselves. We need to remain committed to improved human performance—whatever the mechanism that achieves it. VEs are a collection of technologies that are part of the answer. Let us not assume they are the whole answer any more than are more pixels, better audio, or vibrotactile clothing.

What does the future hold for VEs for training? Who can know for sure, but we know what trajectory we are on now and where that might take us. Clearly, stimuli fidelity is going to improve. There are too many demands (warranted or unwarranted) for more fidelity for this to not be the case. What is less talked about but more critical is the science of learning—cognitive modeling, human behavior modeling, scaffolding, models for learning, real time monitoring of physical, mental, and emotional feedback, and virtual instruction. These combined with high fidelity, affordable VEs clearly head in the direction of *Ender's Game* and the "In the Uncanny Valley." Is there a day when each of us owns our own personal learning machine where we can enter a virtual world that knows us, what we can do, what we need to be able to do, how others have done it, and how to help us improve? Or maybe doing and practicing really are one and the same.

The final chapters of this section discuss the future of virtual environment displays, virtual reality that can be used in both cognitive and physical rehabilitation, personal learning associates that offer a dialogue with the virtual world, and virtual reality displays in the museums of the future.

REFERENCES

Card, O. S. (1991). *Ender's game.* New York: Tor.

Part VIII: Future Visions

Chapter 30

IN THE UNCANNY VALLEY[1]

Judith Singer and Alexander Singer

INTRODUCTION *BY KATHLEEN BARTLETT*

In this chapter, the reader will experience a textual representation of the futuristic fusion of man and machine. Join the ultimate digital narrator, Douglas Craig, a "human-computer diad" (part human/part avatar) as he interacts with a simulated environment (a cataclysmic event of the past). Experience Craig's confusion when his "world" is "reconfigured." Follow along as he searches for the "Control Center" to restore his missing imbedded memory to the "errant search engine of his mind." Puzzle over the meaning of his inevitable conclusion.

This story—the "pseudo-life" of Douglas Craig—challenges traditional definitions of life, of self, of memory, of time, and, ultimately, of reality. By controlling the information available to the narrator, the authors force the reader to identify the essential environmental constructs that contribute to our cognitive perception of (and our subsequent reaction to) the world (our environment) as defined by our easily manipulated senses and our often insufficient memory (experience). Reality in this chapter is only partially constructed, and by the time the narrator (and the reader) put the pieces together, the experience is over (sort of like life).

A renowned Chinese novelist[2] observed hundreds of years ago that "where illusion becomes reality, reality becomes illusion" and that comment describes the most profound paradox embodied in this story, and, indeed, in the continued improvements in the fidelity of simulated experience and the envisioned achievements in augmented cognition (via enhanced memory/experience). As the quality of the simulation, (the illusion) improves, users lose all control over perceptions, experiences, even memory, resulting sometimes in unreliable, unpredictable outcomes. Thus, the "Relativity" of human experience (like the title of the M. C. Echer etching) suddenly becomes questionable and conflicting (surreal or hyper

---

[1]The term originated in robotics, but migrated to the world of computer graphics. It refers to the confrontation with entities whose resemblance to humans either physically or behaviorally is close enough to engender an "uncanny" sense of disquiet and fascination.
[2]Ts'ao Hsueh-ch'in, *The Dream of the Red Chamber* (1715?–1763?).

real). As this story suggests, when the perceived external environment becomes suspect or nonexistent, the internal self is all that remains.

* * * *

Douglas Craig knew his virtual world was being reconfigured. The process in this advanced system was distinctly different from any previous experience. The visual field had gone dark: the dark you perceive when you close your eyes and shapeless afterimages flicker. Once again, phenomena inside his own sensory system were mirrored in the VE system, creating a closed feedback loop, blurring clear-cut separation. The phrase "mind games" had taken on new meaning.

First, he heard the wind, muffled as if through a protective shield. Then he "felt" the folds, pressures, and weight over his entire body of a coverall garment of some kind, responding to the sound of the wind's rise and fall. His audible breathing seemed to come from inside an enveloping helmet; his gloved hands confirmed the shape of the enclosure.

In the after action review (AAR) he was unable to explain why he was certain, before he could see, that he was in a very cold place; but suggested it may have been something that triggered both smell and taste in his sensorium. He was right, of course.

The faceplate gently bounced his body-warmed breath back at his skin. Now, the dark lightened, and as this other world took visual form, he had to grasp wildly at a nearby railing. Far below the metal catwalk he stood upon were a vista of rippling steel blue water strewn with jagged ice flows and an oppressively leaden sky merging into a boundless sea, a polar ocean.

From this point on, the small cloud of his warm breath emerging from the helmet's mouth screen reinforced his sense of the "palpable" presence of lethal cold outside his warmsuit. His exact phrase was that this awareness would now "shadow his reasoning" the way that the "patient's memory fragments" had done.

As he regained his balance he saw that the structure around him was perhaps hundreds of feet above the sea. He tried to make sense of the vast tangle of girders and crossbeams, cables, shafts, and signal lights cutting through the arctic twilight. Mammoth cranes were fed huge structural components by giant choppers: hulking atomic powered freighters built to grind through rare ice flows clustered at floating docks. Wind muffled the soundscape.

A faint vibration under his feet brought his gaze up to see four figures moving toward him along the catwalk. His helmet's sonic enhancers picked up the faint scramble of foreign speech, argumentative, oddly slurred. In the few moments he had to prepare for the encounter he glanced down at the forearms of his warmsuit and was relieved to recognize the universal icons of a long obsolete audible language translator and a GPS (global positioning system) locator with full-zoom outboard imaging. The only reason he had an inkling of their operation was the subcourse he had taken years ago in the evolution of technology controls. He grinned at the memory of sophomores threatening self-immolation if they had to unscramble another screen remote from the early 2000s.

As the four figures came close enough to evaluate, he felt the beginning of unease. A quality in their gait and gestures suggested harsh living, alcohol, and an argument in a babble of dialects he thought were Russian and Norwegian. As the gap between them closed, he backed against the railing so they could pass easily.

As he clumsily fumbled with the translator, he was interrupted. One of the four, a Mongolian woman, whose translated voice and facial movements were out of sync, had turned to him and said bluntly, "Control Central need you . . . (linguistic babble) . . . now."

Craig wanted to put this world on pause, giving him time to think through its underlying logic, beginning with these four: two men, two women. Their body language suggested an impatient split between their ongoing argument and what might be an increasing curiosity about himself. Behind transparent faceplates their work-hardened faces suggested a life on North Sea oil rigs, decades ago. Varying individually, their warmsuits all had multiple geometric patches of odd symbols and colors, both enigmatic yet vaguely familiar. The four did not resemble entities he had ever encountered.

In his second AAR, Craig recalled how much he wanted to ask them about their pseudo lives as avatar projections of advanced students much like himself. Had not he experienced a lifetime of such encounters, in c-games, in years of VE learning?

Inevitably, he had to consider the possibility of a system-induced lucid dream state. Although the pace of events made it difficult to pursue the threads of this conjecture, we know that subsequent events unfolded against the background resonance of these developing thoughts.

Now the four started to move away and he blurted out, "I'm new here. Please. Tell me, where is Control Central?"

The woman who had spoken to him paused as they were moving away; the others slowed, and she glanced back at him as if he might be dim-witted.

In English, heavily accented, "You walk there . . . ," gesturing in the direction from which they had come. She seemed to be deliberately exercising her language skills. "Turn left at . . . post. Follow arm picture after turn."

For a beat he looked puzzled. She came over, vigorously tapped the locator display on his warmsuit's forearm, turned impatiently, and rejoined the others. The haptic physicality of the woman's touch through the warmsuit stayed in his memory as he started away.

The illusory cold outside the protective clothing somehow impelled him to move quickly. As he turned a corner the vast construction in progress revealed itself from another viewpoint.

While one part of Craig's mind tried to make sense of the project before him, another part would continue to search for the underlying rationale of the entire exercise.

He was moving toward some sort of installation in progress on a platform just above the catwalk a short distance ahead. Moving toward the action he continued the internal argument: he was not going to be a neuroscientist guiding surgibots through a patient's brain in some future scenario or some kind of engineer at the pole, decades in the past. As an advanced playerist he felt treated like some Generation D kid on his first imbed high.

Alongside his resentment, the errant search engine of his mind replayed a fragment out of the interior chaos.

Craig flashed back to a moment with his team moving through the patient's hippocampus region: one of the fleeting images that seemed to flicker through the translucent neural holoscape, a girl near a wind farm, playing with an actual live dog. The animal looked like the mutt he had had in childhood, just before his sector excluded organic pets.

The remembrance of his loss suddenly stung him, now, on the polar sea. He gasped, surprised at the impact and shook his head to clear the fluid welling in his eyes.

It is useful to speculate on the causes that disrupted Craig's access to his own, quite good memory of the historic events a quarter century in his past. Remember, he was fully aware that in the final physical preparatory stage his imbed search engine would be temporarily neutralized. In his time, this could be in itself, a somewhat physically traumatic process.

Consider that Craig's natural world of 2058 and the arctic period of 2033 not only spans a gulf of a quarter century, but also encapsulates one of the most turbulent and transformative periods in recorded history. His own time was no more able to organically integrate the range of hierarchical shocks and dizzying potentialities than the generations before and after his own cohort.

Just past the installation rising above the platform he was headed for, he checked the diagrammatic overlay on his locator screen, made a sharp right turn, and almost stopped short in his tracks. Before him was yet another angle of the ongoing construction, but this one set off a chain of synaptic sparks.

There were temporary metal stairways climbing and connecting giant pylons; the whole structure was a display of materials and technologies long obsolete. A handful of workmen hurried up and down the stairs, some carrying gear he could not identify.

The associative process had kicked in and he just hung on, letting it take him wherever it would: a drawing . . . Essen . . . Eckar . . . Escher! M. C. Escher! Dutch! master draftsman; early twentieth century; clinically precise etchings, surreal or hypereal, often convoluted geometric forms; somehow projecting a disarmingly naive sensibility.

The picture he recalled showed humanoid, faceless automata, trudging up and down impossible stone stairways in a timeless, senseless, labyrinthine maze; all perpendicular surfaces hopelessly out of whack with each other. The whole, a mockery of the viewer's instinctual need to locate a fixed position in space/time, aptly entitled "Relativity."

For many of his classmates in the cross-cultural interdisciplinary courses, particularly the scientists, discovering Escher's graphics had whetted their curiosity to explore the two-dimensional fixed visuals of the past; goal achieved. He smiled, remembering the man behind the stratagem: Professor Wellman. Tricky old guy. Using the portal of past art forms, we were guided to explore the evolving discipline of psychohistory, at which point Wellman could repeat his favorite adage to us: George Santayana's, "Those who cannot remember the past are condemned to repeat it" (1905, p. 284). And then Craig's jaw dropped. He checked his forearm screen, fumbled till the date came up: 10/08/2033. Then he let loose a stream of profanity and shouted into the air, "If they hadn't taken out my damn imbed, I might have a clue what the hell I'm supposed to be doing. You set me up to fail and then fail again!" After another stream of invective, he took a deep breath and grew quiet. He was not going to pull the rip cord, not yet.

He checked his locator: 40 meters from Control Central. He turned and started running as fast as his bulky gear permitted. His racing mind barely considered the immensity of the passing holoscape totally integrated with his sensorium.

He focused on possible choices if they had access to full signal contact even in those benighted times. The thought arose, unbidden; my god, in 2033, they were still dying of cancer.

This was the first time Craig had consciously allowed himself entry onto the uncertain terrain of "empathy." The voice of pragmatism argued that all this was "virtual" and why care beyond a final grade rating. Just below full awareness another argument persisted. There was the possibility that this entire experience, including the sense of failure, was somehow more than the sum of its parts.

His full-zoom outboard imaging system signaled, vibrating against his forearm. Checking his screen he was able to unscramble the semiotics: 60 meters away, around a bend, Control Central would be a bulge thrusting into space from the catwalk. His pace slowed as he tried to sort out his chaotic thoughts and begin to focus on an effort to evoke the zeitgeist of 2003, 30 years ago.

Craig had learned of the pandemics tormenting populations in the first three decades of the twenty-first century. Now that he was immersed in the last third of that period, he drove his perspective and historic memory into this very mindset.

Massive quantum computation was still relegated to the status of a laboratory stunt. The Time of Terrors, limited as it was, left national borders convulsed and recovery stunned for another kind of onslaught. It was a time of genetic life extension, full solar/tidal power, and hydrogen-powered locomotion. Profound leaps of understanding in neuroscience and biophysics burst into the late 2040s and 2050s, his own time, with incandescent impact.

He braced himself to bring his sensibilities to that other time, beset by physical trauma and a near frenzy of anxiety: the perceived decline of conventional energy sources and possibly irreversible environmental degradation. He was certain that it had been a large-scale event, somehow related to man-made structures; the same calendar year, 2033, his present screen had indicated. He hoped when he finally entered this Control place, he would find something that sparked the evasive key memory.

Craig now faced the door of Control Central and saw his own image reflected in the WeatherPlex window of the portal. Even with his warmsuit helmet and the glassy surface reflectance, the face was unmistakably himself. Where was the avatar he had carefully built and refined over the years?

We know the full range of his body and brain sensors registered heightened awareness and tension. From later reports we learned that he felt unequal to the task but determinedly rejected the disengage option.

Looking past his own reflection he could make out figures and displays. Before he opened the door he thought, "Nothing will change. The past is immutable and, good as it is, this *is* a simulation." Then he entered Control Central.

On his left he saw, against the sidewall, an open section for hanging cold weather clothing. Deliberately turning his back to the group present, he moved to the clothing area to remove his warmsuit as unobtrusively as possible. Past the awkwardness of strange fastenings, he could not help briefly indulging in the unfamiliar textural and haptic signals in removing the outer garment. Before he could question his actions he was relieved to find he was wearing a one-piece work-site coveralls, like the others present: elastic body-fitted material, sealable pockets; attachment points for gear; basic computer interface stiffened the forearm sleeve; solid brown tone.

The room's air felt overly warm and a bit stale even though the grid-covered vents were audibly working. This was a bare bones construction site work center. Eight people of mixed gender and race moved about in front of the large, spatially projected holoscreens still in use in that time. He thought he must have seen those screens last as a child, at perhaps a techno museum display.

At this point Craig was only minutes from one of the key moments in the exercise, an emotional explosion we will analyze. Consider the critical weight of the cognitive dissonance generated by the unfolding events.

After a moment Craig decided there was an air of latent anxiety in the room. He approached a man off to one side studying a digital clipboard suspended in space, chest high in front of him. A name tag on his coveralls read "Lou Benar."

While Craig was still moving and without taking his attention from the clipboard Benar said offhandedly, "How's it going, Doug?"

Craig realized everyone wore name tags, but he still had to mask his surprise and try to sound both casual and involved. "Came over soon as I could, Lou. Bring me up to speed."

On the verge of responding, Benar reacted to a verbal signal Craig could not discern from a man at the center of the group. The level of agitated exchange had increased among the central group of five people—mostly English in several dialects. Combined with the gestural and voice controls used in their time, a dynamic, relatively integrated interaction between humans and computers was achieved. Craig was strikingly reminded of the gulf between what played out before him and his own time a quarter century later: neural imbeds and (first-stage) quantum computing finally delivered the symbiosis of the human-computer dyad.

As he moved toward the others, Benar said matter-of-factly to Craig, in a tone not requiring response, "Staff was supposed to get the HQ bulletin. Where were you?" He passed a Nordic-looking woman who picked up his head gesture and positioned herself to split attention between the rapidly changing displays and Craig. Her name tag was "Nora Lavrans."

Craig moved just a little closer to her to be heard over the stir around them. Lowering his voice he said, "Please tell me, Nora, what this bulletin is about. Seems to have everybody's attention."

She had the same kind of projected digital clipboard that he had seen before; her attention was now darting between the clipboard and the curved wall pulsing with displays of engineering and geological diagrams, graphic dynamic semiotics. She spoke to one of her colleagues in bursts of monosyllabic techspeak; others were in audio or visually supported contact with remote points. Then flashing him a glance, she responded rapidly in Norwegian-accented English. "Shortly, a red line will connect three regions. That line indicates the borders of the continental shelf that is part of the polar tectonic plate."

Craig would not be able to entirely resolve the conflicting demands for attention, but at least two elements could be sorted out. First, some historic catastrophe was impending and his own organic memory, unsupported by imbed retrieval, had so far been inadequate. Second, the VE experience itself was being distorted by his interaction with participant entities, evincing disturbingly "human" characteristics.

In a few seconds he was parsing those two elements with surgical concentration. His choice of the sequence in which to pursue the two themes was a critical neurological insight, made under temporal and situational constraint.

Craig knew he was distracted momentarily from the display Nora had indicated. His concentration was on the figures around him. Since entering the room he had been making a mental checklist of attributes exhibited by the inhabitants. He ran down the list: the rendering of the clothing's elasticity, fold formation, following every subtle body movement; part of the overall conviction of a solid body inside, subject to gravity; the small imperfections, variations, and asymmetries in limbs, torsos, and hands; the vagaries of hair, skin color, texture, and reflectance; the endless subtle expressivity of faces and gestures, each with its unique rhythm; speech not improved by technology, with its slurs and losses, the product of vocal cords bouncing around the bone and tissue of the human skull. A third form was present: neither avatar nor human.

On entering, he had noticed the debris of food and drink containers and dispensers and a restroom near the open closet wall. Avatars do not use restrooms. He was reminded that his own physiology was temporarily modified by biochemical procedure. Before he could toy with the idea of an improvised Turing test, he became aware that everyone was quietly riveted on the display Nora Lavrans had told him would appear.

Enhanced satellite images of three polar regions, an expanse over 1,600 kilometers, were made contiguous by a glowing red line: the continental shelf bordering the polar tectonic plate. Geographic information hung holographically above the space perspective: WANDEL SEA, north of Greenland; SPITSBERGEN, north of Norway; SEVERNAYA ZEMLYA; north of Siberia. All of them were arrayed along the curve between 80° and 85° North Latitude, south of the vast Angara Basin.

The search engine of Craig's unconscious had been digging inexorably in the dusty files of memory for the triggering stream of images.

As he was framing a question, a rapid series of drone camera angles flashed across the screens, close enough to make out details. The viewpoints revealed a wide causeway, marching to the horizons, great carbon/alloy towers holding it above the ocean. Along one side of the causeway ran a wide maglev system. As the camera angles zoomed closer, he could see structural irregularities and, finally, extensive cracks progressively forming.

The tiny forms of construction crews, like insects at this remove, were scattering in clusters; it was the proverbial "train wreck" in slow motion. The agitation in the group around him was palpable; voices straining to repress panic, moving into emergency mode.

The display produced a new graphic: scattered clusters of pulsing, luminous, concentric white rings along parts of the tectonic plate outline.

Craig leaned closer to Nora and whispered, "The white circles, they indicate seismic shock waves out there, right?" Nora whispered back without turning her head, "Yes. Look at the numbers!" On the screen alongside the concentric rings the numbers climbed; 6.7, 6.9; and climbed; 7.2, 7.8.

Finally Craig could not restrain himself.

Later, he was able to describe his own voice as hoarse, unrecognizable, and sounding like a man waking into *a nightmare.*

He said, "When is this happening? Or is this some kind of replay . . . exercise?" A compact, graying man in the group, named Charles, spun about and yelled in rage, "Where the hell did you come from, you damn fool? You're watching hundreds of people die before your eyes and the biggest engineering project in the world go down the drain and you're asking about replays!" Three of the people near Charles warily moved closer to him.

Craig lost it and erupted with a vengeance, lots of pieces having fallen into place. "Listen to me, damn it! Tacoma Narrows, 1940s, they don't get aerodynamics; Minneapolis, 2000s, they don't get the importance of redundancy; Jintang Strait, 2020s, they don't get prototype c-modeling transition to full size! You've had half a century of c-modeling and seismic analysis and god only knows how many lives and . . . ."

Charles exploded and went for Craig. The three who had been edging close moved fast enough to hold on to him.

Douglas Craig recalled the sequence of events perfectly.

He backed away from the struggling figures. He shut his eyes and shook his head as if to clear away the mind-wrenching scene he still hears: the raging voice, the others placating, reasoning, over the scuffling sounds. It lasts only a few seconds.

And then, there is the unforgettable sound: a low, low frequency roar from the bowels of the earth, making his body vibrate. His eyes snap open. The whole room shakes so hard he fights to keep from falling. People and gear are thrown about. He is slammed back against the row of warmsuits. The lights go out, and they are all shadows against flickering emergency light points. An alarm screams and from somewhere a red light strobes. It lasts forever—32 seconds. Then it is over; the silence and dark is absolute. His breathing and pulse settle. The dark is calming. Selfness is intact.

REFERENCES

Santayana, G. (1905). *The Life of Reason (Vol. 1).* New York: Scribner's Sons.

Chapter 31

TRENDS IN MODELING, SIMULATION, GAMING, AND EVERYTHING ELSE

Jack Thorpe

Every so often it is useful to take a step back from the daily grind of scientific investigation and technology development to look for trends that help trace where our field has been and where it might be going. This is my take.

It appears to me that the major trend in modeling, simulation, war games, and large-scale distributed games, which for the purposes of this chapter I will lump together, is one of *convergence*. This was detectable in the early 1980s following successes in computer networking sponsored by the Defense Advanced Research Projects Agency (DARPA). Networking standards allowed distinctly different computing machines to exchange data at healthy speeds and volume, thus allowing observers to look inside applications running on these different machines and compare and contrast their similarities and differences: We could see inside the application stovepipes[1] and find commonalities, though they often were disguised with stovepipe-specific jargon and iconography. It was not easy, but it was possible.

For the U.S. military, some of the more obvious stovepipes concerned (1) personnel selection and recruitment, (2) initial socialization and basic training, (3) specialty skill training, (4) crew and unit training, (5) doctrine development (operational, strategic, and tactical), (6) structured speculation and analysis about possible future wars, (7) the design of force structure to meet these future threats, including the military systems needed to equip the force, (8) planning (short and long term), (9) mission rehearsal, (10) mission execution (operations), (11) after action review, and (12) construction of the historical record for noteworthy events. Smaller stovepipes sometimes existed within each of these larger stovepipes, for example, the branches and specialties within our military services (infantry, aviation, artillery, surface and subsurface ships, intelligence, and so forth).

---

[1]By "stovepipes" I am referring to the tendency of different functionally oriented groups to talk among themselves with unique terms and concepts, organizations, funding, careers, and so forth, that do not easily lend themselves to communication and collaboration with groups from other "stovepipes." The stovepipe metaphor is often used to suggest parallel but noninteractive enterprises.

Each of these stovepipes tended to live in its own space, had its own keepers and nurturers ("professionals" in these disciplines with professional support associations, trade conferences, and so forth), had its own descriptive concepts and languages, and even had its own funding streams and legislative advocates. For the list above, these might include recruiters and aptitude test makers, sociologists, instructional systems designers, training device manufacturers, instrumented range designers and white hat controllers, operations research and mission route planners, command and control systems designers, performance measurement experts, historians, analysts, and accountants, to name a few.

As information and communications technologies became more prevalent within these various stovepipes, we started to see cracks that allowed us to peek inside to see how someone else's application worked. This was instructive. Most people were expert in just one discipline, so to understand a little about how another discipline was wired together was revealing.

Many examples come to mind. Here are four.

DISTRIBUTED SIMULATION AS A COMMAND AND CONTROL (C2) SYSTEM

Early in the development and testing of large-scale distributed simulations, circa the mid-1980s, an operationally experienced senior officer referring to a network of simulators observed, "What you have here is a command and control system." This was incomprehensible to the simulation engineers working on the project. They knew little of the C2 stovepipe. But after some investigation, it was obvious that a simulation network did many of the same things, and used much of the same technology, as a C2 system: both were essentially a large number of nodes connected in real time sharing significant quantities of multimedia data and interacting in a specific way to achieve a desired end state. In some cases, it appeared that a simulation network actually achieved the end state better and at a lower cost that an iron-built[2] and fielded C2 system. Further, any large exercise (lots of people, units, command elements, vehicles, and other) conducted in a large-scale distributed simulation system *had* to have command and control less chaos result. This could be an external, existing C2 system that the participants brought to the simulation (that is, their own unit's C2 system), or it could be a C2 replication built into the simulation (that is, the essential simulated C2 functions needed to command the forces in the exercise).

ROUTE PLANNING SYSTEMS AND FLIGHT SIMULATORS

A second example resulted from the examination of route planning systems, the computer applications used to create waypoints (x-y coordinates, altitude, time, speed, configuration, and other) for combat aircraft entering and exiting a

---

[2]By iron-built I mean a system rigorously specified by exhaustive requirements documents and constructed by large teams of contractors, often over long periods of time and at significant cost, with the resulting product sometimes underperforming real world needs.

battlefield. As engineers and scientists from the flight simulator stovepipe studied the components of the route-planning stovepipe, it was clear that both used similar algorithms. Both had to have aerodynamic models, terrain models, weather models, and so on. Further, though not designed that way, it was possible to have each interact across the other's stovepipe: a mission could be planned in a route-planning system and flown in a simulator (or groups of networked simulators) for verification and refinement (for example, accommodating pilot preferences, tactics, techniques, timing, and emergency options). Likewise, a mission flown in a simulator could generate an initial set of waypoints that, when converted into a specific format by the route-planning software, could conceivably be entered directly into an aircraft's flight director for route execution.

RECONSTRUCTION OF COMBAT OPERATIONS AND LARGE-SCALE DISTRIBUTED SIMULATIONS

A third example can be found in the DARPA-sponsored reconstruction of the Battle of 73 Easting from the Gulf War.[3] In conjunction with the U.S. Army, DARPA attempted to use simulation tools from the simulator networking (SIM-NET) program to re-create the engagement battle of the 2nd Armored Cavalry Regiment against elements of Iraq's armored Tawakalna Division. The idea was to re-create the battle as if each combatant vehicle had been an entity on a simulator network generating second-by-second vehicle status messages. Since none of the actual vehicles had been instrumented or had been on a network, the paths and behaviors of every vehicle (U.S. and Iraqi) had to be determined, coded into data packets (as if from a simulator update), and then entered into a simulation data stream of the type automatically generated and captured during a SIMNET exercise. This process took 12 months, partly because no single participant in the battle had a complete, detailed picture of what had actually occurred. Once the data stream had been assembled, carefully studied, and verified for accuracy by the commanders who had fought the battle, at least on the U.S. side, the SIMNET system could play it back, allowing observers to view the action from any vantage point, including within U.S. and Iraqi vehicles (for example, gunner and commander reticle views). The lesson for the scientists and engineers involved was as follows: if combat vehicles and individuals could be instrumented and networked as part of their future C2 infrastructure, simulator networking technology could converge with the C2 systems to *capture live combat.*[4]

---

[3]An overview of the battle can be found under "Battle of 73 Easting" in Wikipedia, http://en.wikipedia.org/wiki/Battle_of_73_Easting

[4]A second reconstruction has been completed by a team led by George Lukes at the Institute for Defense Analysis focusing on the defeat of the Taliban in the Mazar-e Sharif region of Afghanistan in the fall of 2001.

Neale Cosby, formerly the director of the Simulation Laboratory at the Institute for Defense Analysis in Alexandria, Virginia, has been a spokesman and advocate for this idea and was instrumental in both reconstructions as well.

REAL WAR AND COMMERCIAL GAMES

A fourth example comes from reconstruction of battles from the current Iraq war, not by military laboratories, but by a commercial gaming company.[5] Kuma Reality Games has produced reconstructions of approximately 100 U.S. operations in Iraq that are sent to subscribers as games they can play, much like any other computer game. The games are run on regular personal computers (PCs). The company selects a well-publicized battle or operation, creates the terrain/feature data base from a variety of commercially available sources, collects details about the incident, often from the troops who participated (according to one news report), and produces and distributes the episode in a short time (sometimes a week). Using modern graphics technology, the rendering of the battlefield environment and combat effects is good. Of interest, it has been reported that these installments can be found being played by Iraqi teenagers in Iraq shortly after they are published. The convergence here: commercially motivated gaming and real world events.

Implications

So these and other examples suggest that many of the stovepipes, because they are increasingly based upon the same or similar information technologies, are converging. Further, this homogenization seems to have accelerated as information and communications technologies themselves moved from "special purpose" to "general purpose," from early computing machines designed to perform narrowly defined tasks (for example, computer-image generators) operated by experts with narrowly specialized skills to today's powerful, mostly consumer level machines and operating systems with robust, built-in networking capability, often browser based user interfaces, capable of being operated by everyday users.

Why Is This Important?

This dissolving of stovepipes by the interventions of common technologies enables new applications and new concepts of operations that in the past were hard to imagine.

As an example, with combat platforms and individual combatants now instrumented with accurate position location information technology[6] and connected via automatically reconfigurable networks, we have a means of capturing some of the details of live combat at an improved level of fidelity and realism. The format of these data, similar to that captured during distributed simulation sessions, allows us to think of both in the same terms. We can use the visualization systems from military simulations or today's commercial gaming systems (for example, PC based graphics; game consoles like the Xbox or PlayStation) to show the details of actual combat operations. Further, we can document combat operations as interactive data streams, living histories rather than written documents, and

---

[5]See www.kumawar.com
[6]This is sometimes referred to as blue force tracking or automated vehicle location.

feed these back into the training of young military personnel via simulation. It is reasonable to imagine a time in the near future where we will have a fully inter-active digital encyclopedia of every battle ever fought, available to every person in uniform, as well as officials in leadership positions . . . and perhaps others.

Such data would be valuable to researchers in a number of areas, such as in "machine learning and reasoning" where one goal is to build personal digital assistants and other tools for commanders dealing with "wicked problems."[7]

This brings to mind the environment that Orson Scott Card painted in his sci-ence fiction classic *Ender's Game* (1991). The protagonist, Ender Wiggins, a teenager selected using sophisticated profiling tools for his potential as a battle commander, undergoes fully immersive training and education using a wide vari-ety of tools to develop his tactical skills, and flexible physical exercise and gam-ing environments to develop his command skills (that is, his development of strong interpersonal relationships with his teammates and/or subordinates). For Card, Ender's assent to becoming a master tactician and effective commander is an all-consuming, continuous effort, without respite.

For Ender, any differences among the spectra of representations of reality are nonexistent: the convergence is complete. The information and communications technology that he interfaces with blurs the distinction between the real world and its many representations, all of which he can interact with and manipulate. It provides him the fluid command environment for collaborating with his team-mates. In the end, he puts in another day in simulated battle without realizing he is actually commanding a real battle. But because technology has allowed this complete convergence, it does not matter. It is all the same.

It seems to me we are on that path, fueled by the emergence of common infor-mation and communications technologies, and as we look at trends in modeling, simulation, and gaming, we can trace a path that takes us into a completely differ-ent landscape of futuristic, revolutionary applications. Here is one view.

BATTLE SCHOOL (UNDERGRADUATE AND GRADUATE) AND THE BATTLEPLEX

What would Orson Scott Card's battle school notion look like with near- and mid-term scientific and technological advances? I like to think of three interre-lated thrusts. Many of the specific components are already here today or shortly achievable, but others are farther off.

Battle School (Undergraduate)

The core curriculum of the battle school (undergraduate) is focused on leader development. Its goal is the maturing of leaders and leader teams (multidimen-sional/multiorganizational) for dealing with future conflicts and very large

---

[7]This means ill-defined, complex, dynamic problems of the type routinely faced by commanders in chaotic, nonlinear situations.

operations (for example, humanitarian assistance/disaster relief). It has five technology components, many of which require scientific advances:

a. A strategic syllabus for lifelong learning—adaptive learning tailored to each learner;

b. Commander's personal reference library—100,000 volumes;

c. Interactive encyclopedia of all battles and major operations;

d. Worldwide access to online mentors: anytime, anywhere;

e. Families of decision exercises.

Battle school "students" all are equipped with the battle board, a la *Ender's Game,* a tool that accompanies them throughout their careers. It is the main way they port into their battle school worlds.

Battle School (Graduate Level)

This is the part of battle school for advanced studies. It also has five components:

a. The designer's workbench—where new technologies and concepts are evaluated against past, present, and future worlds.

b. Extreme scenarios—a laboratory for thinking the unthinkable using unconventional, creative techniques;

c. Reconstruction lab—development of advanced tools and skills for reconstructing past battles and distributing them into the interactive encyclopedia;

d. Iterative history—developing the capability to interdict the historical record, manipulate events and decisions, and create alternative futures;

e. Cultural immersion—virtual travel to all points of the world for immersion into current, past, and future histories.

The BattlePlex

The BattlePlex is a complex of practice fields, coaches, trainers, "sports medicine," labs, media, and stadiums, where initial learning and skill acquisition is practiced, specific missions are rehearsed, real world operations are conducted (via the BattlePlex), and senior leaders (especially political) can observe, learn, and understand.

All three of these thrusts are interconnected. Advanced studies develop tools and content for core leader development, skills are practiced and honed in the BattlePlex, and real operations conducted there, captured live, are fed back into the core and advanced studies activities.

Can we do things like this? Given the continuing trend toward convergence, I think the answer is yes.

REFERENCES

Card, O. S. (1991). *Ender's game.* New York: Tor.

Chapter 32

TECHNOLOGICAL PROSPECTS FOR A PERSONAL VIRTUAL ENVIRONMENT

Randall Shumaker

BACKGROUND

Short-term predictions for the kinds of technology important for virtual environments (VEs), say two to five years into the future, are fairly safe, but not really all that interesting. Longer-term predictions of technological advances, perhaps 10 to 20 years into the future, have a somewhat spotty, generally humorous history. Some examples of this can be found in the nice little book *Yesterday's Tomorrows* (Corn & Horrigan, 1984), which has collected predictions in many domains: flying cars, all kinds of cities of the future, and even intelligent robots are shown. Now that we are in that future time, or long past it, we can enjoy the naiveté of those predictions. I am still hopeful on the flying car, though. One of my goals is to try to avoid bringing too much pleasure to future readers by using a reasonable methodology and by avoiding specific technical delivery predictions. My focus will be probable capabilities, however implemented.

So, how can we make predictions reasonably far into the future that are not doomed to being either completely wrong, or otherwise useless, and why try? The "why" is relatively easy to answer. While short-term predictions are safer, and can be of some value in tactical planning, they are not especially useful in creating a long-term strategic vision. There is also some reason to believe that a reasonable trend prediction may actually be self-fulfilling by creating a viable vision and time scale. The key is the reasonableness based on known laws of physics and economics. Consider Moore's law for growth in the number of logic elements on a microcircuit for an excellent example of this principle in action. Making reasonable projections that are qualitatively accurate predictions and do not violate the aforementioned laws of nature and economics are my goals here. My objective is to define the sort of features that could be expected to be available in multimodal virtual environments for home use, a personal VE.

Another lesson I have learned from a long career in information technology is that even in cases where long-term predictions are reasonably correct technologically, assumptions about how these advances will be used and their long-term impact have not proven too useful. The discussion here will be confined largely to considering the technologies for implementing personal VEs, with only modest speculation about the most obvious applications that this development will enable. There will be some discussion of what I hope will be less obvious applications.

APPROACH AND BASIC PREMISES

A great deal of VE literature appropriately focuses on human requirements and the degree to which technology can meet them now and in the near future, perhaps five years ahead. Stanney and Zyda (2002) provide a fine example of this useful literature. This chapter covers the many technical, psychological, and evaluation issues inherent in VEs in their current and near-term forms. Typical of many VE publications, this chapter expresses great confidence in the continued growth of technologies. Unfortunately, due to those previously mentioned laws of nature and economics, this trust may not be completely justified, particularly when the horizon may be 10, 15, or 20 years ahead. For this time frame it may be most useful to reverse the question: Can we specify just how much capability might ultimately be required to stimulate the senses of a human being in all necessary modalities at the maximum rates that can be processed by a human? Given this information, estimates can be made concerning if, and when, that might be possible for each modality. This information will also put us in a position to decide whether we really need, would want, or can afford to provide maximum stimulation capability.

While stand-alone applications for entertainment and education are potentially important applications for a personal VE, many really interesting applications will involve interacting with other virtual environments and the individuals or groups within them. In order to bound the problem a bit, the focus will be on applications where interaction among people and avatars within the personal VEs will be limited to perhaps a few tens of individuals. This assumption allows us to reasonably safely predict the inter-VE communication requirements that might be needed in a home environment. Another important issue is cost. While laboratories and shared facilities will invest substantial resources to achieve effective capability, most individuals cannot or will not. We have an effective example to guide us in deciding how much the majority of individuals will spend to acquire a high end interface technology that they desire, but do not require: large-screen, high definition television (HDTV) with home theater audio. My goal is to make reasonable estimates of when, if ever, an individual might be able to acquire a personal VE at a cost of $2,500 in current year dollars, about the median price for a high quality, large-screen HDTV with an associated high end audio system today. My short-term prediction follows: within a very few years home entertainment and personal computing will be one commodity or at least will be so tightly coupled that the distinction will not be important. Given this, the fiducial point for projections that follow will be in multiples of "PC-

2008" capability, the computing system that anyone can buy today for about $1,200. Such a typical machine from a high end vendor would contain a dual-core processor running at 2.2 Ghz (gigahertz) clock rate, 3 gigabytes of memory, a 500 gigabyte disk drive, a high end graphics card, wireless LAN 802.11g (nominally 54 megabits per second), at least 100 megabits per second wired networking, and possibly gigabit Ethernet capability. I will have more to say about external connectivity later.

This middle to high end PC will also include a 22 inch LCD (liquid crystal display) monitor, capable of full HDTV video display, and a reasonably good printer. For a few more hundred dollars a quad-core processor, as much as a terabyte of disk space, and an HDTV tuner could be added. All in all, this very capable machine might have cost more than $100,000 ten years ago. Note some calibrating information: the qualification about a maker of high quality computers is important because good design and good components are necessary to harvest the theoretical power of such computers. Note also that the clock rate is not really a good indicator of computer performance, but most PC manufacturers are silent about computing power that users might actually be able to harvest. In fairness, computer performance benchmarking is very workload dependent, so it is hard to provide accurate figures. For comparison, many very high end scientific computers use similar processors with a typical per-processor throughput of perhaps 10 to 20 billion floating-point operations per second (gigaflops). This is a lot of computation, but personal PC mileage will vary significantly, very much to the downside. And finally, almost no generally available software for personal computers can take advantage of the second core in our current machine, let alone use the four in the quad-core machine. This is expected to change though, and I am counting on this for building my personal VE.

CAPABILITIES THAT MIGHT BE POSSIBLE FOR A TOTALLY IMMERSIVE PERSONAL VE

Wonderful data for understanding the human as an information processing entity exists (McBride, 2005); however, it is in a form that requires some interpretation to derive appropriate technical parameters for fully "exciting" the human sensor systems. For purposes of this discussion, I will attempt to provide rough order of magnitude (ROM) estimates of the engineering parameters, not specific values. Also necessary in shaping this analysis are a number of assumptions about how the personal VE will be implemented and what it might cost to be a practical possibility. For ethical and practical reasons I have assumed that the user interface will be entirely external to the body in the time frame of interest. This may not always be the case; cochlear implants have been very successfully applied in dealing with some hearing impairments, and various other implantable interface devices have been investigated with some success. For pragmatic reasons though, unless we are attempting to overcome a specific personal disability, I do not believe the average person will consider implantable interface devices in the time frame I wish to consider.

VISION

Vision is our highest bandwidth communication channel and the channel we think of first when planning a VE. It is also our most heavily overused channel in creating such environments, in many cases the only one. There are several aspects of importance here: data delivery to the human visual system, bandwidth needed to supply the delivery system, and processing necessary to generate visual content. McBride (2005) says that the human eye binocular field of view is 200° wide by 135° high from central fixation, and we can infer that perhaps a worst case equivalent resolution would be 80–100 megapixels. The issue is clearly more complex than this number would imply. For example, the fovea is perhaps only 1 megapixel, with peripheral areas much less. If we could track the fovea and provide only high resolution imagery, there the total number of pixels needed would be very substantially less, but requires a lot more complexity for high performance eye tracking, communication, and dynamic image generation. We will still need tracking, even if we provide full-image resolution for everything in the immediate visual field, but at much lower resolution and rates. I do not know exactly how we might supply those 80–100 million pixels. Such technologies as direct retinal writing and advanced head-mounted displays may be able to achieve the equivalent relatively sooner and much less expensively than full high resolution displays. Whether we would actually want or need to build such full visual capability everywhere in the visual field is questionable. In any event, my hypothetical personal VE user will likely opt for less, but still pretty spectacular, capability. For reference, a 1080P (1,920 × 1,080 pixels, progressive scan) HDTV, the highest resolution expected to be commercially available in the United States, has 2,073,600 pixels nominally refreshed 30 times a second. The bandwidth needed to achieve this in raw format would be about 3 gigabits per second.

In practice, using currently approved compression technology and broadcast bandwidth, 720P (1,080 × 720 pixels) is the standard, providing 921,600 pixels and requiring an uncompressed bandwidth of 143.18 megabits per second. The compression algorithms used allow this to be transmitted at 37–40 megabits per second. Commercial broadcast experience has shown that most users are quite happy with compressed data at 22 megabits per second and are generally unable to notice the difference. By an interesting coincidence, if we had a way to directly deliver the signal appropriately divided between the fovea and the peripheral vision areas of the eye, 720P HDTV would fairly well fully excite human vision. A high performance eye tracking system coupled with a very large field of view display or direct retinal write system would be required to make this possible. There is much more that could be said about the differences between progressive (P) and interlace scan, various available frame rates, and the human ability to effectively use it. Lacking direct retinal excitation, we will be more interested in the trade-off between visual resolution and field of view, required computing power, local storage needs, and communication capability for creating a personal VE. The performance numbers for 720P HDTV are adequate for this discussion.

Obviously outstanding video is a key capability for a personal VE; however, a lot of research over the past 10 years has shown that adding additional sensory channels can greatly improve the sense of reality, immersion, and enjoyment of the experience. Moreover, these additional channels can heighten the perceived visual effects or reduce the required visual capability for creating an effective experience. This will be a consideration when we inevitably have to compromise on capability for cost or other practical reasons.

AUDIO

McBride (2005) also provided outstanding detailed specifications for the human auditory system for those who wish to delve more deeply. For creating a personal VE we can take a less comprehensive view and say that we will consider audio compact disc (CD) standards as adequate to provide the sense of audio presence that is required from a transducer standpoint. Separate excitation of the two ears in a range of 20 to 20,000 Hz (hertz) is accomplished by using a sampling rate of 44.1 KHz (kilohertz) and 16 bits of resolution. This yields a data rate of 705.6 kilobits per second per channel. In round numbers, a standard CD can record 700 megabytes yielding 80 minutes of stereo audio uncompressed.[1] A high quality audio headset or a more costly speaker system should work well as the transducer. Note that a high performance speaker system will also involve additional audio channels. Simply transmitting stereo audio from some real space or a fixed virtual space into a personal VE could be accomplished without additional processing other than perhaps dealing with relative changes in position. Generating an audio "soundscape" for a virtual space, the kind of rich audio environment we encounter in normal life, may involve quite a bit of preliminary computation and significant dynamic calculation that is very dependent on the complexity of the audio situation to be portrayed. In any event, our canonical reasonably high performance personal computer can accomplish this task now. For this channel we will also assume the ability to generate speech. This will require additional computation, but no additional bandwidth.

TACTILE, CHEMICAL, AND VESTIBULAR SENSES

I am including discussion of these capabilities, but as a practical matter these are less critical information channels to be provided in a personal VE for the kinds of applications I will describe later. Moreover, for tactile, in particular, while the total bandwidth involved is relatively small, the number, spacing, and types of tactile and thermal sensors that humans possess are likely to confound practical total synthesis for a very long time. Limited but useful capability may come from progress in direct stimulation for dealing with nerve injuries;

[1]Note that communications information is usually expressed in bits, and computer information in bytes. There are 8 bits in a byte; however, there are usually additional bits needed in communication systems for framing, error correction, and other overhead functions, so the throughput relationship is not just a simple ratio.

however, I do not believe that most unimpaired people will opt for direct stimulation. Where needed, tactile capability for a personal VE is probably limited to low fidelity physical devices until direct nerve stimulation becomes feasible. This may not be as important a limitation as it appears. Experience with some low resolution stimulation devices has shown promise for some special applications and should prove useful for dealing with some visual, auditory, and vestibular handicaps as well.

Pioneering work in sensory substitution by Bach-y-Rita and Kercel (2003) show that auditory and tactile analogs for vision and vestibular sensing can be effective and may be rapidly incorporated into the communication channel repertoire of an individual. Even the tongue has been shown to be a viable channel for providing reasonably high resolution information to people with vestibular problems (Tyler, Danilov, & Bach-y-Rita, 2003). These and other novel interface concepts will be of interest for some applications of personal VE discussed later. For olfactory stimulation, it is known that the sense of smell can add significantly to the sense of presence (Brewster, McGookin, & Miller, 2006); fairly simple technology to implement it is available, but I do not believe this will be an important element of a personal VE except perhaps for some kinds of entertainment. Similarly, I believe that most people will want to stimulate their senses of taste with real rather than virtual stimuli, and it is not clear what sort of application might require this capability. This said, perhaps some limited ambient olfactory capability might be an interesting adjunct to improve a sense of presence.

Vestibular and proprioceptive senses provide our sense of balance and postural position. Providing effective stimulation of these senses is important for some applications, such as flight simulation, locomotion in a VE, and for practicing physical skills. For widespread personal use, however, I believe that the main issue will be finding ways to avoid significant sensory discrepancies among the visual, auditory, and vestibular scenes; otherwise, our system will have to have simulator sickness placards prominently displayed.

HUMAN COMPUTATIONAL CAPABILITY

Among the things that would be nice to know is how much computational capability the human brain has to apply. This is of prime interest to psychologists of course; they are interested in precisely how the brain operates. I would like to know for a different reason: when, if ever, might computers become able to interact with humans as a reasoning peer or, more realistically, a reasonably responsive avatar. Such an avatar would be really nice to have in a personal VE for such applications as games, every kind of educational use, and as a personal companion or assistant. I have no idea at the moment how to build the software for this, or even if it is theoretically possible, but understanding the magnitude of computational power that might be required may still be useful. Not too many years ago the answer would have been very sobering—we were very far away. How far away are we now? McBride (2005) has some numbers, most of which are implementation specific to the human brain, a "wetware" computer. This text provides some guidance, but is a bit difficult to convert to computer-oriented numbers. Fortunately, someone has done this kind of analysis for us. Hans Moravec (2003) considered this very problem and provided numbers from which we can work. He projects that monkey level intellectual performance might require 10 million MIPS.[2] Moravec also projects that it would require 300 million MIPS (300 tera operations per second, or 300 TOPS) for human level performance. The 30 to 1 ratio of human to monkey capability is arbitrary, but we will work with these numbers, maybe having to settle for a monkey level assistant, but with good verbal skills. In any event if the main goal is to produce a pleasing, reasonably responsive, socially adequate, automated avatar, the amount of deep intelligence required might not be that significant. After all, monkeys function well in a rich social environment and can solve many interesting problems in finding food and dealing with threats. However, they are not too good at calculus and other high level reasoning functions.

Also there is a positive for our prospects for creating social competent avatars: human and monkey brains have a lot of responsibilities and concerns that an intellect-only artificial colleague does not need. For example, maintenance of life support functions, concerns about finding food, finding a mate and reproduction, coordination of motion, and most kinds of peripheral sensory processing will not be necessary. Similarly the reliability of the individual computing elements of computers is better than the biological equivalent, particularly when we consider a working life before replacement of about three to five years, so we can expect to get by with a bit less raw power and redundancy than the brain has, giving us a cushion in our estimates. The largest supercomputer listed in November 2007 was an IBM Blue Gene capable of a peak speed of about 600 TeraFLOPS (a measure of a computer's speed), with a memory of 73,728 gigabytes. We may have the raw computing power to achieve Moravec's vision, except, of course, for the huge cost and knowing how to actually build the required software.

There is another element of Moravec's prediction; How many MIPS might be purchased for $1,000. His projections were based on personal computers available in 2000, extrapolated from 1995 trends. My reading of his chart shows an estimated 10^9 MIPS for $1,000 by 2025; this is 1,000 TeraFLOPS, not too far from today's fastest computer. There have been some significant processor implementation changes for personal computers since 2000 that warrant another look regarding this issue.

CAPABILITIES THAT WE WILL WANT TO PROVIDE

Everything so far has been a prelude to better understanding what might constitute the requirement and capabilities for a VE that an individual might

---

[2]MIPS is a million instructions per second, a common unit of computer performance. This measure includes all computer instructions, whereas MFLOPS, another common measure, refer to only floating point arithmetic instructions. Unfortunately, MIPS and MFLOPS are not easily interchangeable, and the relationship varies depending on the computer architecture. Since I do not know what mix of instruction types might be needed to build a high performing avatar, I will use these values as loosely interchangeable to suit my ROM arguments.

purchase for personal use. For practical reasons and to keep these speculations from becoming too much like science fiction, the year 2020 is my target time frame. This is far enough ahead to be interesting and require a certain amount of guesswork, but near enough to allow reasonable technical speculation.

High performance vision and audio are the primary VE modalities we must have, but how much is reasonable and usable? Chemical and tactile capability may be useful, but tactile, in particular, will be hard to implement in a totally virtual environment. Tracking body motion, gaze, and facial expression detection will probably be useful or even necessary, but how much physical mobility will be needed? Having high performance avatars would be very nice, but how much should be expected? Which elements of these are likely to be worthwhile or even technically feasible? And finally, what are we likely to be able to afford within our proposed budget and time frame?

NOTES ON MY APPROACH TO PREDICTION

Physics, other laws of nature, and the known laws of economics provide useful guidance in making reasonable predictions. Some needed technologies are within sight of fundamental limits that impact our expectations. Fortunately these limits are in many cases orders of magnitude beyond our present capability and at high enough performance levels that they are probably more than enough to keep us very happy in 2020. I will be making use of a nice tool for qualitative physics and mathematics, the ogee curve, coupled with and driven by those laws of nature and economics mentioned. Taken together these have historically proven useful in explanation and prediction in many fields. An ogee, or S-shaped curve, appears very frequently in nature, social systems, mathematics, architecture, and art. It represents growth, possibly geometric or exponential, that is ultimately limited by a lack of some resource or a limitation in some crucial feature. In nature, the limiting factor might be the availability of water, or some nutrient, or competition for some resource, or perhaps a physical limit, such as bone strength. In economics, it might be a shortage of a commodity or limitation in some other resource, such as transportation or saturation of a market. In technology, such limitations are often due to physical properties of available materials, heat dissipation capability, or size constraints in manufacturing technology. In any event this phenomenon is so ubiquitous that it is fairly safe to choose it as a prediction paradigm. The key feature is the phenomenon of interest grows with an increasing rate initially, then with a decreasing rate until it becomes asymptotic to some ultimate level.

For our purposes, consider three regions of the curve: early, mid-life, and mature. In the early phase the growth rate is actually increasing over time, in mid-life the rate of rate growth changes from increasing to decreasing, and in the mature phase the growth rate is asymptotic to zero. For a medium- to long-range prognosticator, this is a useful tool in that with relatively few data points we may be able to determine which part of the curve is in effect. For technologies where we may know something about the ultimate limiting factors, we may still be able to make useful predictions about how far away the limits are without knowing specifically how we might get there.

WHAT I CAN HAVE IN MY 2020 PERSONAL VE: THE GOOD, THE BAD, AND THE DISAPPOINTING

The Good

This discussion will provide expected capabilities, modulated by technological and economic reality. Cave-like immersive virtual environments are nice for some purposes. Because of space and cost considerations, and considering likely available sources of content, I do not think this is the way a personal VE primarily used for recreational and social purposes will be implemented. The most likely format will be a wraparound screen for individual or small group use, or more economically, a head-mounted display with head and perhaps eye tracking. Either should also provide hand, eye, body joint, and facial expression tracking. Looking at costs for LCD computer displays in 2007 (about $300 for a 22 inch $1,680 \times 1,050$ pixel display) and noting that we are probably in the early part of mid-life for these technologies, we can reasonably predict that we can have an $8,000 \times 2,000$ pixel display arranged to subtend perhaps 160° of horizontal field for about the same cost or less. This might be physically large, covering a curved wall, or closer for a more personal display, such as looking out the large windscreen of our personal VE navigation machine. (The vehicle metaphor helps considerably with visual coverage requirements, locomotion issues, and expectations for haptic interaction with the environment.) With a head-mounted display and appropriate tracking, we could achieve 360° capability with about the same effective field of view for the same cost. The high quality, current-day graphics in our hypothetical PC-2008 can drive two 22 inch displays already. It is not much of a reach to project the roughly eight times capability improvement at constant cost in 13 years.

I expect to have postural tracking, hand position tracking, and facial expression and eye tracking available for our personal user as inexpensive peripherals by 2020. Some of this capability exists now; for example, the Nintendo Wii interface can detect acceleration in three dimensions and has proven effective for simulating physical interaction in sports games. Facial expression detection and posture tracking are available in the laboratory now, with performance improving rapidly and costs dropping dramatically. Projecting a 10-year-in-the-future cost of perhaps $200 for such an interface is fairly safe. I do not anticipate the need for wide-area physical locomotion technology in a typical personal VE both for cost reasons and the observation that most people's entertainment and social technology is generally done comfortably seated or perhaps standing within a small area. This is not likely to change. High quality spatialized audio will clearly be available, the economical version headphone driven, the higher end area surround version costing significantly more. This will be integral to any combined computer and entertainment appliances available in 2020, the grand unification of entertainment and computing already being well under way.

The end of growth in computing power has been long predicted, but for the past 50 years, just as the rate of performance growth started to flatten out, a new technology has been available to keep the curve moving. This cannot go on forever, of course, and basic physical rather than technological limits are appearing on the horizon. Computer technology still has a way to go both in individual processor performance and more recently through physical replication (multicore technology and multiprocessor computers) and better processor utilization (multithreading architectures). For example, the IBM Blue Gene mentioned earlier uses more than 100,000 processors. Dual- and quad-core and multiprocessor personal computers are available today, and 8, 16, and 32 core, and so forth, processors are clearly on the horizon. An experimental 80 core processor was demonstrated during 2007 and thousand-core processors have been discussed. Various sources claim that computing hardware has demonstrated more than 10^{14} improvement in what $1,000 in constant dollars will buy in computing capability. (This is not a factor of 14, but 14 orders of magnitude, an almost unbelievable number—unduplicated in any other human technological endeavor.) By my estimate this capability is in the latter part of the middle of the maturity curve, but still has 5 orders of magnitude of potential growth left in the next 20 years or so. If true, these 5 orders of magnitude are enough to buy a 600 TeraFLOP human-throughput-capable computer for $1,000 inflation adjusted in 2027. By 2020 we should easily be able to afford high grade multi-"monkey level" computing power for the same price.

Storage seems to be actually a bit further from limits than computing power, perhaps mid-curve in growth. Currently a gigabyte of RAM (random access memory) costs about $30, and a terabyte of disc $250. Assuming a very conservative 2 orders of magnitude increase at constant cost by 2020, this would give you 0.1 terabyte of RAM and 100 terabytes of disc for under $300. What we could possibly need this to hold and whether such information would best be held locally are other questions. On the other hand, limitations in availability of very high bandwidth connectivity for individual users may justify local caching of large amounts of information. More on the reason for this follows.

The Bad, or at Least not as Good

Even modest-size college campuses and businesses have access to optical backbone networks rated at 10 gigabits per second today, and many personal computers have gigabit Ethernet capability. Optical networks have been demonstrated that provide as much as 1.6 terabits per second connectivity using wave division multiplexing on a single fiber pair, and of course fiber links have many such pairs available. All seems very promising for essentially unlimited communication capability, except perhaps for a personal VE user, individual users on large networks, and especially for mobile users. A typical small business currently might have a T1 connection, nominally rated at 1.5 megabits per second (Mbps), and a home user with "high speed" connectivity might have a nominal 7 Mbps connectivity. Most have significantly less. This is both an economic and pragmatic issue involving the time and cost to change basic infrastructure. Current home and small business connectivity is not particularly inexpensive either, at least by computing capability standards. A T1 line ranges from $250 to $1,000 per month, while the "high speed" home connection is typically $40 per month. There are plans afoot to provide fiber to individual houses, but, of course, it will take a long time to completely rewire the nation and the world. Cost and delivered bandwidth have not been clarified either.

I am less hopeful for multiple order of magnitude bandwidth improvements for individual home users and widespread use in businesses, although a factor of 10 or so is very reasonable by 2020. Part of this is due to the "last mile" problem, or delivery to the end user, and part is due to the shared nature of network distribution where multiple users share each channel. Wired connectivity within a building or house now is generally 100 megabit Ethernet, or perhaps gigabit Ethernet. These networks are shared, of course, and the number of other users and what they are doing can significantly reduce available bandwidth, and, of course, these cannot achieve full rated speeds for technical reasons. Even so, for text, still images, Web browsing, and low resolution video these data rates are satisfactory most of the time. Fiber bandwidth is, of course, significantly higher, but not too widely available, and unlikely to reach the average home users for 15 or 20 years. For distribution of a few hundred HDTV channels to many tens of thousands of users one way, current entertainment cable and fiber connectivity is fine. For hundreds of thousands of sessions of a few tens of users interacting, the situation is not so promising, particularly that last mile. I hope to be surprised by how fast universally available, genuinely high bandwidth connectivity is available; however, realistically, we will probably have to rely on workarounds, such as local caching, high levels of compression, and avatars that mirror or mimic human body language rather than live video for personal use for some time. This latter may not be a bad thing for some applications that I will discuss later where photo realistic imagery is not required and may even be undesirable.

Wireless networking is even more problematic. There are really two issues of interest here, connectivity on the go and untethered connectivity indoors. Radio frequency bandwidth is a limited commodity, and we cannot make more of it. While some more frequencies are becoming available with the move from analog to digital television broadcast, there is tremendous competition for it for cellular telephony, entertainment delivery, and all kinds of business, government, and military applications. For these reasons I do not foresee a great increase in the bandwidth, perhaps a factor of 10, that an individual outdoor mobile user might hope to have available. Indoors, and untethered, the situation is a bit better, but there are still a number of laws-of-nature limitations. Frequency congestion is still a critical issue, but for short-range line of sight communication, such as in a house or building, it is much easier to spatially share available channels.

Current WiFi wireless uses a protocol called, not very descriptively, 802.11 (pronounced eight oh two dot eleven). There are various flavors of this labeled with a suffix letter, for example 802.11a, 802.11b, 802.11g, 802.11n, and so on.

These range in data rate from 11 megabits per second (802.11b) to 248 megabits per second (802.11n), but, of course, their typically delivered data rate is realistically about a quarter of the nominal rate; and since the media are shared among local users, individual user data rates are further reduced. These rates are actually pretty good, though, and the highest of them could allow untethered, multichannel two-way video for a modest number of local VE users. The bad news here is that there probably will not be too much more radio based bandwidth ever available for frequency allocation and user health reasons. Also, many of the frequencies are shared with other services, so interference can be a problem. Perhaps local optical line of sight methods will become available that could significantly increase this, but I am unaware of any current development. In any event, do not count on many orders of magnitude increase in connectivity in the time frame considered here.

The Disappointing: Software

Conventional software, the kind that constitutes operating systems, devices drivers, office suites, and the many applications in conventional use, has come a long way, but painfully slowly compared to hardware advances, and not too efficiently either. Entertainment software, such as games, is often even complex and costly to produce on a per line of code basis because of heavy demands for fine graphics and complex motion. A large piece of conventional software, such as an operating system,[3] may constitute 50 to 100 million source lines of code (SLOC) that are handcrafted by teams of hundreds of individuals. Currently the operating systems for the personal computers that will be the backbone of future VEs cannot take full advantage of the multiple core processors available now, although techniques for doing so exist and are in use in high performance computers today. For our personal VE, I do not envision the need for fundamentally better operating systems and office suites; however, I do expect this software to disappear from direct user view. The days of everyone needing to be a system administrator or having to know what the "C drive" does need to pass into history. This process is under way already, as the computer becomes the brains behind multimedia home entertainment. Think about the iPod as an early example of how to hide underlying complexity while providing sophisticated behavior.

What else is there? Well, building the software that will permit the computers underlying our personal VE to change to be a collaborator with which users can seamlessly interact to solve problems or seek information, and deal with in the same general manner as in human-human social interactions is a major challenge. We are still in the early days of making this a reality. A key part of this is being able to use verbal and nonverbal communication channels effectively in both

---

[3] Apple's Mac OS 10.4 is estimated to be 86 million SLOC, and Windows Vista is estimated to be about 50 million SLOC. This is among the reasons why memory growth has been driven to such large numbers. Their sheer size and complexity alone explain the reliability issues so often seen. Fortunately this is improving nicely; by 2020 who knows how stable commercial operating systems will become.

directions. Good speech recognizers and excellent speech synthesizers already exist and will continue to improve. If a user were to have the gesture, posture, and facial feature peripherals I expect a personal VE to have, more effective and more natural dialog between humans linked by VEs would be possible,[4] and these would also provide excellent cues to be used for moving human-computer interaction to the more abstract level of communication that I envision. Note that recognizing and generating speech is not the same thing as understanding language and generating cogent responses. This latter level of software is the kind on which we need to devote some significant energy.

MODELING HUMAN BEHAVIOR AND CREATING ADAPTIVE, RESPONSIVE AGENTS

The field of artificial intelligence (AI) has had its heady times of great expectations, lows of disappointment at how hard this undertaking proved to be, and a resurgence with more modest goals and much better computing and information resources upon which to draw. We are beginning to see good progress in cognitive modeling and the simulation of certain individual and group behaviors. Maybe one day we will have a workable biologically plausible model of the human mind that could drive a companion avatar, probably not by 2020 though. If and when we do, it might well take a petaFLOP (10^{15} operations per second) computer or so to run the billions of SLOC needed to implement it. Do we really need such a full-up model to be useful? I believe that most human-human social, business, and recreational interactions do not require anywhere near full human cognitive capability. This is a huge advantage to us in normal life; who wants to have to use all of their cognitive capability all the time? It also means that we can set objectives for social avatars and more natural human-computer dialog at much more reasonable levels. I do not know precisely how to create this social software yet, but there is very promising work in the literature, and some really important supporting technologies already exist.

One of the problems with which early AI researchers had to deal was how to compile the huge amount of information that humans draw upon in decision making and in daily life. A great deal of research went into capturing expert knowledge, and many of the techniques developed then have become standard practice in computer science today. An important misapprehension at that time was the view that general human intellectual performance could be captured using logic or some variant as a general model. What was not as successful was capturing "common sense," the many pieces of important and not so important data that humans make use of in ordinary discourse, and development of good methods for piecing these together. Well, today the compilation of a lot of this general discourse-supporting information has been done, continues to

---

[4] I am assuming here that live video may not be the best way to interact in a VE for bandwidth and other practical reasons. I believe that avatars reflecting human body language and facial expressions of real humans, or computer-generated avatars with reasonable body language and verbal skills, will be the most effective way to build such systems.

improve and expand, and is available rapidly and free. Consider Wikipedia (wikipedia.org) as an example of a repository for huge amounts of reasonably vetted knowledge, mostly in compact narrative form. Consider Google (www.google.com) as a link to huge amounts of information ordered in sequence of likelihood of being of interest. Both of these are constantly updated and expanded by people—learning and adaptations are intrinsic and free to end users. These even support multiple languages. If we are interested in less factual and structured data to support discourse, opinions, and viewpoints on almost every subject of interest to humans, we have blogs. At least in theory these could provide a dialog avatar a rich source of considered analysis, opinion, ignorance, prejudice, and idle musings. The key, of course, is sorting through and applying this information. This is where I believe some more research should be focused.

So, it is not too risky to predict that by 2020 users will be able to conduct an effective and satisfying social dialog with an avatar for tens of minutes on such topics as the weather, current events, travel planning, and information access. Just do not expect too much intellectual depth,[5] nor is this really necessary for many of the applications I envision as the most widespread. The hypothetical high end monkey computational capability mentioned before will be more than adequate to implement this level of interaction, and others have provided the storage for the huge information base required to carry on a responsive and believable dialog.[6]

WHAT COULD BE DONE WITH A PERSONAL VE?

Aside from the obvious applications for VEs, such as entertainment, virtual travel, education, and above all, games, I believe that whole new classes of augmentation, assistive, and quality of life applications will become feasible and perhaps, with an aging population, very desirable or even necessary. First let us consider some obvious uses, then the less obvious, but potentially more interesting ones.

Second Life (http://secondlife.com) is a current example of an online virtual world that is not a game, but a venue for social interaction and even real world commerce. Even in current form with primitive avatar motion and text interaction, a great many experiments are being done on what such environments might have to offer in the future. Enhanced avatars, and using a personal VE as the user window into a virtual world, should provide a highly effective means for learning, social interaction, and commerce. Virtual worlds have even been proposed as venues for research in social science as realistic microcosms of human experience and decision making (Bainbridge, 2007). A recent article in the *Washington Post* (Stein, 2007) discussed a number of other applications, including the use of a virtual world in therapy and as an accommodation medium for physical and emotional limitations. Some businesses have built private virtual meeting rooms,

---

[5]I have been in many meetings and social gatherings where none of the human participants appeared to employ even a modest level of cognitive prowess, yet the events were considered to be successful.
[6]Some early examples of such discourse agents already exist; visit Ramona! at www.kurzweilai.net

claiming better results than for audio or even video conferencing, and there are even commercial showrooms for products and off-site college campuses in *Second Life*. A recent pilot study in using a virtual classroom for teacher selection (Dieker, Hynes, Stapleton, & Hughes, 2007) showed high user acceptance of intelligent avatars in a VE and the potential for a personal VE as a future tool for delivering immersive experiences. This system had multiple avatars that displayed many of the high level social interaction features that I have proposed to automate. The jury is still out on how effective these limited VE experiences really are for commercial and educational purposes, but a number of companies now provide excellent tools for developing business- and training-oriented virtual worlds.[7] However, the widespread acceptance and use of virtual worlds for purely social interaction, even in their present limited form, are indicators that there may be significant unexploited potential for other useful purposes.

A personal VE as the user portal into such systems could enable many new uses, enrich existing ones, and remove some of the technical impediments for individuals, facilitating broader usage. High performing social avatars and wider rates of human participation would improve virtual world capabilities immensely. An emergent application has been the use of virtual worlds in therapeutic applications and as a means for mitigating social barriers for people with disabilities, mobility limitations, and some type of psychological impediments to full social interaction. *New Scientist* magazine produced a series of articles during 2007 showing examples of *Second Life* as a mechanism for dealing with some of these issues. Two of these articles are of particular interest; the June 27 issue focused on *Second Life* as a vehicle for people with Aspergers' spectrum disorder to more effectively interact with others, and the August 22 issue dealt with physically handicapped individuals participating in virtual world activities on an equal basis with nondisabled individuals. While still anecdotal, there is growing evidence that virtual worlds have an important role to play in providing social and business access to people across space and as mitigation for various physical and social limitations. In the case of Aspergers, the simplified social signals and even the inherent delays are part of what make these environments effective. Similarly, physical handicaps need not be apparent or an inhibitor in a virtual world. A personal VE that allows a better sense of immersion, multisensory interaction, or even discourse involving synthetic entities can only enhance this capability.

One key application I foresee is social accessibility for the elderly who may have mobility issues, but still wish to engage in a rich social life with friends, shop, consult with physicians, or engage in mentally stimulating activities. Conversational avatars may also serve as companions who never get bored hearing repeated stories or discussing politics. Companion animals have served some of this function in the past, but have generally not been able to hold up their side

---

[7]During a recent demonstration of a virtual world for training decision making in a high impact environment, in this case disaster response, it was noted that an avatar of one of the participants was recognizable as representing him, but his avatar was much slimmer and more agile. The participant noted that while he did not have time to exercise, he had his avatar run five miles a day.

of the conversation. Such conversational avatars might even be able to conduct ongoing cognitive assessments and provide reminders to patients that may be helpful in health maintenance. The current generation of elderly has not generally embraced the Web and e-mail as fully as might be hoped, although this is changing. I do not expect that to be the case with members of the next generation, having been users of personal computers, cell phones, ATM machines, MP3 players, and purchasers of large-screen HDTVs. In particular, the disappearance of the computer itself as a separate entity that has to be learned, debugged, and generally tolerated for the benefits it provides should facilitate the transition to multimodal entertainment, communication, and information appliances. This process is already well under way.

REFERENCES

Bach-y-Rita, P., & Kercel, S. (2003). Sensory substitution and the human-machine interface. *TRENDS in Cognitive Sciences, 7*(12), 541–546.

Bainbridge, W. (2007, July 27). The scientific research potential of virtual worlds. *Science, 317*(5837), 472–476.

Brewster, S., McGookin, D., & Miller, C. (2006). Olfoto: Designing a smell-based interaction. *Proceedings of the SIGCHI Conference on Human Factors in Computing Systems* (pp. 653–662).

Corn, J., & Horrigan, B. (1984). *Yesterday's tomorrows: Past Visions of the American Future.* Baltimore, MD: Johns Hopkins University Press.

Dieker, M., Hynes, C., Stapleton, C., & Hughes, C. (2007, February). Virtual classrooms: STAR Simulator. *New Learning Technologies 2007,* Orlando, FL.

McBride, D. K. (2005). The quantification of human information processing. In D. K. McBride & D. D. Schmorrow (Eds.), *Quantification of human information processing* (pp. 1–41). New York: Lexington.

Moravec, H. (2003, October). Robots, after all. *Communications of the ACM, 46*(10), 90–97.

Stanney, K., & Zyda, M. (2002). Virtual environments in the 21st Century. In K. Stanney (Ed.), *Handbook of virtual environments, design, implementation, and applications* (pp. 1–14). Mahwah, NJ: Lawrence Erlbaum.

Stein, R. (2007, October 6). Real hope in a virtual world. *Washington Post,* p. A1.

Tyler, M., Danilov, Y., & Bach-y-Rita, P. (2003). Closing an open-loop control system: Vestibular substitution through the tongue. *Journal of Integrative Neuroscience, 2*(2), 159–164.

Part IX: Military and Industry Perspectives

Chapter 33

THE FUTURE OF NAVY TRAINING

Alfred Harms Jr.

BACKGROUND

With the U.S. Navy investing billions of dollars annually to educate and train its broadly skilled and widely dispersed forces, it is essential that relevant learning content, captivating presentation methods, ubiquitous delivery options, and responsive support systems be efficiently employed to ensure that our sailors have near-continuous, ready access to the most effective learning experiences possible. This mandate becomes even more compelling as the navy continues to reduce in size to an active duty force of 320,000 or less, with minimally manned crews and increasing requirements for multiskilled, cross-trained professionals.

Today, one area of focus should be virtual environment learning scenarios as they offer uniquely powerful, creative, and adaptive opportunities for best preparing sailors to successfully serve in ever-changing, increasingly challenging, and potentially risky settings. From high end, full-scale, multiuser, immersive applications to low cost, individual, part-task trainers, virtual environment learning experiences are generally more exciting, impactful, fun, and effective than traditional education and training methods. However, the use of virtual world environments should not be viewed as a panacea for all learning requirements. Rather, they are but one tool in what must be an engaging, tailorable, and incredibly flexible set of blended learning solutions to satisfy the needs of the twenty-first-century sailor. When determining how to best provide any learning opportunity, it is instructive to review the crucial role of education and training in the long-term success of the U.S. Navy (and similarly, all of our armed forces). Once education and training imperatives are understood as cornerstones of individual excellence, successful team development, and operational force primacy, one can smartly consider how to best use virtual world environments to enhance future learning experiences.

The U.S. Navy has been the world's preeminent maritime force for well over half a century, and whereas some may debate the reasons why, there are three certainties that highlight the U.S. Navy's unquestioned supremacy on the high seas and around the world. First and foremost, today's navy is the best in the world due to the wonderful people who fill its ranks from the most junior seaman recruit to the highest ranking admiral. These marvelous volunteers are remarkable in many ways, but most notable are their far-reaching talents, courageous dedication to mission, steadfast trustworthiness, selfless teamwork, and unparalleled commitment to a life of service. Motivated, disciplined, loyal, and hardworking, these amazing professionals are and always will be our nation's most significant "asymmetric advantage" in any military action, and they will long be our "surest guarantor" of military success if called to fight. No matter what the situation or circumstance, our servicemen and servicewomen routinely provide the pivotal difference between success and failure whether operating in seemingly simple, incredibly complex, or unimaginably demanding scenarios around the globe. In the end, it is the human, and not the machine or technology, who provides the winning margin.

That is not to say that the numbers, availability, and capability of our nation's military and other security equipment are not exceedingly important. In fact, most experts would agree that the second principal reason for America's military excellence is due to the fact that our fighting forces are generally well-outfitted with modern, world-class equipment. Furthermore, they are adequately resourced to consistently maintain and regularly upgrade these substantial capabilities. Multiple times in our history, the enormous manufacturing capacity of America's industrial complex has enabled our forces (and various allies) to continue fighting when our enemies simply could not keep up. Ultimately, we outproduced and overwhelmed the enemy with virtually unlimited force structure. Likewise, our nation has for many years aggressively fielded top-of-the-line technological capabilities for its fighting forces, further ensuring consistent and predictable success on the battlefield. Generally speaking, our fighting forces have the necessary tools to fight early, win decisively, and return home safely. This fact does not imply that combat or daily operations are now or ever will be risk-free, as has been so graphically highlighted by the recent years' casualties experienced in Iraq and Afghanistan. However, no opposing force in the world can match the quality of the people and the equipment routinely fielded by American forces, and together this combination has been key to our military's many successes over the years.

The third and perhaps most dominant factor ensuring the exceptional performance of our military forces has been the nation's unparalleled commitment to educate and train all personnel, officers and enlisted alike. The finest people, outfitted with the finest equipment, will not alone produce an effective fighting force; extensive and continuing education and training are absolutely necessary to fully develop and exploit the enormous human skills and equipment capabilities resident in our forces. It is this third factor that truly distinguishes the U.S. Navy (and sister services) from other military forces around the world.

CHANGING REQUIREMENTS

Given that the U.S. Navy is the world's best and that all three dimensions of excellence—personnel, equipment, and education and training—play a key role in end-game success, observers might proffer the broken logic that questions "if it ain't broke, why change a thing?" One would think that when living in a world with almost incomprehensible and still accelerating rates of change, the answer should be readily apparent. Frankly, with so many things changing at unprecedented rates, and an institutional proclivity toward rapid change in most operational scenarios, it is hard to fathom why there exists such inertia opposing change in learning approaches. It is also disappointing that some leaders "just don't get it" and seemingly undervalue (ignore) education and tolerate suboptimal training regimens designed generations ago. Are our ranks filled with such talented and adaptive people—great patriots committed to success at any cost— that we have become too comfortable with the status quo and not been forced to change? I hope not, for just as status quo operational tactics will eventually get our sailors killed as the enemy continuously adapts for advantage, so will status quo learning approaches fail (or unnecessarily falter) as global realities and mission requirements rapidly change.

The navy's legacy, mass-production approach to learning was extremely well suited for last century's "force on force"—"fixed enemy" mindset (World War I through the Gulf wars); however, the twenty-first century will demand a broader, more complex mission set than ever before experienced. From large naval platforms with large crews having significant individual specialization and rating-specific tasks to smaller mission-configurable platforms with smaller crews of multitalented individuals requiring extensive cross-training, it is imperative that today's learning environment be precisely relevant, highly tailorable, readily adaptable, and pervasively available to meet ever-changing threats and mission requirements. Whereas last century's navy faced well-defined threats generally dominated by our superior forces, technologies, and supporting infrastructure, today's navy operates in a more complex operational environment where the collective threat of unpredictable and often unknown situations can be metaphorically described as a "thousand daggers" versus a "single, savage thrust"! This changing environment, coupled with the realities of increasing competition for talent and increasing costs for manpower, makes it imperative that navy education and training efforts be creatively adopted, wisely managed, and predictably effective in producing sailors who can consistently "outthink," "outlearn," and "out-adapt" the enemy. This imperative becomes even clearer with time as virtually no enemy is likely to revert to any traditional force attrition warfare model in the near or foreseeable future. Rather, our great navy will continually face a pervasive, shadowy mix of professional warriors, hard-core criminals, determined extremists, and zealous bunglers, all capable of delivering unmistakable horror in unimaginable places with unthinkable consequences. We must prepare for this new world and not wallow in the relative comfort of the past ... a past that we essentially dominated, but one that has become increasingly less relevant.

FUTURE POSSIBILITIES AND DIRECTION

So, how does the navy ensure that the education and training pillar of operational excellence continues to provide the winning margin of performance in the future? I contend that adequate numbers of wonderfully talented and dedicated men and women will continue to serve their country as members of the armed forces, taking their turn in helping preserve the blessings of freedom and opportunity for all. Likewise, I am convinced that our nation will continue to provide our forces with the finest equipment in the world and adequately resource the outfitting, upgrading, and modernization of our equipment inventory. Finally, I am convinced that we can achieve, enhance, and sustain a world-class learning environment that will interest, challenge, and richly prepare our sailors for their demanding responsibilities in the future. Although the task is daunting in many respects, there is less mystery than many imagine in crafting a successful way ahead. Boldly executing necessary changes to existing policies and programs will be neither painless nor cost-free; however, failing to understand, value, and embrace needed change in the ways we educate and train naval personnel will assuredly result in the diminished growth, development, and accomplishment of our most important "force multiplier"—our people!

Transitioning from legacy learning programs to an approach fully leveraging proven science of learning principles and human performance considerations will help ensure our people pursue and master relevant content in an efficient and effective manner. More specifically, we have the capability today to accurately analyze and define actual performance requirements with high specificity and then build learning opportunities tailored to relevant performance objectives. To neglect this fundamental step in ensuring that our education and training efforts are properly targeted on specifically known and thoughtfully projected performance requirements would be both foolish and costly in today's constrained resources and high stakes environment. A more serious misstep would be a conscious decision to endorse a "good enough" mentality toward navy learning as it exists today, and worst of all would be the utterly disingenuous, "head in the sand" approach used by those who neither value the importance of learning nor understand its direct linkage to our readiness status. We would never approach the people and equipment pillars of military readiness in this manner; rather, we always assess needs and capabilities with laboratory precision, valiantly maneuver to attract and retain the very best people, and procure and maintain the very best equipment. Similarly, we should never shortchange the science, support, and pursuit of what is arguably the most differentiating pillar of the readiness equation—educating and training our people.

The use of virtual world environments and other supporting simulations will enhance the learning experience of all our sailors by providing discriminating realism, contextual fidelity, and dynamic and interactive participation, confirming repetition and controlled assessment of cognitive, affective, and psychomotor skills, all in a cost-effective manner. This type of learning experience will become more and more critical with expanded multimission tasking, increased cross-training requirements, and a growing likelihood of short-notice, unpredictable, and unconventional mission scenarios. The expanded use of virtual world environments will enable more meaningful learning experiences whether pursuing (1) simple "exposure" events for improved awareness,

(2) repetitive "practice" events for enhanced competency, or (3) one-of-a-kind, "final rehearsal" events designed to both test and assess the learner's ability to outthink, outlearn, and out-adapt the enemy. In addition to enhancing individual or team performance, these special learning environments can also provide a reliable, dynamic, and affordable methodology to determine attainment of qualification standards or certifications with high levels of specificity, standardization, fairness, and repeatability. Finally, without virtual world learning environments, there may be no practical or affordable way to expose, educate, and train our people to successfully operate in those infrequent, remotely located, potentially unsafe, and highly volatile or overtly dangerous situations we know our forces will encounter. The benefit to our forces to even be minimally exposed to these types of scenarios before actually encountering them will prove to be invaluable in terms of individual confidence and performance, thereby enhancing the likelihood of mission success and personnel safety.

Whatever the mission, having the benefit of repeatable and varied learning experiences will unarguably enhance the end-game performance of most individuals, again measurably increasing the likelihood of mission success and personnel safety. Legacy learning methods alone are simply no longer adequate to prepare our forces to engage, counter, and defeat a determined enemy. Furthermore, legacy learning approaches typically do not facilitate real time, in-depth assessment of performance for feedback and evaluation purposes, thereby limiting learning and performance improvement for most individuals. Both of these situations are unacceptable shortcomings given the upside potential of today's learning options, especially when considering the pervasive threat and unforgiving scenarios that many of our forces will face with increasing frequency in the coming years.

Beyond providing more relevant, better tailored, and more easily assessed learning experiences for our people, virtual environments can clearly be more attention grabbing, more engaging, more stimulating, and, in some cases, more pedagogically sound than traditional learning approaches. Anything done to spark, broaden, and sustain an individual's desire and passion for learning is critically important, especially in an era where self-driven, independent, lifelong learning will become more and more the norm. Yes, self-administered, self-assessed learning will become commonplace in the years ahead, and these experiences must be relevant and rewarding in order to motivate our people to learn the knowledge, skills, and abilities necessary for survival and success in the twenty-first century.

The use of virtual environment experiences will also enhance important "self-discovery" aspects of learning. Although sailors are expected to essentially master the content of mandatory topic areas, virtual environment learning experiences are easily designed to offer additional, positive aspects of learning. These learning skills include widespread use of information-seeking, information-analysis, and information-synthesis tools, real time exercise of adaptive problem solving skills, and routine implementation of collaborative planning and decision-making skills. Finally, whereas traditional learning approaches tend to be individualistic by nature and often rooted in either the textbook author's or classroom instructor's perspective, virtual world experiences are typically more inclusive, offering the learner opportunities to gain familiarity with multiple and diverse perspectives. Then, using his or her own knowledge base and personal experiences, the learner can better differentiate between fact and opinion, draw balanced conclusions, and more successfully engage both our allies and our enemies around the world.

CONCLUSION

Virtual learning environments can provide relevant and rewarding learning experiences and, in many cases, offer far more innovative and exciting learning scenarios than in any traditional setting. If we are to be fully responsive to the needs, interests, and strengths of our sailors, and if we are truly focused on maximizing their growth and performance, it is hard to imagine that the U.S. Navy would not exploit the many advantages of virtual world learning environments throughout its education and training programs. As stated in the opening paragraph, however, there is no "one size fits all" solution; rather, a judicious blend of traditional and virtual learning environments will be needed in the future. In that future though, virtual environment learning experiences can and should play a large role in helping prepare our sailors maintain the competitive edge necessary to remain the world's greatest navy and successfully fulfill their incredibly important role as guardians of our grand republic.

Chapter 34

THE FUTURE OF MARINE CORPS TRAINING

William Yates, Gerald Mersten, and James McDonough

> What ought one to say as each hardship comes? I was practicing for this, I was training for this.
>
> —Epictetus

ROAD MAP FOR TRAINING

The U.S. Marine Corps Training Modeling and Simulation Master Plan (TM&SMP) was signed in January 2007 (Trabun, 2007) and provides a road map for developing training technology requirements for fully implementing training modeling and simulation (M&S) in support of the Marine Air-Ground Task Force (MAGTF). In broad terms, the Marine Corps relies on simulation to enhance training across the continuum that begins with classroom instruction and culminates in live-fire and live-maneuver exercises. The long-term vision for Marine Corps training is that simulation will be a transparent window from the live environment to a virtual tactical environment that will provide marines from the riflemen to the MAGTF commander with faithful representation of the battle space in which to hone their skills.

Simulation training in the MAGTF of the future will begin with entry level training for marines and officers in their basic skills and primary military occupational specialty, for example, marksmanship training and vehicle operator training. After graduating from the entry level training pipeline, the marines in operating units will use simulations to train for the performance of collective tasks within their organic capabilities and also with other units. One example of such collective training would be exercising virtual close air support linking a forward air controller or joint terminal attack controller (JTAC) to a pilot in the cockpit of a high fidelity flight simulator. As a culmination of the training continuum, the many components of the MAGTF will employ live and virtual simulations together in a common scenario conducted inside the wrapper of a constructive simulation prior to deployment. The capstone pre-deployment exercise includes elements of intelligence and a geographic frame of reference that are tailored to prepare the deploying MAGTF for the specific mission for which it is to deploy. While a MAGTF is deployed, simulation will be embarked with it and connected via distributed networks to the higher headquarters and a tactical exercise control group that will support the MAGTF commander's requirement for rapid scenario generation for mission planning and rehearsal while embarked and under way.

USE CASE FOR A SIMULATION-ENABLED INTEGRATED TRAINING EXERCISE

Consider the case of a future training exercise integrated across the domains of live, virtual, and constructive training and spanning the echelons from the individual marine on the battlefield to the MAGTF commander's staff. A typical series of events might begin with a squad of marines in the 1st Marine Expeditionary Force Battle Simulation Center at Camp Pendleton, California, training in a desktop or laptop based virtual tactical decision simulation (TDS) on a patrol in a geospecific representation of a real town in a foreign country. In the course of the patrol the marines encounter a citizen of the country in which they are operating and converse with him in his native language while using appropriate cultural behaviors. The citizen of this foreign country would be represented by an artificially intelligent avatar of a human being that is indistinguishable from other human avatars controlled by the marines. This artificial intelligence (AI) person would speak the native dialect and react in a culturally appropriate manner to the actions (kinetic, verbal, and nonverbal) of the marines. In the course of the conversation with this AI entity, the marines learn of a potential high value individual (HVI) who is thought to be located in a compound outside the town. This information is relayed via a virtual tactical radio link from the marines who are physically located at the simulation center to a regimental combat operations center (COC) in the field at Camp Pendleton.

The staff in the COC receives the information on the location of the potential HVI. The suspected location of the individual is correlated from the map location in the virtual area of operations to a grid location in the training area of the continental United States training base where the COC is set up in the field. A decision is made to launch an unmanned aerial vehicle (UAV) to reconnoiter the location. Marines using a Raven-B laptop control station simulated by a virtual and constructive simulation fly a virtual UAV sortie over the area of interest and gather real time video that substantiates the presence of an HVI at the location gleaned from the conversation with the civilian in the TDS.

A decision is made to attack the compound where the HVI is located with tactical air and then follow up with a live team of marines to conduct sensitive site exploitation. The sortie to attach the target will be launched from a simulated ship offshore, but flown virtually from flight simulators at Marine Corps Air Station Yuma (Arizona). Terminal control will be provided by a JTAC on the desert floor at Twentynine Palms (California). The voice communication between the JTAC in the field and the pilot in the simulator would be via a radio frequency to voice over the Internet protocol bridge. The JTAC would designate the target with his laser designator and the intelligent targetry would sense the laser and relay the location of the laser designation wirelessly through the range telemetry system into the simulation.

As the pilot of the virtual close air support sortie comes within visual range of the target in the simulation, he will see a representation of the laser on the target that is actually taking place 200 miles away at Twentynine Palms. The pilot transmits to the JTAC "wings level," and the JTAC turns his head to the sky to visually acquire the aircraft. Through his lightweight, mini-head-mounted, mixed-reality display (receiving simulation-injected position-location information, aka position-location information data for the aircraft) the JTAC sees the virtual representation of a Joint Strike fighter against the blue background of the sky and announced "cleared hot." The pilot in the flight simulator released virtual precision-guided bombs that impact the location of the target.

The simulation tracks the bombs to the target and, upon impact, stimulates the range targetry to activate pyrotechnics visible to the JTAC. Alternatively, if the JTAC is wearing a lightweight see-through mini-head-mounted, mixed-reality display, there would be an augmented reality explosion visible to the JTAC. The JTAC reports the bomb damage assessment, and the virtual sortie returns to the ship. The constructive simulation that serves as the "wrapper" for this exercise adjudicates the effects on the target and "killed" the target entities affected by the attack. Meanwhile, a live squad of marines isolates the target area and checks for dead and survivors. They discover a badly wounded individual represented by an anatomically correct dummy that displays wounds consistent with the concussion and fragmentation of being in proximity to a bomb blast. The corpsman employs lifesaving measures, and a request for casualty evacuation is sent via radio to the COC. Documents and other relevant information are collected from the site and conveyed to the COC, and the exercise continues.

The hypothetical case described here is used only for illustration of how the elements of live, augmented, virtual, and constructive simulation might be integrated in a scenario that provides highly realistic and valuable training for marines from the corpsman, to the JTAC, to the pilot, to the staff in the COC, to the marines operating in the TDS at the simulation center. Whether orchestrating such a training scenario is worth the overhead in planning and synchronization is a decision made by the commander, but creating the training infrastructure necessary to facilitate such an event is the intent of the TM&SMP.

INVESTMENTS AND BENEFITS

The justification for M&S infrastructure is tied directly to MAGTF capability lists and MAGTF requirement lists. The decision to make an investment in a simulation for training is based on an analysis to determine whether a simulation offers an economy of risk, time, or consumable resources (fuel, ammo, and sweat) compared to alternative venues for training, for example, live fire and maneuver. After a simulation application is identified that offers the potential for efficiently training marines, the candidate system must be vetted to ensure that it is truly effective at imparting the desired skill to trainees. At the same time, the system is evaluated to ensure that it does not impart negative training.

One of the most important potential efficiencies of simulation training is realized by distributed training over a network that connects marines at geographically distant locations. Operational tempo combined with the cost of transportation of troops and equipment makes it very expensive to conduct live training exercises that span the entire MAGTF. Distributed simulation facilitates marines remaining at their home station, but interacting in a virtual environment (VE) with their counterparts across the command element, ground combat element, air combat element, and combat service support element. Robust distributed simulation exercises require a relatively high bandwidth network backbone dedicated to training. The Marine Corps presently participates in distributed training conducted over the Joint Training and Experimentation Network (JTEN) administered by the Joint Forces Command Joint Warfighting Center. Training conducted using the JTEN must be of a joint nature and be under the umbrella of the Joint National Training Capability. The Marine Corps is also beginning the task of federating some of its training simulations to operate in the Navy Continuous Training Environment (NCTE), which is similar to the JTEN, but was created specifically to conduct training of U.S. Navy and U.S. Marine Corps mission essential tasks.

While training in a joint and naval service context is essential the JTEN and the NCTE do not provide a readily available distributed training network on which the Marine Corps can conduct Title X training. For this reason the Marine Corps is studying the requirement for a training network to enable training exercises between dozens of Marine Corps installations that span the force to include the reserve component. At the present time this future capability is being referred to as the Marine Corps Training and Experimentation Network. A high bandwidth training network is a key enabler for the Marine Corps' vision of the future training capability.

CAPABILITY GAPS

The TM&SMP identifies seven science and technology (S&T) "long poles in the tent," aka S&T long poles, for achieving the objective capabilities in training simulation. These S&T challenges are identified gaps that exist in the current training capabilities of the Marine Corps. Research into potential solutions for these technological challenges is of keen interested to Training and Education Command (TECOM) Technology Division, as well as such organizations as the Office of Naval Research and the Defense Advanced Research Projects Agency. Proposed solutions to these challenges coming from industry or academia must be an open source approach to both software and hardware. The Marine Corps will not invest in proprietary technology that is not interoperable or for which full interface design documentation is not available. The following is a list of the

S&T long poles paraphrased and quoted from the TM&SMP:

1. Rapid generation of high fidelity three-dimensional (3-D) terrain databases to include contour, vegetations, hydrography, and man-made structures and equipment.

2. Targets for the Live Virtual and Constructive Training Environment that are present and interactive from the perspective of all participants in a training environment whether they are participating in a live, virtual, or constructive capacity. Targetry on a live-fire range must "come alive" by a stimulus from a virtual or constructive simulation. When a virtual representation of a target engages in a simulation, the effects on the target must be visible to observers in the live realm. For example, marines in the field instrumented with position-location information and target sensor telemetry would be notified and react appropriately if the location at which they stand is engaged by a fire for effect in the constructive simulation. Tactical command, control, communications, computers, and intelligence (C4I) systems will be stimulated by simulations that communicate via simulation-to-tactical gateways providing data feeds to the commander and his staff that are indistinguishable from operational data feeds.

3. Representations of man-made structures, particularly those in an urban environment, must be faithful replications of actual buildings and objects. These virtual structures must incorporate realist physics models of their construction so that they respond appropriately to kinetic actions, such as breaching of a wall by a vehicle or destruction of the building by ordnance. To be useful in the context of mission planning and mission rehearsal, it is a requirement that these virtual urban structures be created rapidly by the training unit.

4. Live forces must be tracked via position-location information and those tracks represented in virtual and constructive simulations both indoors and outdoors. The position-location information systems cannot be dependent on line-of-sight telemetry. Virtual representations of humans, both computer generated and those representing position-location information feed from live troops, must fully implement human anatomy motion-tracking and display. Physical movements, such as hand and arm signals, must be fully articulated in the virtual environment. Human models must exhibit appropriate cultural, emotional, religious, and ethnic responses to the stimulus from the training environment.

5. Accelerated learning science is a requirement to assist marines in rapidly assimilating knowledge on how to perform complex tasks. Specific S&T gaps include the following:

   a. Foundations of learning applied to complex tasks. Cognitive load theory and instructional efficiency must be extended to complex tasks.

   b. Training interventions triggered by neurophysical markers of learning and cognition.

   c. Principles of expertise development and strategies tailored to continual proficiency models, beyond today's simple novice to expert techniques.

      Artificially intelligent opposing and friendly forces should sense the proficiency level of the trainee and scale the difficulty of the training to a level appropriate and optimal for learning and skill acquisition.

6. Experiential learning technologies that provide an experience similar to the experience of completing the live Mojave Viper pre-deployment capstone exercise

conducted at Twentynine Palms to forces at their home stations or deployed in a virtual environment. The content of the learning experience must be tailored for the individual marine's learning aptitude and base of knowledge.

7. The injection of political, military, economic, social, infrastructure, and information nonkinetic effects into operational level staff exercises is needed to support all elements of national power, future operations, and long-term assessments. This capability must not add to the number of support personnel for a training event. Currently this aspect is war-gamed in a seminar format combined with the simulation effects of a kinetic exercise.

IMMERSIVE TRAINING

Recent experience of marines in combat, training in preparation for combat, and the considered opinion of senior operating force leadership are the basis for the Marine Corps' investment in VE training technology. The short "dwell times" spent at home-station bases between deployments compel the Marine Corps to develop immersive training to augment and improve upon live training capabilities. Virtual training at a less than immersive level facilitated by the deployable virtual training environment will continue to be an important component of training. What follows is a detailed discussion of a research initiative into the higher end of virtual training environments.

The Marine Corps is investigating a fully immersive infantry training environment that blends live participants using real weapons and equipment with augmented/virtual reality in order to provide an experience of sufficient realism that it results in a suspension of disbelief. The infantry immersive training environment seeks to (1) replicate as closely as possible the effects and conditions of the battlefield and (2) allows a dismounted infantry squad, platoon, or battalion to effectively train in the wide spectrum of tasks necessary to execute the full range of military objectives.

To sufficiently stimulate the senses of trainees to the point of immersion, this trainer must synthesize the "fog of war" through a cluttered, confusing, combatlike environment. As marines assimilate into the environment they will experience stress and fatigue and encounter more chaos and randomness (fewer linear events). The increased realism and interactivity of the training will begin to synthesize the exposure that comes from a first firefight and the nonkinetic events leading up to the baptism by fire. This training environment must stimulate all the senses and overwhelm marines into believing they are experiencing a "real, life-like" firefight.

The mixed reality training experience should include such elements of realism as the following:

- ✓ Getting smacked in the face by branches and getting a face full of spider webs while on a security patrol;
- ✓ Conducting an extended security halt on the snow north of the Arctic Circle;
- ✓ Encountering a snake that falls out of a tree on the patrol;
- ✓ Finding the lost members of a squad-sized patrol at night in heavy vegetation;

✓ Having the enemy detect and open fire on a patrol if the marines do not exercise noise and movement discipline.

Current tactical training environments are in general too sterile. By having wounded who scream, dead who smell and decay, civilians, and so forth, we can make the experience more challenging and realistic. All who are about to go into a firefight wonder how they will act and how will they handle fear or the initial panic of combat. The immersive training environment must answer those questions in the mind of the marine being trained.

PAYING IN SWEAT VERSUS BLOOD

In early air-to-air combat over Vietnam, navy pilots achieved a kill ratio against North Vietnamese MiG jets of only two to one. A careful study showed a dramatic seasoning curve increase for pilots after combat. Forty percent of all pilot losses occurred in their first three engagements; however, 90 percent of those who survived three engagements went on to complete a combat tour. In 1969 the navy began a program that sought to provide a pilot his first three missions risk-free. Top Gun pitted novice airmen against a mock aggressor skilled in North Vietnamese aerial tactics. Combat was bloodless yet relatively unfettered. Uncompromising instructors recorded and played back every maneuver and action. The results were dramatic. From 1969 until the end of the air war, the navy's kill ratio increased sixfold.

The Marine Corps is investigating immersive training for the ground infantry equivalent to the navy's Top Gun for pilots. Some of the impetus behind developing the technologies for the creation of home-station and/or deployed immersive training environments is to significantly enhance that first combat experience and provide a realistic, life-like experience as close to a first fight combat experience as possible. Although the focus is on surviving the first kinetic engagement with the enemy, the simulation environment will also encompass nonkinetic experiences of interaction with host nation civilians and local government officials. Training for the escalation of force will be a critical element of the objective virtual training capability.

While realism is paramount, the objective training capability must also be scalable to accommodate training for the entire force. The most sophisticated and realistic virtual training environment is not a solution to the Marine Corps' requirements if it is not affordable in sufficient capacity to train every rifle squad in the Marine Corps. The training environment must be interoperable with joint training simulations, especially in the exchange of C4I data feeds. The objective environment must replicate not only the employment of organic equipment, but also supporting arms and sensors.

WHY IMMERSIVE MIXED REALITY?

The goal of virtual and mixed reality training is to make the experience at home station more like a Mojave Viper experience. Immersive and mixed reality does not provide the entire answer, as we cannot fully replicate such emotions as fear and panic; however, the more senses we can stimulate in a virtual training environment to deliver the required suspension of disbelief, the more completely we can achieve the goal of learning by "living" the experience and the closer we can get to that first firefight experience and the emotions of fear and panic on the battlefield.

The true benefits that we see from immersive mixed reality are as follows:

1. Provide a more realistic and engaging environment in order to allow leaders to make decisions in a near-combat environment experiencing those stressors and stimuli that are possible to be re-created in simulation. This consequence-free environment will provide these leaders with a reservoir of experience from which to base their future decisions in combat.

2. Create a revolutionary training environment where marines can interact with not only live players, vehicles, and aircraft (real when available and virtual/augmented otherwise), but also accurately generated virtual entities in both an urban environment and in open terrain. The potential cost savings and flexibility are tremendous. One example where this capability is clearly seen is close air support conducted in an urban environment. In the current area of operations (AOs), leaders are forced to make targeting decisions based on the threat, as well as the surrounding environment. The decision to engage a target and with what weapons systems in order to eliminate the threat, as well as minimize collateral damage, can be a challenge. We must present marines with this situation in a dynamic realistic training environment. With augmented reality, we could now have the observer look down on a real village populated with both live and virtual (augmented) role-players, both civilian and hostile, and based on what he sees he could then direct either a real (if available) or virtual aircraft to deliver virtual precision-guided munitions onto the target and observe the augmented effects.

3. Another potential benefit that the use of this technology will provide is the ability to quickly change not only a scenario, but also the environment. In an interior building this could be a literal change of the climate, but more importantly it means a change of the people that the marines will encounter in this environment. Instead of having to go out and hire a completely new set of role-players from a different part of the world, it simply requires loading a different set of entities and modifying their behaviors to match the current threat seen in the new AO. In addition, as the system is used the interactions of the virtual players can be updated based on current intelligence.

Fully immersive mixed reality training will enable the Marine Corps to quickly adapt training to prepare marines for any emerging threat. Virtual (augmented) environments blended with live training offer the potential of both improved realism and cost savings compared to a live-only approach in which all of the space and structures must physically exist. As the live elements of training remain constrained by location, the virtual (augmented) elements can change the context and tailor the experience to the demands of the mission.

REFERENCES

Trabun, M. A. (2007, January 18). *U.S. Marine Corps Training Modeling and Simulation Master Plan.* Quantico, VA: U.S. Marine Corps, Training and Education Command, Technology Division.

Chapter 35

THE FUTURE OF VIRTUAL ENVIRONMENT TRAINING IN THE ARMY

Roger Smith

VE GROWTH STIMULANTS

There is a rich history in researching and developing virtual environments (VEs) within the military. The simulator networking (SIMNET) program of the late 1980s and early 1990s demonstrated the deep value of virtual environment applications (Miller & Thorpe, 1995; Davis, 1995; Singhal & Zyda, 1999). Twenty-five years later there have been significant advances in this area, but there remains vast unexplored potential in this field. There are potentially hundreds of valuable applications to real military operations in logistics, command and control, situation understanding, and information fusion. In both the commercial and the military worlds, the power of VEs is significantly enhanced by the growing availability of digital data in every industrial and government domain. In a world where reconnaissance photos are captured on physical film, there is little that computation and VE can do to enhance this information. Once those photos become digital, it is possible to analyze, fuse, integrate, and morph them so that they become the visible skin of a VE. As most information about the world becomes digital, it creates opportunities to generate higher levels of understanding and new advantages over competitors. As the world has become networked, digital data have also become globally accessible so that digital photos from every continent can be viewed in real time anywhere in the world. As network bandwidth, computational power, and VE algorithms advance, there will be a point at which these images can be stitched together into a seamless three-dimensional (3-D) map of the entire world and navigated in real time. These data will include digital images, sound waves, weather patterns, population densities, personal locations, radio frequency spectrum, financial transactions, and dozens of other specializations.

From a military perspective, most situations of interest are geographically based. In the past, our technologies have limited our ability to construct information into a geographic form similar to the world from which it was collected. The

VE is a new and powerful alternative to textual, graphic, and other paper-oriented representations that have dominated our decision making for centuries. Today's leaders, managers, and engineers are very comfortable communicating information that has been structured in the form of graphs and tables. The next generation will be just as comfortable structuring information into unique VEs and exploring those collaboratively as a means of understanding and manipulating the world.

COMMERCIAL LEADERSHIP

Sometime in the late 1980s there was a tipping point (Gladwell, 2002) at which commercial industry took the lead from government laboratories in advancing computer technologies. The explosion of consumer-grade computing power led to a corresponding explosion in software applications that could exploit this power. One of these growth areas was the computer gaming world that created such products as *Quake* and *Unreal* and an annual harvest of new competitors presenting the best VE rendering available at consumer price levels. This civilian market will continue to drive research and development into VEs and the creation of ever-more beautiful and immersive worlds in which to interact with information and other people (Smith, 2006; Dodsworth, 1998).

Just as e-mail and instant messaging have replaced the telephone as the leading medium for personal communication, and the Web has replaced the library as the leading repository of information, VEs will replace the textual Web page as the primary medium for shopping, socialization, and exploration. VEs can capture both the contextual relationships of hyperlinks and the proximity relationships of geographic collocation. Some form of VE will become the context within which online digital information is organized, significantly extending the linked, flat Web pages that convey this information today. People who are browsing through data will be able to discover related items that are geographically close to each other just as they do when browsing in a physical library or bookstore. Such applications as Google Earth, *Second Life,* and World Wind are beginning to illustrate this future. Imagine a World Wide Web in which all personal information is tied together in a single context. For example, a social network of friends live as 3-D avatars in a VE apartment where favorite video clips are streaming on one wall and the contents of an online encyclopedia are lying on a coffee table. Further, in a VE there is no reason that the apartment has to look anything like a physical habitat; it could be a giant garden, forest, cloud city, or ant colony. The information that people need and enjoy may grow like flowers in the garden all around them, their colors and sizes representing currency, importance, source, or other key attributes.

Most commercial VE expressions are uniquely personal, playful, and civilian, but the technologies behind them are seriously powerful. Like the radio and the semiconductor before them, these technologies are not limited to entertainment, business, or national defense, but can be applied equally to each domain. The commercial world will be the source from which advanced VE technologies spring and the foundation from which military applications are built.

Though computer-generated VEs are primarily visual, there may be other alternatives to loading information into the human mind. Direct neural stimulation may allow information to enter without going through the eyes. Technology that enables a blind person's mind to "see" is similar to that required to generate a VE directly in the mind. The advantages of this approach are beyond current understanding. A neural image may be superior to a standard visual scene. It may create a new sense of the data that are contained in the world, effectively enhancing the human ability to perceive rich mixtures of data within a VE.

Further a field is the possibility of creating or enhancing the VE through the use of chemicals. It may be possible to chemically stimulate the brain to construct useful representations of information. The 1960's experiments with LSD (lysergic acid diethylamide) cast a dark shadow over these kinds of experiments, but new research into chemically enhancing athletic and soldier performance is bringing these ideas back into vogue. Just as caffeine can enhance alertness and reaction time, other chemicals may improve understanding of information that is part of combat operations or that drives training for life-threatening missions.

VE APPLICATIONS

The term "serious games" is often used to describe the application of game technologies to military or industrial problems. This has been a useful term, but it will become archaic as the distinction between game technology and nongame applications fades away. Computer chips and graphics cards are not referred to as "entertainment chips" or "serious graphics cards." They are just tools for constructing useful applications. The same will occur with serious games. All industries will have VEs that meet their needs, just as they have specialized computing and communications devices today (Bergeron, 2006; Lenoir, 2003).

Since 1992, the military has identified its simulation tools as live, virtual, or constructive. This delineation has highlighted the computational and conceptual limitations in representing both breadth and depth in a VE. "Virtual" refers to the use of simulated objects by real humans, and these systems have typically represented small areas with few objects at relatively higher levels of detail. "Constructive" refers to the use of simulated objects by simulated people, and these systems have represented very large areas and many objects with relatively less detail. In the years since these definitions were standardized, advances in computation have enabled the creation of many systems that combine one or more domains. Further advances in computation, communication, and conceptualization will allow us to stretch the boundaries of these domains so that there is little difference between them. In the future, constructive and virtual will refer only to the view that is being presented to the human or to an artificial intelligence, not to any inherent limitation of the models that are driving the virtual world.

There have been three distinct generations of "constructive simulation," and perhaps future VEs will create a fourth. The first was the use of sand tables and miniature figures, essentially a scaled representation of the real battlefield. The second was the paper board game that allowed greater abstraction and additional rigor in the rules and mechanics of behavior. The third was the computerization of the war game that extended the algorithms to the limits of the computer rather than the limits of a human player (Allen, 1989; Perla, 1990; Dunnigan, 1993). Advances in VEs will enable the creation of a constructive simulation that is just as detailed as any virtual simulator if so desired. It will employ aggregation and abstraction as a useful metaphor rather than as a core design limitation driven by limited computational power.

In the "live" domain there will be VEs embedded in real equipment just as two-dimensional map displays exist in equipment today. These VEs will be integrated into the control screens and head-mounted displays that are currently portals into flat, disassociated, two-dimensional data. Rather than seeing the battlefield from a top-down, two-dimensional view, the operators will be able to see it in three dimensions from any angle that they find useful. This is a hugely powerful paradigm and carries so many potential options that the challenge will be in determining where the valuable views lie, not in rendering and animating them for the operator. In this world, there will be little difference between the objects that come from a simulation and those that exist in the physical world. All of them will be seamlessly integrated into a VE.

ARMY MISSIONS

VEs are supplemented with physical and cognitive models, software management and control tools, and external interfaces to operational devices to create simulation based training systems. As the nature of the army mission has changed, simulations and VEs have been challenged to represent new missions, new threats, and new tactics that capture the essential elements of the real world and can be used to teach this reality to humans. We have emerged from four decades of a Cold War in which most military training focused on large combat operations that occured on specified battlefields where all participants were expected to be combatants. More recent missions have focused on small units in an urban environment where they must perform humanitarian operations, search and reconnaissance, facility defense, and combat operations all on the same day. This has created a situation in which our VEs and simulations are expected to represent a much more diverse set of objects and interactions. These can no longer be "combat only" models of the world. The focus of current and future missions appears to be on much smaller areas, making it both possible and desirable to deliver very high levels of detail in the area of operations. This detail calls for a VE that can re-create combat operations in a single city block, but also allow personal communications with the populace to build an understanding of the societal factors surrounding the military operations. These factors will trigger important actions and reactions as the simulation progresses. Many of the current simulation models focused on immediate action and immediate consequences. In most cases, these actions and/or consequences are discrete and do not influence actions between objects in the future.

While the military simulation community has been wrestling with models of information processing and human reaction, it has just begun to explore the richness of person-to-person relationships and their influence over different groups within a population. There is a great deal of "soft social science" that needs to be incorporated into VEs in the future. Accurate physics models of weapon penetration and aircraft lift remain important, but a useful understanding of the urban battlefield is driven by human interactions, motives, and group dynamics. In the past, military simulation systems have been able to focus on the universal and verifiable behavior of the physical world. But models of personal relationships and group behavior are highly cultural, social, and geographical. Huntington (1996) has suggested that all future competitions will be based on seven unique cultures that have emerged in the world: Western, Orthodox, Latin American, Muslim, Hindu, Sinic, and Japanese. Rather than a bipolar world threatened with traditional combat, we live in a more complex world in which the confrontations may be focused in the political, military, economic, social, infrastructure, or information domains and involve seven different and powerful cultures. VEs that are able to represent such a diverse world accurately and effectively will be a significant challenge and a significant focus in the future.

ADVANTAGES

VEs that are created electronically, biologically, or chemically all present significant advantages for military operations and training. They create an improved space for accessing, absorbing, understanding, and applying information. These are all information based terms that create a pattern very similar to the observe, orient, decide, and act loop that was first proposed by Col. John Boyd (Coram, 2004).

The advantages to be gained are so significant that VEs will continue to grow in importance and in the breadth of their application. Specialized versions of VEs will be used for hundreds of different applications, each with a unique focus, but built on a core set of technologies. As the limitations of computer and communication technology fall away and our level of expertise in creating and manipulating these environments increases, VEs will appear in all types of consumer and military systems to aid people in making better decisions and taking more appropriate actions. VEs combine technologies that have been maturing in the training, entertainment, computer science, and communications domains for several years and have reached a point at which they can be adopted by hundreds of commercial and government organizations.

REFERENCES

Allen, T. (1989). *War games.* Berkeley, CA: Berkeley Publishing Group.

Bergeron, B. (2006). *Developing serious games.* Boston: Charles River Media.

Coram, R. (2004). *Boyd: The fighter pilot who changed the art of war.* New York: Little, Brown, & Co.

Davis, P. K. (1995). Distributed interactive simulation in the evolution of DoD warfare modeling and simulation. *Proceedings of the IEEE, 83*(8), 1138–1155.

Dodsworth, C. (1998). *Digital illusion: Entertaining the future with high technology.* New York: ACM Press.

Dunnigan, J. (1993). *The complete wargames handbook: How to play, design, and find them.* New York: William Morrow.

Gladwell, M. (2002). *The tipping point: How little things can make a big difference.* New York: Little, Brown, & Co.

Huntington, S. P. (1996). *The clash of civilizations and the remaking of world order.* New York: Touchstone Press.

Lenoir, T. (2003). Programming theatres of war: Gamemakers as soldiers. In R. Latham (Ed.), *Bombs and bandwidth: The emerging relationship between information technology and security* (pp. 175–198). New York: The New Press.

Miller, D. C., & Thorpe, J. A. (1995). SIMNET: The advent of simulator networking. *Proceedings of the IEEE, 83*(8), 1114–1123.

Perla, P. (1990). *The art of wargaming.* Annapolis, MD: Naval Institute Press.

Singhal, S., & Zyda, M. (1999). *Networked virtual environments: Design and implementation.* New York: ACM Press.

Smith, R. (2006). Technology disruption in the simulation industry. *Journal of Defense Modeling and Simulation, 3*(1), 3–10.

Chapter 36

FUTURE AIR FORCE TRAINING[1]

Daniel Walker and Kevin Geiss

Aviation pioneer Wilbur Wright (1900) stated, "It is possible to fly without motors, but not without knowledge and skill." Our vision for the future emphasizes the capabilities of operators, not simply the hardware that confines them. Future air force training will be driven by expected operational requirements involving personnel extensively connected to their weapon systems, other operators, and coalition forces in a global environment. From a research perspective, we observe weapon systems that are increasingly capable and complex. Reflecting these advances, the future of air force training is live, virtual, and constructive (LVC): "live" personnel and equipment, "virtual" simulated adversaries and environments, and "constructive" computer-generated entities.

OPERATIONAL CONTEXT

Operational Roles, Policy, and Doctrine

Transformation, evolution, adaptation—the operational roles of air forces—are adjusting along with the nature of conflict. The Department of Defense (2006) Quadrennial Defense Review reemphasized the necessity and value of transforming training to account for the shifts from conventional or symmetric conflicts to asymmetric and unconventional engagements that go beyond traditional kinetics based operations and now focus on such areas as cyber warfare and humanitarian operations. Robust training systems must accommodate future weapon systems along with the makeup and tactics of future adversaries in diverse global operational contexts. The exact makeup of adversaries 20 years hence is unknown, but we do know that technology will advance the capabilities of our forces, as well as those of our adversaries.

To enable effective operations, training methodologies require incorporation of advances in both technology and doctrine. In this context, air force personnel participate in military operations through a variety of weapon systems beyond

---

[1]Disclaimer: This manuscript reflects independent views of the authors and is not an official opinion of the U.S. Air Force. Approved for public release WPAFB 08-0027.

inhabited aircraft to include autonomous and semi-autonomous aircraft, space, missile, and ground systems. These systems are further functionally integrated with special operations, stability operations, and information operations. Distributed mission operations are discussed elsewhere in Volume 3 (see Andrews and Bell, Volume 3, Section 1, Chapter 8) and reflect a key evolution in the operational framework.

System Attributes and Capabilities

Military weapon systems continue to separate individual operators from ultimate mechanical events. Pilots no longer push a stick connected to a wire for manipulating a wing aileron, but rather they manipulate electronic interfaces sending digital commands to control uninhabited vehicle systems. The essential competencies required for such tasks may differ. However, for some systems, physical separation is mirrored by cognitive integration that imbeds humans in technological systems. The manner by which work is divided between human and machine is increasingly complex. A recent National Research Council report (2008, p. 30) asserts that for "today's aircraft" it is now impossible to precisely assign "the percentage of responsibility to humans or machines." Thus, careful analysis is required to determine for which tasks the human must be trained and how the human is integrated into the virtual environment (see Barnett, Volume 3, Section 1, Chapter 3).

Air force weapon system technologies are becoming so diverse and powerful that training, testing, and skill maintenance will increase demands on training and simulation systems. Ackerman (2006) describes one such weapon system, the F-35 joint strike fighter (JSF). JSF targeting capabilities utilize substantial sensor and information fusion, including electro-optical targeting and scanned array radar. The JSF tracks all aircraft within a 10 mile radius and integrates information from 1,000 independent scanning radar arrays, which may be tracking unmanned aerial vehicles (UAVs), missiles, or moving ground targets. To support the currency needs of the JSF operator, training systems must provide innumerable variants on key dimensions (for example, weather, adversaries, and weapon systems). The JSF ultimately requires pilots to take on the additional duty of "chief information officer." Through interacting with other JSFs, one aircraft has the ability to perform a mission by relying substantially on information provided by a second aircraft (Ackerman).

Resource Constraints

Two main resource constraints are driving greater implementation of virtual and constructive simulations for training. First, with weapon system capabilities, such as the JSF, it is difficult to put enough real assets in play to fully train a pilot. There is neither sufficient airspace nor enough capable live adversaries routinely available to enable training operations for the pilot of such aircraft, and certainly not for a whole squadron. Second, military operations are costly and fuels are precious commodities. Using funds or fuel for training rather than operations becomes a difficult decision. The cost-benefit/effectiveness analysis presented by Moor, Andrews, and Burright (2000) indicates that simulator based aircrew training is a valid alternative for the development of training strategies and requirements in light of these resource constraints.

APPROACH

Creating the training tools and strategies to improve warfighter performance using LVC operations demands development in five areas: competency based assessment, performance measurement, continuous learning, cognitive modeling, and immersive environments.

Competency Based Assessment

Consistent with other armed services, air force personnel receive primary training via traditional formal instruction courses. Advanced and continuation training is often administered using different approaches. For example, in the Ready Aircrew Program (RAP) discussed by Colgrove and Bennett (2006) aircrews train with a frequency and event based system to maintain proficiency through specified numbers and types of events.

One consequence of a RAP for performance improvement or maintenance is its limited assessment capabilities. The assessment is conducted in two often uncorrelated parts. The primary assessment is conducted by tracking event numbers and frequency. Personnel could be deemed not mission ready by virtue of completing too few events or by exceeding a predetermined period between events. The second part of the assessment is a subjective evaluation of crew member mission competency. If the required events are not performed well, or the crew member appears incapable of succeeding in a designated mission, a supervisor could disqualify him or her. Simply performing the required events in the appropriate time period may be indicated as satisfactory training. A crew member might be deemed mission ready without any linking to qualitative assessment, since poor performance is not tracked by this method. In practice, these subjective assessments are not regularly conducted, and other methods, such as supervisor observation or self-reporting, are required to validate a need for further training.

An alternative for aircrews is to use a competency based system versus simply accomplishing a required number of events. Competency based assessment requires detailed mission essential competency[2] (MEC) evaluation, which is being instituted for many aircrew combat specialties. The MEC process determines the knowledge and skills, not just tasks, required for proficiency in a mission. Research presented by Colgrove and Bennett (2006) showed that MEC based training produces favorable results. For example, one aerial defense

---

[2]The phrase mission essential competency, mission essential competencies, and associated acronyms have been service marked. Air Combat Command, Air Force Research Laboratory, The Group for Organizational Effectiveness, Inc., & Aptima, Inc. are the joint owners of the service mark.

scenario study of MEC based training effectiveness showed 63 percent fewer enemy bombers reached their target, 24 percent more enemy fighter aircraft were killed, and friendly aircrews suffered 68 percent fewer simulated mortalities. Ensuring aircrews can perform such skills requires improving the measurement system.

Performance Measurement

Technological advances, fielded and under development, provide promise to capture the objective metrics to enable meaningful evaluation and tailored training. One advancement is evident in simulator based training; high fidelity simulation testbeds collect over 750 different performance parameters every 50 milliseconds. This dense data environment provides one component of a performance evaluation and tracking system. Schreiber, Watz, Neubauer, McCall, and Bennett (2007) describe this system as an emerging set of performance measurement strategies and tools to support competency based continuous learning. It includes subject matter expert observer assessments using behaviorally anchored grade sheets and objective measurements based on data from simulation or live operations.

Robust and extensively instrumented live training would enable data collection similar to simulator environments. Important parts of this future environment for collecting objective performance data are available within the weapons system, but not generally transmitted on instrumented training ranges. Efforts are under way to develop live, virtual, and constructive techniques to capture those data, such as internal cockpit switch positions. Gathering comparable performance data from virtual and live experiences will enable seamless training for aircrews irrespective of domain. When procedures are in place to gather detailed performance data in all training events, then it will be possible to more efficiently tailor training to specific individuals rather than the "one size fits all" approach of many continuation training regimens. For further discussion of aviation training, see Schnell and Macuda, Volume 3, Section 1, Chapter 12.

Continuous Learning

A recent Defense Science Board report (Department of Defense, 2003) recommended that traditional schoolhouse training be replaced with continuous training employed on-site with the individual. LVC environments introduce the concept of the transparent venues with an added opportunity that such tools could support both training and operations, allowing personnel to take advantage of nonmission time for training. The continuous learning strategies we foresee go beyond simply "on-the-job training" and should become a standard feature of military systems.

Conventional job based training reflects learning during the course of normal duties rather than a situation where the operator is unable to discern the training events from normal mission events. Admittedly, even laboratory experiments have not achieved completely seamless integration of simulation and operations for complex weapon systems, yet the value of continuous training and performance assessment is apparent. Hancock and Hart (2002) discuss one simplistic example of the integration of training, competency assessment, and operations. The Transportation Security Administration uses the Threat Image Projection software program where the performance of individual screeners in detecting weapons and explosives by X-ray imaging is evaluated continuously. This approach also allows the system to integrate up-to-date intelligence on specific threats. Likewise, the power of constructive simulations would allow training system designers to incorporate the latest information (for example, threats or terrain data) into training.

Cognitive Models

Cognitive models for replicates and imbedded tutors are additional elements for enhancing mission-effective performance training. Cognitive model products are projected to shape service training. One approach is to develop models for performance prediction. Research models can account for the effect of training frequency on models of memory and may allow commanders to predict performance for specific training regimens. Jastrzembski, Gluck, and Gunzelmann (2006) propose that these predictions could then be used to determine effective application of limited training resources while having the greatest impact on improving individual crew member performance.

Ball and Gluck (2003) present one pathfinder effort for advancing computational replicates, the development of a Predator UAV pilot computational model. The researchers first created a synthetic task environment (STE) tool to simulate operation of the Predator aircraft. The STE includes aircraft performance simulation and three synthetic tasks: basic maneuvering, landing, and a reconnaissance problem requiring sensor positioning over a target within given constraints (for example, wind, cloud cover, and flight path restrictions). In this STE, various cognitive models were developed in an effort to replicate human performance in dynamic and complex tasks. As this foundational work is expanded, future training strategies will include models of synthetic adversaries and allies.

Well-developed models will provide a richer training experience than current rule based constructive simulations. Synthetic adversaries and allies will continue to be an important part of air force training for a number of reasons. As discussed above, modern weapon systems need large, complex scenarios to fully exercise their capabilities. Live adversaries and allies are expensive and less available due to shrinking force structure. Also, peacetime training restrictions (for example, range, space, and speed) decrease the effectiveness of live adversaries when matched against our most advanced systems. As Gluck, Ball, and Krusmark (2007) contend, computational replicates, when fully developed and deployed, offer greater flexibility as they can be modified more cheaply and perhaps more effectively than hardware-intensive live weapon systems.

Live   Virtual   Constructive

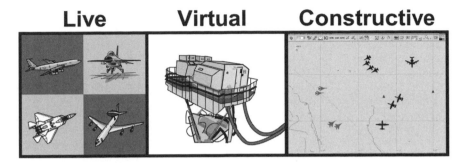

Figure 36.1.   This graphic depicts three elements of future air force training technology systems—live, virtual, and constructive.

Immersive Environments

As weapon systems continue to diminish the barriers between human and machine, training systems must follow suit. Continued advancements in training technology toward LVC environments (see Figure 36.1) can provide enhancement of immersion through sensory fidelity. Maximizing this fidelity by using more operational equipment may obscure the perception of an active training environment versus an actual mission. Mixing live and simulated entities in the same domain can challenge the situational awareness of participants, although preliminary research has discovered effective mitigation techniques. For instance, Hughes, Jerome, Hughes, and Smith (Volume 3, Section 2, Chapter 25) discuss aspects of integrating terrain data in simulations. Other concerns relate to operating with differing security levels, simulation hardware, and fidelity requirements. Governments and industry will have to continue to work toward standards for data protocols and multilevel security in order to realize an effective coalition immersive environment.

CONCLUSION

We have seen increasingly immersive operational environments, such as the merged information stream for the JSF pilot. Because of the data sharing capabilities of the JSF (sensor data and information provided from one platform to another), the pilot may never actually see the target before or after weapon deployment. Thus, in a training scenario, a virtual adversary may be inserted that cannot be distinguished from a live asset. Conceptually, what we are describing is a convergence of perceived experience: a training environment that increasingly incorporates both constructive and real entities, and likewise, a real world activity that is integrated with simulation and training-specific tasks. The future of air force training will be enabled by continued advancements in live, virtual, and constructive environments.

REFERENCES

Ackerman, N. D. (2006, October). Strike fighter partners with pilot. *SIGNAL Magazine.* Retrieved November 11, 2007, from http://www.afcea.org/signal/articles

Ball, J. T., & Gluck, K. A. (2003). Interfacing ACT-R 5.0 to an Uninhabited Air Vehicle (UAV) Synthetic Task Environment (STE). *Proceedings of the Tenth Annual ACT-R Workshop and Summer School.* Pittsburgh, PA.

Colegrove, C. M., & Bennett, W., Jr. (2006). *Competency-based training: Adapting to Warfighter needs* (U.S. Air Force Research Laboratory Publication No. AFRL-HE-AZ-TR-2006-0014). Retrieved October 30, 2007, from http://handle.dtic.mil/100.2/ADA469472

Department of Defense. (2003). *Defense Science Board Task Force on Training for Future Conflicts* (Final Report).

Department of Defense. (2006). *Quadrennial Defense review report.* Retrieved October 30, 2007, from http://www.defenselink.mil/qdr/

Gluck, K. A., Ball, J. T., & Krusmark, M. A. (2007). Cognitive control in a computational model of the Predator pilot. In W. Gray (Ed.), *Integrated models of cognitive systems* (pp. 13–28). New York: Oxford University Press.

Hancock, P. A., & Hart, S. G. (2002). Defeating terrorism: What can human factors/ergonomics offer? *Ergonomics in Design, 10,* 6–16.

Jastrzembski, T. S., Gluck, K. A., & Gunzelmann, G. (2006). *Knowledge tracing and prediction of future trainee performance.* Paper presented at the 2006 Interservice/Industry Training, Simulation, and Education Conference, Orlando, FL.

Moor, W. C., Andrews, D. H., & Burright, B. (2000). Benefit-cost and cost effectiveness systems analysis for evaluating simulation for aircrew training systems. In H. F. O'Neil & D. H. Andrews (Eds.), *Aircrew training and assessment* (pp. 291–310). Mahwah, NJ: Lawrence Erlbaum.

National Research Council of the National Academies. (2008). *Human behavior in military contexts.* Washington, DC: National Academies Press.

Schreiber, B. T., Watz, E., Neubauer, P. J., McCall, J. M., & Bennett, W., Jr. (2007). *Performance evaluation tracking system.* (Available from AFRL/HEA, 6030 South Kent Street, Mesa, AZ 85212)

Wright, W. (1900, May 13). Personal letter to Octave Chanute.

Chapter 37

FACTORS DRIVING THREE-DIMENSIONAL VIRTUAL MEDICAL EDUCATION

James Dunne and Claudia McDonald

A PERFECT STORM

A perfect storm of adverse factors compels the search for innovative means to provide experiential learning rooted in critical thinking, not only for degree based education, but also as continuing education for medical practitioners. Research and development of three-dimensional (3-D) virtual learning platforms coordinate an interdisciplinary response to these adverse factors, which include widespread medical error, imminent shortages of medical personnel, a shift driven by demographics in the nature of medical care, rapid evolution in military medicine, and challenges posed by mass-casualty terrorism.

The Institute of Medicine of the National Academy of Sciences estimated in 1999 that 44,000 to 98,000 people die annually from avoidable medical errors and that costs associated with these deaths ranged from $17 billion to $29 billion (Kohn, Corrigan, & Donaldson, 1999, pp. 1–2). The report distinguished between latent errors, whole classes of mistakes waiting to happen due to defects in the complex system of medical care, and active errors, individual miscues, most of which stem from system defects. The report calls for systemic reform in order to solve latent errors rather than enhanced vigilance to reduce errors at the point of contact. In essence, according to the report, fixing the complex system of medical education in America will reduce the overall number of errors (Kohn et al., p. 146).

The institute's call for broader and deeper training and continuing education carries with it a degree of difficulty compounded not only by "a broadening array of topics," but also by other factors, including shorter hospital stays and residents' workweek, which are reducing clinical training opportunities and expertise development (for example, Verrier, 2004, p. 1237); and baby-boom retirements from academic faculties and other demographic factors, which are creating looming shortages of medical personnel, especially physicians and nurses (Rasch, 2006, p. 3).

Modeling and simulation are generally recognized in the medical community as remedies for the shortcomings of degree based health-care education curricula that do not provide sufficient clinical experience. A U.S. Food and Drug Administration panel, for example, "has recommended the use of virtual reality simulation as an integral component of a training package for carotid artery stenting," according to the *New England Journal of Medicine* (Reznick & McRae, 2006, p. 2667). Reznick and MacRae conclude, "Given the advances in technology and the accruing evidence of their effectiveness, now is the time to take stock of the changes we can and must make to improve the assessment and training of surgeons in the future" (p. 2668).

SHIFTING SANDS: WARFARE AND TERRORISM

The shape and nature of military conflict has changed dramatically, from traditional "theater" engagement to what Bilski et al. (2003, p. 814) call "the ever-changing nonlinear battlefront." A new warfighting tactic developed since the mid-1990s, expeditionary maneuver warfare, has driven military medicine to develop the forward resuscitative surgical system (FRSS), mobile, highly trained teams of surgeons and paramedical personnel who establish trauma treatment facilities within 10 miles of enemy contact (Chambers et al., 2005, pp. 27–30). Such forward-area medical intervention is necessary because expeditionary tactics and warfighting often increase evacuation times to major medical facilities; moreover, surgical practice in the FRSS differs significantly from that of institutional settings wherein surgical care seeks not only to keep patients alive through "damage control," but also to effect definitive treatment of injuries.

Terrorism has evolved to become a military tactic, as well as means of attacking and destabilizing civilian populations. The use and development of improvised explosive devices and other "weapons of opportunity" (Ciraulo & Frykberg, 2006, p. 943) create rapidly evolving generations of blast wounds, including soft-tissue damage and amputations, for which most military and civilian physicians are not prepared by clinical experience.

Terrorism also has struck the United States. Studies since the 1995 bombing of the Alfred P. Murrah Federal Building in Oklahoma City, Oklahoma, and the 2001 jetliner hijackings with devastating effects in New York City and Washington, D.C., show that most physicians are not well prepared for mass-casualty incidents (for example, Treat et al., 2001; Galante, Jacoby, & Anderson, 2006). There have been significant expenditures since 9/11 in training first responders, but there has not been a corresponding provision for training physicians to deal with mass casualties stemming from a terrorist attack.

Major training issues for civilian and military medicine flow from these developments as global conflict rooted in terroristic attacks looms on the horizon. Routine training required for effective FRSS operations, for example, is simply not available through most military medical institutions (Schreiber et al., 2002,

p. 8). Mass-casualty incidents are rare enough in the United States that few physicians have clinical experience in dealing with the triage, let alone treatment, of complex blast injuries (Ciraulo & Frykberg, 2006, p. 948). Few civilian medical facilities have either the equipment or clinically trained staff to handle a major bioterroristic attack, the victims of which may not begin appearing at hospitals and clinics for weeks after their exposure (Treat et al., 2001, p. 563).

Developing clinical education in virtual space is consistent with the military's long-standing commitment to simulation as an effective part of its training mix. Zimet, Armstron, Daniel, and Mait (2003) observe:

> With considerable assistance from the electronic game and entertainment industry, coupled with virtual reality environmental trainers, training systems correspond with actual combat to an unprecedented degree. Training software now is embedded in actual equipment, allowing continuous training on station. In addition, warfare itself has moved from the mostly physical to the mostly mental demands of information management and decision-making; thus, virtual training particularly approaches operational conditions in information age warfare.
>
> (¶36)

A NEW LEARNING PARADIGM: VIRTUAL MEDICAL EDUCATION

Virtual medical education research and development calls into play expertise in the fields of medicine, medical education, computer science, software engineering, physics, computer animation, art, and architecture to collaborate with the commercial gaming industry in producing research based virtual learning platforms based on cutting-edge computer gaming technology.

Virtual medical education is an initiative for improving the assessment and training of future medical and health-care personnel. Three-dimensional virtual environments can be used for clinical learning by all health-care disciplines to supplement traditional didactic materials and methods by providing iterative clinical training that poses no threat to patients, even as it enhances critical thinking. Such learning platforms also take into account the remarkable development of computer technologies as tools for teaching the "Net Generation" born since 1982, of which 89.5 percent are computer literate, 63 percent are Internet users, and 14.3 percent of whom have been using the Internet since age four, according to the U.S. Department of Commerce (2002, p. 43).

Virtual medical education also can hone clinical skills for medical cases not usually encountered in actual environments, for example, avian flu epidemics and attacks of bioterrorism. Clinical experience for these events must be simulated to be learned; in theory, 3-D virtual space provides the most effective means for delivering asynchronous, iterative, clinical training, anytime and anywhere as an in-depth complement to traditional didactic materials.

Three-dimensional virtual learning platforms, armed with intelligent tutors utilizing artificial intelligence to monitor user performance, can provide immediate feedback in the form of instructional material simultaneous with the user's encounter with a virtual case. Case-editing systems with user-friendly interfaces can provide instructors with flexibility to design cases consistent with an institutional setting and educational goals consistent with problem learning and the procedural requirements of credentialing agencies. Virtual environments, moreover, are customizable and can be authored, as cases are, to be consistent with site specifics.

The development of three-dimensional virtual simulation comes at a moment in the history of American health care when the paradigm is shifting from cure to care as the population ages. The current system is built on the concept of cure, but that does not reflect current U.S. demographic data and society dominated by aging baby boomers living longer than any previous generation due to pharmaceutical intervention, surgical advances, and the successful treatment of chronic diseases. U.S. medicine is evolving a new concept of care that has implications for medical education and underscores the need for virtual learning space. Medical practice in the future will become more complex as the concept of care for older patients matures and technology continues to advance. Virtual simulation of such complexity makes possible the kind of education and training that will rise to these challenges.

Coalescing adverse factors in health-care education—fewer clinical opportunities, less time for clinical training, and declining medical and nursing school admissions—compel the search for alternative means not only of degree based education, but also continuous education for practitioners. Virtual medical learning platforms may provide cross-disciplinary expertise and resources required to meet these looming health-care crises.

Virtual medical education research may have implications for disseminating cutting-edge medical knowledge to economies and cultures throughout the world that have not developed sufficient infrastructure to provide adequate clinical experience in traditional curricular formats. Virtual medical learning platforms are conveniently deliverable by various electronic media, playable on sufficiently powered and configured computers, and quickly utilized through the application of user-friendly tutorials and training routines.

THE WAY FORWARD: CENTERS FOR VIRTUAL MEDICAL EDUCATION

Sophisticated learning in virtual space is, so far, just a theory. To be credible, virtual medical education must be based on rigorous research and testing to establish its validity and reliability. Centers for virtual medical education, such as one proposed at Texas A&M University–Corpus Christi, would provide cross-disciplinary expertise and resources to educational, governmental, and business entities engaged in meeting looming health-care crises with three-dimensional virtual learning platforms that are iterative, providing unlimited, repeatable clinical experience without risk to patients; portable, for training anywhere there is a computer; asynchronous, for training anytime; and immersive, providing first-person experience leading to critical thinking and practical knowledge.

Virtual learning platforms must be grounded in research findings and equipped with tools and generators that enable clients to author their own cases and create their own scenarios within a variety of virtual environments. The platform in development at Texas A&M University–Corpus Christi, for example, is being rigorously researched and developed and extensively tested for reliability and validity, which may be expected to yield a product for delivering curricula with confidence for medical and other health professions.

Virtual medical education research must be continuous as products are refined with successive generations of computer electronics, especially game-development technologies. Sophisticated learning platforms currently are pushing commercial developers to produce true-to-life images not previously achieved for entertainment purposes. Total fidelity in replicating physiological and pathophysiological states in virtual space is key to the success of these technologies as pedagogical and training tools, for only then can students' clinical experiences in the learning platform be truly immersive and true to life.

Virtual medical education also must develop an entrepreneurial dimension through collaboration with other entities that will generate revenue to support continuing research. Developing three-dimensional virtual learning platforms is expensive, requiring from-scratch funding of no less than $30 million. The good news is that, as sophisticated learning platforms are developed, subsequent refinement will not require reinventing the wheel. Virtual medical education researchers will license their proprietary software to others for far less than it would require to stake development from scratch. Over the long term, case-development costs may be expected to decrease due to economies of scale, even as clinical cases become more complex and the demand for sophisticated visual fidelity pushes beyond current gaming industry standards; however, in the short term, it is not unlikely that case-development costs will remain relatively high for the foreseeable future.

In the beginning—and we are at the very beginning of developing these learning tools—the U.S. military may be key in funding research programs leading to the development of valid, reliable, sophisticated learning platforms. The research and development field of virtual medical education would be a pool of training resources for military medical training, professional certification and credentialing, professional development, graduate medical education, and improved joint-force military deployment. By their very nature, such training products will produce efficiencies of operations and economies of scale for joint military medical activities as a source and distribution point for clinical training materials transmitted electronically to joint operating forces anywhere in the world.

CONCLUSION

Three-dimensional virtual learning platforms are the right thing at the right time in medical education—technological development meeting critical needs and generating pioneering research. Virtual medical education will coordinate a diverse array of academic disciplines in an interface with government and business interests toward understanding and validating the dynamics of learning in virtual space, which will benefit not only medical education, but higher education in general. Students will learn, but more than that, new opportunities and career paths will evolve as the theoretical benefits of learning in virtual space are better understood and its capabilities—and limitations—become clear. The gaming industry recognizes that so-called "serious games" are on the growing edge of this dynamic sector, but it will take collaboration with the medical academic community to substantiate claims that learning platforms in virtual space can provide valid and reliable educational strategies for high level critical thinking and clinical skills.

Contributor: Ron George, Texas A&M University–Corpus Christi

REFERENCES

Bilski, T. R., Baker, B. C., Grove, J. R., Hinks, R. P., Harrison, M. J., Sabra, J. P., et al. (2003). Battle casualties treated at Camp Rhino, Afghanistan: Lessons learned. *The Journal of TRAUMA, Injury, Infection and Critical Care, 54*(5), 814–822.

Chambers, L. W., Rhee, P., Baker, B. C., Perciballi, J., Cubano, M., Compeggie, M., et al. (2005). Initial experience of U.S. Marine Corps forward resuscitative surgical system during Operation Iraqi Freedom. *Archives of Surgery, 140*(1), 26–32.

Ciraulo, D. L., & Frykberg, E. R. (2006). The surgeon and acts of civilian terrorism: Blast injuries. *Journal of the American College of Surgeons, 203*(6), 942–950.

Galante, J. M., Jacoby, R. C., & Anderson, J. T. (2006). Are surgical residents prepared for mass casualty incidents? *Journal of Surgical Research, 132,* 85–91.

Kohn, L., Corrigan, J., & Donaldson, M. (Eds.). (1999). To err is human: Building a safer health system. *Committee on quality of health care in America.* Washington, DC: National Academy Press.

Rasch, R. F. R. (2006). Teaching opens new doors. *Men in Nursing, 1*(5), 29–35.

Reznick, R. K., & MacRae, H. (2006). Teaching surgical skills: Changes in the wind. *New England Journal of Medicine, 355*(25), 2664–2669.

Schreiber, M. A., Holcomb, J. B., Conaway, C. W., Campbell, K. D., Wall, M., & Mattox, K. L. (2002). Military trauma training performed in a civilian trauma center. *Journal of Surgical Research, 104,* 8–14.

Treat, K. N., Williams, J. M., Furbee, P. M., Manley, W. G., Russell, F. K., & Stamper, C. D. (2001). Hospital preparedness for weapons of mass destruction incidents: An initial assessment. *Annals of Emergency Medicine, 38*(5), 562–565.

U.S. Department of Commerce, Economics and Statistics Administration, National Telecommunications and Information Administration (2002). *A nation online: How Americans are expanding their use of the internet.* Retrieved May 11, 2007, from http://www.ntia.doc.gov/ntiahome/dn/anationonline2.pdf

Verrier, E. D. (2004). Who moved my heart? Adaptive responses to disruptive challenges. *Journal of Thoracic and Cardiovascular Surgery, 127*(5), 1235–1244. Retrieved May 17, 2007, from http://dx.doi.org/10.1016/ j.jtcvs.2003.10.016

Zimet, E., Armstrong, R. E., Daniel, D. C., & Mait, J. N. (2003). Technology, transformation, and new operational concepts. *Defense Horizons, 31.* Retrieved May 4, 2007, from http://www.ndu.edu/inss/DefHor/DH31/DH_31.htm

Chapter 38

VIRTUAL TRAINING FOR INDUSTRIAL APPLICATIONS

Dirk Reiners

The goal of this chapter is to provide the reader with an overview of industrial applications of virtual reality in general and virtual training specifically and to discuss the issues involved for industry to employ virtual training and how these can be overcome in the future.

Even though the initially envisioned applications areas for virtual reality (VR) were in the entertainment and scientific, as well as military and medical, realms, industrial users were quick to try and see how to take advantage of the potential of the technology. Ressler (1994) already lists a number of prototype and research applications focused on industrial uses of VR technologies. While design review and visualization have been the focus areas for industrial applications, training has always been a core area of interest for industrial users.

However, actual productive acceptance of virtual reality has been rather limited. This chapter looks at reasons for this phenomenon up until now and how the landscape will change for training in VR in the future.

SPECIAL CASE: DRIVING SIMULATORS

Industry has been an avid and successful user of specific kinds of virtual environments for training for a very long time, in fact, long before the term was even created. These systems were just not called virtual environments (VEs), but simply simulators, primarily for such vehicles as cars, planes, boats, and others. Many different kinds of driving and flight simulators have been and are in productive, daily use both on the side of the manufacturers, for example, as driving simulators to evaluate and train on such driver support systems as antilock brakes, as well as the users, such as airlines for pilot training.

They share some of the characteristics of other VE training systems, so some of the issues described in the later parts of this chapter apply to them, too, but due to their wide availability and development, somewhat independent from the rest of the virtual environment continuum, they are not a part of this chapter.

APPLICATION AREAS

There have been many prototypes and developments for employing VE technology in an industrial context. Virtual worlds have significant advantages in industrial training in several areas.

A major advantage is that training can be done without actual access to the physical facility for which the training is taking place. This is useful for facilities that have not been built yet, allowing the creation of a trained team to be ready by the time a facility becomes operational, for example, training an assembly crew before the factory that things will be assembled in has been built. It is also useful to train for facilities that cannot be taken out of production for training, for example, training a painter without having to shut down a full paint booth, for cases where training needs to be done in a separate geographical location from the final facility, or in cases where the object of the training is not available in the training location, for example, because of expense or space reasons. Examples include maintenance for vehicles and aircraft (Kaewkuekool et al., 2002; Wenzel, Castillo, & Baker, 2002). Other advantages of virtual training can be reduced resource use and faster reconfiguration to provide training for different scenarios.

Even if the facility is available, using it for training could endanger the trainee, for example, operation of large machinery (Sanders & Rolfe, 2002), chemical plants (Nasios, 2001), or nuclear reactors (Kashiwa, Mitani, Tezuka, & Yoshikawa, 1995; Mark, 2004). Other scenarios that cannot effectively be trained in real life are realistic emergency situations, as by their very nature they can threaten the trainee (Nasios, 2001).

DEVELOPMENT OF INDUSTRIAL USE OF VIRTUAL REALITY

There are a number of different reasons for the slow adoption of VR into productive use in industry.

Cost of Entry

In the early days of VR (before 2000), projection and computing systems were a major investment, up to and above the million-dollar mark. High end Silicon Graphics, Inc.'s (SGI's) graphics supercomputers were the only practical option for driving a VR setup. Head-mounted displays were either of low quality, such as 640×480 pixels with less than 24 bit color resolution, or very high cost, while low quality projection systems capable of displaying stereo images were not available at all, leaving only high end, expensive projectors. An additional, very important cost factor was the size of the installations. Computer systems were the size of a large refrigerator, with corresponding noise and air conditioning requirements. Cathode ray tube based projectors were not much smaller and needed regular calibration by specialized staff to maintain good quality images. Many of these installations would not fit into a regular office building floor and needed special construction, which posed a significant extra expense when looking at introducing them.

This effectively prevented anybody but very large companies from acquiring a system for experimentation and prototyping applications. Unless there was a very clear and present return of the substantial cost involved, it was next to impossible to argue for buying a virtual reality system.

As a consequence only large companies with correspondingly large research budgets and large prospective benefits considered and invested in virtual reality. For most of the 1990s, especially in Europe, this meant car and other vehicle companies.

They were and still are in a business situation that has large potential benefits to be gained from the use of virtual environments both for training and more general design and planning purposes. Competition in the field is fierce, and time to market is an important factor for success. Being able to use a virtual model of a product or a virtual environment simulation of a scenario can significantly reduce the number of real models and prototypes (major time and cost factors) that need to be built, which can pay off even a high end virtual reality system fairly quickly.

Software Availability

A major hurdle for industrial adoption of virtual environments was and still is the availability of adequate and effective application software. The potential industrial users of VE technology are not software developers, and systems designed for software developers that are used very successfully in a computer science centric environment are not immediately useful in industrial applications. Larger companies worked with universities and hosted researchers or Ph.D. students to develop specialized systems for them, which also helped alleviate the need to have their own hardware setup. Some of these developers became part of the companies and continued to do specialized development, but overall having to develop software is a major deterrent.

Thus for a long time VE usage was limited to large companies that could afford to fund specialized software development in an innovative field with a small number of available developers, mostly from the research community. Most companies did not want to fund the development of general software suites that could support various application areas of VE technology; very narrowly focused solutions were much less expensive and more effective. Actual uses were mainly limited to research prototypes or applications with a very limited scope.

Productive deployment was and is hindered by support issues. Research organizations and universities are great partners for developing prototypes and new technologies, but productive use needs constantly available support, user service, documentation, and continued development. Research organizations are not set up to provide these services, and commercial software vendors were not able to see enough business due to the high cost of entry and the small number of customers.

Data Creation and/or Conversion

The data conversion and processing needs of applications vary widely. On the one end of the spectrum are applications that have only one scenario, like most driving simulators. In that case, a lot of manual work can be put into model preparation, as it has to be done only once; therefore, the cost can be amortized over many uses of the system.

Especially in the early days, preparing a model for running a virtual environment was a major effort. Graphics systems could handle only fairly small models, on the order of 100,000 triangles or less, and still maintain the update rates and latency necessary for achieving immersion. Even with many optimizations, such as level of detail, that focus this triangle budget on the visible portion of the screen, that is a severe limitation. Creating a convincing virtual world within this limitation required strong simplification, especially if the original data came from a computer-aided design (CAD) system and was designed for constructing real objects, thus containing a large amount of detail that is irrelevant for a virtual environment, such as the threads of the screws. Typical CAD data can easily exceed millions of triangles, requiring a reduction by orders of magnitude to provide satisfactory performance. It was also not a trivial endeavor to export data from CAD systems, as there were no standardized formats that could be used for exchange. Every CAD system supported only its proprietary format, requiring a specialized exporter converter to get access.

Another important aspect of model preparation is assigning material characteristics, colors, and textures in an effort to simulate the natural appearance of the object. The graphics hardware could process only very simple lighting calculations fast enough, so creative approximation was needed to generate convincing effects, and a lot of experience in assigning these parameters was necessary to get the best results.

This data preparation effort is a major problem for application from the other end of the spectrum for the ones who need to work with up-to-the-minute current data, such as reviews of the current design state or assembly/disassembly training and simulation. For these applications any manual intervention beyond model selection is a problem, and even automatic conversions that take longer than a few minutes limit the usefulness severely.

CURRENT STATUS

While virtual veality is still a rather small area compared to the size of the computer industry, VR systems benefit significantly from developments in other areas, changing the playing field, especially for interested parties from an industrial background.

Cost of Entry

The times of Silicon Graphics graphics machines are long gone. The competitive drive in the race for a better gaming personal computer (PC) has led such manufacturers as nVIDIA and ATI (now AMD) to develop highly integrated, extremely powerful graphics systems that fit on a PC card and cost a fraction of an SGI machine. Such standard manufacturers as Dell, Inc. and Hewlett-Packard Company can deliver a machine capable of displaying scenes consisting of several million triangles with sophisticated lighting and shading models at interactive and/or immersive rates. The cost of such a system is only slightly higher than a regular desktop PC, and, in fact, many CAD designers use just such a system in their daily work. This makes it possible to run an immersive VE system from standard components that integrate well into a standard information technology (IT) infrastructure and that do not need major investments to be acquired. They also do not have unusual requirements as far as power and air conditioning are concerned, making their installation very easy.

The display side has not evolved quite as much. High end stereo-capable projectors and screens are still large and expensive, but it is now possible to use low end boardroom projectors with simple passive filters to create entry level systems that for most practical purposes look very similar to a standard meeting room.

This allows even small companies to set up a VE-capable environment without unduly large effort. The hardware side of the equation is ready for widespread adoption.

Software Availability

In the long run, in-house software teams are not a sustainable solution for most companies, given the speed of change of the software and hardware environment and the growing user requirements. The amount of effort required to keep an internal software system for virtual environments up-to-date is just too high for companies for which the software is not the core business.

Higher level software systems that hide the complexity of programming applications behind graphical user interfaces promise to reduce the barrier of entry into application development. Many systems have a node- and route based structure that provides basic building blocks that can be connected using graphical tools. However, for most of the industrial users, especially the many smaller companies that have not had contact with the technology at all, the learning curve even of those systems is not a viable option.

Therefore, the availability of off-the-shelf applications or low cost configuration and/or specialization is a necessity for a much wider adoption of VE technology in a larger variety of industries. It is an obvious hen and egg problem, as the substantial efforts involved in creating specialized applications need to be offset either by high prices or larger numbers of customers to be a viable business. A number of small- and medium-sized companies are competing in this market with reasonable success, offering turn-key solutions for a number of problem areas.

A compromise that has been attempted with good success by some companies is to create very specialized applications with very limited but valuable functionalities either as configurations for a high level system or special programs based on existing libraries. This could be a successful path for wider spread use in industry, but it depends on the availability of development manpower at affordable pricing. At this time, software availability and usability remains a critical shortcoming in widespread adoption of VE technology in industry.

Data Creation and/or Conversion

Thanks to the rapid developments in graphics hardware capabilities, the requirements for models to be suitable for virtual environment systems have been relaxed quite significantly. The precision of CAD models has increased from the early days of VR, but not at the same rate as the performance. Therefore, current CAD models for small- to medium-scale objects can be used pretty much as they are in a VE system, reducing or eliminating the need for simplification. For large objects, such as airplanes or whole factories, or for mechanically complicated objects, such as an engine with all construction details, nonsimplified models are still too complex. But thanks to advances in automatic software simplifiers that remove visually irrelevant components, usually only a small amount of manual work is necessary.

It has also become easier to get data out of other systems. Exporters for such standard file formats as VRML/X3D or Collada are common in many construction and/or simulation systems. If those are not available, a common fallback is STL, the stereolithography format, which is a trivial triangle format that loses a lot of structural information, but gives access to the geometry and is supported in virtually every geometric construction system.

Given the growing need for high quality images coupled with the wider availability of high speed graphics hardware, more and more construction systems support the specification of high quality surface characteristics. These are not necessarily used by every designer, but having them available helps increase awareness. In addition, the grown hardware capabilities allow the direct use of realistic lighting models, alleviating the need for a large amount of experimentation to achieve a desired look. Instead, an automated or semi-automated assignment of surface characteristics to objects can be done.

In conclusion, while model conversion and availability for VE systems is not a totally obvious and automated process yet, it has been significantly simplified and does not pose a serious deterrent to introducing VE systems into an industrial context anymore.

FUTURE DEVELOPMENTS

A lot has happened since VR was first prototyped and used in industrial applications. Cost and other barriers of entry have been significantly lowered, and things are going to get only better in the future.

Cost of Entry

There does not seem to be an end in sight for the evolution of graphics card performance. They keep getting faster and more powerful, at a stable price level, supporting running VE-style interactive/immersive three-dimensional (3-D) graphics application on almost every available computer system. New developments, such as the PCI-Express standard for expansion cards, will support putting more graphics cards into one system, allowing larger, multiscreen or very high resolution projection systems to be driven from a single off-the-shelf PC.

Three-dimensional-capable displays are just now entering the mainstream. Three-dimensional movie theaters are becoming ubiquitous, and Texas Instruments together with Samsung is pushing 3-D-capable TVs into the market, at prices that compete very well with regular TVs. This noticeably reduces the cost of entry into 3-D displays and introduces a wider audience to them, increasing acceptance and, after that, demand for them.

Software Availability

The problems for providing software are getting smaller due to increasing numbers of customers and a growing market, but it is still a major issue and a deterrent to really wide adoption. The existing companies will slowly but surely expand their offerings, but there is a market for new companies that can provide solutions for industries that see potential at the now reduced price points.

CONCLUSION

Industry has looked at virtual environments for a long time, as there are a number of application scenarios that can clearly benefit from VE technology and provide significant savings both in time and cost compared to traditional methods. In the beginning many technical limitations were barriers for exploratory or even productive entry into the field.

Thanks to many developments in hardware and software many of these barriers have been lowered or will be lowered in the near future. A limiting factor that still exists is the availability of turn-key software solutions for quick and seamless introduction into new businesses. There is a need for more providers of specialized know-how that can help companies quickly create practical, working solutions for new application scenarios and industries.

REFERENCES

Kaewkuekool, S., Khasawneh, M. T., Bowling, S. R., Gramopadhye, A. K., Duchowski, A. T., & Melloy, B. J. (2002, May). *Using virtual reality technology to support job aiding and training.* Paper presented at the Industrial Engineering Research Conference, Orlando, FL.

Kashiwa, K., Mitani, T., Tezuka, T., & Yoshikawa, H. (1995). Development of machine-maintenance training system in virtual environment. *Proceedings of the 4th IEEE International Workshop on Robot and Human Communication* (pp. 295–300). Piscataway, NJ: Institute of Electrical and Electronics Engineers.

Mark, N. K. (2004, November 25–26). *VR-system for procedural training and simulation of safety critical operations in relation to the refuelling at Leningrad NPP.* Presentation at the NKS Seminar on Nordic Safety Improvement Programmes, Halden, Norway.

Nasios, K. (2001). *Improving chemical plant safety training during virtual reality.* Unpublished doctoral dissertation, University of Nottingham, Nottingham, United Kingdom.

Ressler, S. (1994, June). *Applying virtual environments to manufacturing* (Rep. No. NISTIR 5343). Gaithersburg, MD: National Institute of Standards and Technology.

Sanders, S., & Rolfe, A. C. (2002). The use of virtual reality for preparation and implementation of JET remote handling operations. *Fusion Engineering and Design, 69,* 157–161.

Wenzel, B. M., Castillo, A. R., & Baker, G. (2002). *Assessment of the Virtual Environment Safe-for-Maintenance Trainer—VEST* (Rep. No. AFRL-HE-AZ-TP-2002-0011). Mesa, AZ: Air Force Research Laboratory.

Chapter 39

CORPORATE TRAINING IN VIRTUAL ENVIRONMENTS

Robert Gehorsam

Interactions in virtual training worlds can be much richer, deeper and more realistic than with existing computer based techniques. In addition, using virtual training worlds for corporate training has such advantages as decreasing training cost by lowering trainees' travel and lodging expenses, providing a wide range of flexibility for training schedules, and improving the motivation of trainees.
—Accenture Technology Labs

Historically, virtual environments have been most useful for training and practicing procedurally oriented skills when the risk of failure in the operational environment is high and alternative methods are either prohibitively expensive or, at the other end of the spectrum, ineffective in providing appropriately immersive experiences to the trainee. Aviation, disaster response, hazardous materials handling, and military training have typically been the sweet spots for the application of virtual environments. Furthermore, these applications have tended to target the single user (such as fighter pilots) and rely on expensive, specialized, location-specific hardware and software systems. Finally, these training applications focus on human interactions with complex physical or instrumentation systems—the environment itself is a key antagonist in the training scenario.

The rise of massively multiplayer online games and social virtual worlds has transformed this paradigm, providing an opportunity for new cohorts of professionals to derive benefits from virtual environments that were unforeseen by traditional virtual training developers, as well as by game designers. Fundamentally, these consumer technologies use game design along with graphic, artificial intelligence, and networking technologies to focus on the *identity, social,* and *interpersonal* aspects of virtual environments. In essence, rather than focusing on the human-environment interaction, they focus on the human-to-human interactions within the environment. The *avatar*—the virtual representation of an individual in a virtual environment—is the focus. Rich, multiuser virtual environments are now available to anyone with a standard personal computer (PC) and an Internet connection, and this will ultimately spread to mobile devices.

Beyond these technology trends, it is an often-repeated truism that a generation of employees is now entering the workforce whose first experience of software is not the spreadsheet, but the video game. *America's Army,* albeit a military application, is a key example of how contemporary organizations believe they must communicate to their prospective workforces. It stands to reason, then, that corporations are now looking at how games and virtual environments might play a productive role in the workplace. And when they do, they see training and e-learning as areas of primary interest.

The global knowledge management market is informally estimated by Claire Schooley, Senior Analyst for e-Learning at Forrester Research, to be $195 billion per year, which includes spending on technology, course development, and formal and informal learning (personal communication, April 2008). Today's global corporations have a seemingly endless demand for training and e-learning that can be satisfied only with a range of learning modalities: internally, there are sales, management, and leadership and technical training needs; externally, there are customers and partners to be educated and trained. The global corporation is thus faced with several challenges for developing effective learning strategies: how to overcome the financial and time costs of bringing people to learning centers and how to foster cohesion when the workforce is distributed, mobile, time-shifted, and comes from diverse cultural backgrounds.

Virtual environments—and specifically persistent online virtual worlds—are essentially a model. Dr. Byron Reeves, the Paul C. Edwards Professor of Communication at Stanford University and Faculty Director of the Stanford Media X Partners Program notes, "MMORPGs [massively multiplayer online role-playing games] mirror the business context more than you would assume. They presage one possible future for business—one that is open, virtual, knowledge-driven, and comprised of a largely volunteer or at least transient workforce" (IBM Corporation & Seriosity Inc., 2007, p. 7). As a model, then, virtual worlds provide a potentially limitless environment for learning and training. The premise for the successful deployments of virtual worlds in enterprise-oriented training is that these worlds can (a) replicate the full range of everyday and extraordinary situations for employees, (b) provide the necessary support for various modalities of learning, (c) do so in a manner that is easy to deploy, operate, and learn, and (d) provide either superior training or more cost-effective training . . . or both.

However, because of the traditional resistance of corporate information technology (IT) departments to nonstandard desktop applications, and an equally traditional "cultural" bias against any software that seems "game-like" in the serious workplace, the use of virtual environments is still in its earliest stages in corporate settings. Some of the early work explores the use of virtual environments for collaboration, some for marketing, and some for training. At this point in time, studies and pilots are under way in a number of different industries, notably health care, energy, and, not surprisingly, technology. What is most striking is the broad range of applicability suggested by avatar-enabled virtual environments, from general purpose management training through specific industrial uses to individual professional development, these immersive environments show the potential to deliver training and learning through a range of modalities, including team based training, mentoring, formal curricula adapted to virtual environments, and even individualized training. This chapter provides a series of snapshots of how some of the more innovative learning-oriented companies are utilizing virtual worlds to develop new training and learning capabilities and what these early efforts might presage for the future.

In 2003, as massively multiplayer online gaming was becoming a mainstream phenomenon and Linden Lab and There.com were launching *Second Life* and *There,* respectively, Accenture, through its Technology Labs, began exploring a general purpose, horizontal application of virtual worlds—the use of distributed immersive environments for management and leadership training to solve critical problems in a collaborative manner. In a 2003 paper "Using Virtual Worlds for Corporate Training," published as part of the *Proceedings of the Third IEEE International Conference on Advanced Learning Technologies* (ICALT'03), the authors describe how virtual worlds are well suited toward solving the problem of delivering "synchronous interactions among distributed trainees" (Nebolksky, Yee, Petrushin, & Gershman, 2003, p. 412).

The authors describe how the virtual training world provides for three distinct categories of users: facilitators (the observer/controller equivalent in military exercise), subject domain experts, and the students themselves. While the first two categories of users provide context and content, the students, who may or may not be co-located, are represented as three-dimensional (3-D) avatars on computer screens, which are the communications interface between the participants. Students may be assigned to play their actual real-life role (for example, vice president of manufacturing) or be asked to switch roles to provide them with new perspectives. For example, the vice president of marketing may be asked to take on the head manufacturing role in order to learn about the operating constraints that manufacturing experiences as it tries to deliver to marketing specifications.

Accenture identified 10 skills necessary for leadership development, ranging from such basic skills as communications and planning through more demanding skills, such as envisioning the future and intelligent risk taking. They designed a fictional scenario well suited for the immersive, visceral qualities of 3-D environments based on the polar expeditions of Sir Ernest Shackleton, a scenario that demanded—and challenged—both team formation and leadership skills. The participants then proceeded through a series of episodes with a "narrative" arc that involved scenario description, plan, plan execution, complication, and debriefing.

Relative to live role-playing exercises that previously required collocation of participants and a paucity of immersive experience, Accenture concluded that using virtual worlds for corporate training "had such advantages as decreasing training cost . . . giving the flexibility for training schedule, and improving the motivation of trainees" (Nebolksky et al., 2003, p. 413).

Interest in corporate uses of virtual worlds for organizational training was further echoed by SRI Consulting in a major 2007 study of virtual worlds and serious gaming, in which the company noted that "training and education are likely to be the most mature virtual worlds applications outside games and social world" (Edmonds, 2007, p. 50). Citing the long history of custom-built virtual environments as training tools for military and government organizations, it concluded that organizations would "continue to demand specific environments that they are free to control themselves and host on their own intranets" as opposed to using public virtual worlds for organizational purposes (Edmonds, p. 50).

Nevertheless, robust, publically disclosed deployments of private virtual worlds for corporate training have been few and far between, due to reasons cited above. One early adopter organization has been pharmaceutical giant Johnson & Johnson, which in 2005 began using virtual environments to address the onboarding and familiarization challenges of new employees.

Johnson & Johnson (J&J) is emblematic of today's large, globally distributed companies. It has over 119,000 employees, distributed among 250 operating companies in 57 countries, and generates over $53 billion in annual sales. It has both enormous consumer brand recognition, as well as a technology research and development–centric culture. According to surveys conducted by Fortune, Forbes, and other media, it is one of the world's most admired—and diverse—companies. Furthermore, it has achieved consistent sales growth for 75 consecutive years. How does a company of this scale maintain such a high level of performance?

One critical component of success is through J&J's e-University, an online, traditional Web based portal comprising over 75 distinct schools organized by region, function, and operating company. The schools offer thousands of individual courses, integrated into a learning management system. However, one functional unit of J&J, the Pharmaceutical Research & Development (J&JPRD) unit, concluded that a more immersive virtual environment was required for certain training tasks. J&JPRD consists of 10 research centers located in North America, Europe, India, and China engaged in the development of new pharmaceutical products and conducting clinical trials. These efforts require high levels of collaboration between diverse groups.

J&JPRD has developed an extension of its e-University presence, known as 3DU. As reported in a case study prepared by Brandon Hall Research, the unit felt the need to "shrink the physical world to allow for engaging interactive learning techniques . . . to enhance retention and knowledge transfer among the global employees of J&JRPD" (McKerlich, 2007, p. 17). The environment, based on Proton Media's Protosphere tool, re-creates physical classroom environments and provides individual employees the opportunities to interact via avatars (with speech and text), exchanging documents, conducting classes, and engaging in informal learning opportunities. Unlike typical synchronous teleconference environments, 3DU is "always on," meaning that any employee anywhere in the world can log in anytime to interact with others. Furthermore, the integration with J&J's learning management system provides self-paced learning within the virtual environment as well.

As noted earlier, one barrier to adoption in the enterprise is the cultural resistance of IT departments or even management to enable game-like applications on corporate intranets. Even when this barrier is overcome, other cultural barriers can exist within the workforce. Technology apprehension by older, less tech-savvy employees came to the forefront of J&J's implementation challenge, and, thus, a technology-training curriculum had to be introduced to enable culturally and demographically diverse populations of employees to not only use the technology itself, but to understand how to communicate and interact with other employees in an avatar-mediated environment (McKerlich, 2007, p. 18).

Private e-learning-oriented virtual worlds, created from commercial off-the-shelf platforms, deployed behind firewalls, and integrated into corporate learning management systems, are an important and perhaps central trend in creating the milieu for the widespread adoption of virtual worlds for training, but not the only. Members of today's global, mobile workforce have individual professional development needs that exist separately from the specific training regimens of individual enterprises. Almost certainly, one of the most in-demand skills the workforce needs is language proficiency. In particular, learning English as a foreign language (EFL) is a multi-billion-dollar industry, with demand expected to rise to 2 *billion* English language learners worldwide by 2010 (Graddol, 2006, p. 101). According to informal research conducted by Paideia Computing, a technology based EFL instructional company, today, the vast majority of learning English occurs in live classroom or private tutor settings, with less than 20 percent technology enabled, mostly through CD-ROM based media.

The efficacy of using multimedia and game based language training tools is increasingly accepted, and the U.S. Department of Defense's *Tactical Iraqi* program has, in fact, spawned a commercial endeavor to extend the reach of the capability beyond U.S. forces. Now, however, some companies are looking at how to employ virtual worlds to address the insatiable commercial demand for English as a second language in the Asian market.

Paideia Computing is a company that utilizes Forterra Inc.'s OLIVE virtual world platform to develop English language curriculum for the Asian market. The solution is not intended to be delivered behind a company's firewall for training purposes, but rather as a public, subscription based service for individual and small-group training. In addition to the new form of delivery (a public, special-purpose virtual world), Paideia introduces several other innovations relative to the use of virtual worlds for enterprise-oriented training.

Language acquisition is a skill best learned through practice in context. Paideia quickly understood that a virtual world provides the optimal blend of both: an on-demand virtual environment that can re-create the physical settings most language students are going to need in both acquiring and practicing conversational and reading capabilities. Unlike a classroom setting or a textbook, the virtual environment provides the student with the actual experiences of, for example, being at an airport, in a restaurant, and in a work meeting. And unlike CD-ROM based instruction, the virtual environments could be populated not just by artificial characters, but by other role-players, students, and teachers.

The distributed nature of a virtual world platform also enables native-speaking or otherwise qualified teachers to be available from any location in the world to any student located anywhere in the world, which in turn provides a capability not available to other single-user technologies, such as CD-ROM voice interaction. While there have been considerable advances in voice recognition technology to enable a synthetic instructor to understand and respond to a human, the integrated voice over Internet protocol capabilities of a virtual world platform enables high fidelity multipoint communication between a teacher and any number of students, no matter where they are located. Furthermore, the integrated recording and replay capabilities of the platform enable the teacher to review with any number of students over the network key lessons and performance assessments.

That said, with over 1,000,000,000 students studying EFL, it is easy to imagine the significant business scalability problems in providing personal tutors for every student online. So while virtual worlds excel and differentiate themselves from other game-like environments, Paideia has recognized that a full language training solution involves not just virtual environments that provide teacher-to-user training, but integration with learning management systems, the development of appropriate synthetic characters that can either provide instruction or be "background" characters in a scenario, and full data analytics to support the optimization and evolution of the system. The result is a learning environment that is part virtual world, part social network, part game, and part classroom.

In conclusion, we see that while the uses of virtual environments for training in enterprise are still at their earliest stages, a range of applications, learning modalities, and deployment methodologies are available. It seems clear that there is no "one size fits all." Unlike government programs, where requirements frequently result in technology being built from the ground up, corporations will rely on commercial off-the-shelf platforms, subsequently modified for specific organizational needs, to satisfy their requirements. Nevertheless, some barriers will continue to exist and need to be overcome by internal evangelism and the as-yet-to-be-published success stories in a range of lighthouse deployments. Technologically, the limited graphical capabilities of desktop PCs and the limited capacities of internal networks to handle the increased traffic requirements will challenge widespread adoption in the near term. Culturally, the potential resistance of IT departments and management to "game" deployments on networks will play an inhibiting role, and the not-uncommon phenomenon of user apprehension relative to a new, dynamic technology can be anticipated and answered. However, the potential benefits and cost savings relative to travel and other overhead expenses show high promise. Most importantly, the entry into the workforce of a young, technically sophisticated population, growing up on games and virtual worlds at home, presages a demand for a similar high quality immersive experience at work.

REFERENCES

Edmonds, R. (2007). *Virtual worlds.* Menlo Park, CA: SRI Consulting Business Intelligence.

Graddol, D. (2006). *English next: Why global English may mean the end of 'English as a foreign language.'* London: The British Council. Available from http://www .britishcouncil.org/learning-research-english-next.pdf

IBM Corporation & Seriosity Inc. (2007). *Virtual worlds, real leaders: Online games put the future of business leadership on display* (A Global Innovation 2.0 Rep.). Available from http://www.seriosity.com/downloads/GIO_PDF_web.pdf

McKerlich, R. (2007). *Virtual worlds for learning: How four leading organizations are using virtual environments for training* (Analysis Report). Sunnyvale, CA: Brandon Hall Research.

Nebolksky, C., Yee, N., Petrushin, V., & Gershman, A. (2003). Using virtual worlds for corporate training. *Proceedings of the 3rd IEEE Conference on Advanced Learning Technologies—ICALT'03* (pp. 412–413). Los Alamitos, CA: IEEE Computer Society. Available from http://csdl2.computer.org/comp/proceedings/icalt/2003/1967/00/ 19670412.pdf

Part X: Next Generation Concepts and Technologies

Chapter 40

VIRTUAL ENVIRONMENT DISPLAYS

Carolina Cruz-Neira and Dirk Reiners

The display has always been a critical component of virtual environment (VE) systems. This is not surprising, as humans take in 80 percent of the information about the environment through their eyes; therefore, presenting a convincing version of the virtual environment to the eyes is a necessary step for full immersion. Displays have also shaped the public image of virtual reality (VR) to a large extent; people wearing weird contraptions on their heads or, at the minimum, funny glasses are a staple element of many movies, and that is not actually an inaccurate description of the reality in most labs.

A LITTLE BIT OF HISTORY

The initial idea for VE displays originated from Ivan E. Sutherland's (1965) vision for "The Ultimate Display" that enabled users to enter and control a computer-generated world. Sutherland's (1968) Sword of Damocles was the first head-mounted display (HMD) and, therefore, marked the beginning of a new field. Fisher, McGreevy, Humphries, and Robinett (1986) at NASA (National Aeronautics and Space Administration) went a step further integrating HMDs with three-dimensional (3-D) sound, voice recognition, voice synthesis, and a DataGlove (Zimmerman, Lanier, Blanchard, Bryson, & Harvill, 1987). A few years later, the first commercial HMD, VPL Research Inc.'s Eyephone, became available. The early 1990s clearly marked the acceptance of VEs, with the introduction of the CAVE (cave automatic virtual environment) by Cruz-Neira, Defanti, Sandin, Hart, and Kenyon (1992), the virtual portal by Michael Deering (1993), and the responsive workbench by Krueger and Fröehlich (1994). These systems helped establish the field of virtual reality by introducing the novel, yet pragmatic, use of proven, familiar projection systems to create immersive displays. The real validation for projection technology as an accepted form of VR happened when General Motors Corporation (GM) installed the first CAVE in the industry in late 1994. GM pioneered applications of VR in the area of vehicle design and virtual prototyping. A few years later, the oil and gas industry "discovered" VR, getting a significant number of projection based immersive systems across its different branches and groups. By 1997 there were over 50 projection based VR systems operating worldwide in academia, research, and industry.

Projection systems solved several limitations of HMDs by providing a physical space that could be shared by multiple users (although only one could be tracked) and that allowed the blending of virtual and real space, including the user's own body. But they also introduced their own limitations, in particular, issues related with the use of projectors, such as multiple projector calibration (color, convergence, blending, and so on) and the need for large and dark spaces.

Thus displays have always been and will continue to be an active area of research and an important aspect of future growth and development for VE systems. To reach high levels of acceptance, VE displays need to have high quality, in the best case the quality of the human visual system, and they need to be affordable. Given that these are conflicting goals, there are many sweet spots in the continuum that can be exploited, and current developments in commercial off-the-shelf (COTS) components enable future devices to raise the quality bar without raising the price.

Given the enormous flexibility and the wide range of capabilities of the human visual system, quality can have many aspects in the context of displays. Volume 2, Section 1 in this handbook provides a detailed discussion of the human visual system and its relationship to display design. For the purposes of the discussion in this chapter, resolution, field of view, stereo presentation, and color and brightness precision all can play important roles in whether a display is merely good enough or able to get users to suspend their disbelief and get immersed in the virtual environment.

NEED COTS COMPONENTS TO GROW

A recurrent problem with virtual reality equipment and especially displays is that they tend to be very expensive, as they are usually built in very small numbers and have high quality and precision requirements. As a consequence, only a few people can afford them, which drives the price even higher to the point that nobody is able or willing to afford them, and the company disappears.

The most promising solution to this problem is to use COTS components as much as possible, such as projectors, liquid crystal display (LCD) panels, or interactive devices from gaming. To some extent that has always been the case, as it is economically unfeasible to develop such specialized components as LCD panels for VE displays. The disadvantage is that this limits the possible capabilities to whatever the rest of the commercial market needs at the time. These capabilities might not match the needs of VE displays exactly, opening a space for ingenious engineers to find ways to push the components far beyond that for which they were designed.

HEAD-MOUNTED DISPLAYS

Head-mounted displays have been one of the defining components of virtual environments since the very beginning. Making them work well and deliver high quality visuals is a very challenging problem, which is why early HMDs had very bad quality and/or usability.

An HMD combines a large number of challenges: as it is rigidly attached to the head it needs to be very light to allow comfortable use over time. It also needs to cover a large field of view to avoid distracting tunnel-view effects. Because the displays are close to the eye, the pixels need to be very small to avoid blocky-looking images.

Especially the wide field of view, in combination with limited resolutions of LCD panels, has been a significant problem in HMD design, and for a number of years very little development and/or advancement could be seen. In the early years the focus was on wide field of view systems (50° to 60° horizontally) with low resolutions (in the order of 640 × 480 pixels), which led to displays that allowed the identification of every individual pixel (or subpixel), which severely limited the realism of the displayed images. Later the focus shifted to higher resolutions (up to 1,280 × 1,024 pixels), but partially at the cost of field of view (30° to 40° horizontally), leading toward many HMDs that left the impression of wearing black blinders at all times. In the recent past this has changed, and two new developments give hope for a resurgence of HMD display systems.

The HMD system designed and developed by Bolas and McDowell (2006) targets the field of view problem through the use of two LCD panels per eye and widespread optics. The result is a display that provides a good field of view, but with noticeably limited resolution outside of the direct forward view direction. The benefit is a fairly light design that can be used very comfortably and that can be built at a reasonable cost.

The second approach introduced by Sensics (2008) tries to cover all the bases by combining a large number (up to 24) of LCD panels to cover the whole field of view. This approach has the potential to achieve very high resolution everywhere in the field of view, allowing the user to look around freely. The cost is a fairly high weight, requiring physically strong users or an intelligent counterbalance scheme. It is also a very challenging problem to design the right optics to hide the discontinuities, including the geometric, as well as color and brightness discontinuities, between all the LCD panels and to make them appear seamless to the user. Solutions exist, but they require fairly precise calibration for and by the user.

A critical component of all current HMDs is small, high resolution LCD panels. They used to be very specialized components with very limited use, which led to few options and high prices. Given the rise of higher and higher resolution cell phones, as well as such portable devices as ultralight PCs, such as Sony's Vaio UX50, the need for high resolution in small displays is rising, helping to add interest among display manufacturers to provide more products in this market. This increased availability and quality will support increased quality for HMDs pretty directly. Both systems with few panels and widespread optics, as well as multipanel tiled systems, will benefit, and there are market opportunities for both designs. The resulting expectation is that we will see some resurgence of HMD developments and technology, reviving a market that had become rather stale.

MONITORS

In the beginning monitor based systems, also known as fish tank VR, such as the one described by Ware, Arthur, and Booth (1993), were pretty common. They provided an inexpensive entry way into virtual environments, as the common cathode ray tube (CRT) monitors were capable of displaying active stereo images directly, alleviating the need to buy specialized display systems. This changed dramatically when LCD panels became the standard display, to the point that they have essentially replaced CRT monitors 100 percent. LCDs are not fast enough for active stereo, removing the ability to directly display stereo images and all but removing monitor based VR systems. New developments in stereo-scopic display for regular TVs can open new avenues here, where Samsung has introduced plasma displays that can display the necessary high refresh rate to drive active stereo glasses. These are designed to be used as home TVs and, therefore, are too big for desktops, but they can serve as an alternative to large monitors or small projection screens.

On the LCD side the quickly increasing resolution of current panels, routinely reaching $2,560 \times 1,600$ pixels or more, has made it possible to actively use tech-nologies that have been around for a long time, but which need very high resolu-tion displays for high quality results. These technologies are lenticular or parallax-barrier screens for autostereoscopic display.

These approaches both trade resolution for stereoscopic display by redirecting a subset of the available pixels into different directions away from the screen. By placing the eyes into the right position so that one eye sees a subset of pixels that is not visible to the other eye, stereo display is possible. The difference between the two methods is that lenticular displays use a lens sheet to redirect the pixels into different directions, while parallax-barrier displays use an array of black stripes that hides parts of the screen for certain locations in space. The quality of these displays depends on the resolution and on the number of separate images that can be redirected, which influences how large the area is from which correct stereo can be seen, as well as how many users can see individual views of the 3-D scene. Autostereoscopic displays add a new quality to the display space, as they allow immersive display without any special glasses of any kind, which also allows multiple viewers. This can help in gaining acceptance among a larger audience spectrum and help make stereoscopic and/or immersive displays more widely used in regular office settings.

An alternative to the old active stereo, which LCD panels do not support and will not support in the near future, is passive stereo using polarized glasses. Through the use of two coupled LCD panels it is possible to create polarized stereo images directly, without having to create two images separately and polar-izing and overlaying them. These displays are becoming available commercially now, such as the iZ3D monitor (2008), and at competitive prices, as they are being targeted at the game-playing public.

Somewhere between monitors and projection systems is an alternative technol-ogy to the previously mentioned stereoscopic plasma displays, which are stereo-scopic digital light processing (DLP) displays. These are projection TVs that have been put on the market by different manufacturers at price points that are very close to regular TVs. They are projection systems, but at a small scale, and they support regular active stereo signals. They also form a basic component of larger-scale projection displays (see below).

Research and development for monitors for VE displays had been fairly dormant for some time. New developments in commercial applications for stereoscopic displays for regular TV and game audiences help to make them more available and affordable, and also make them a more usual occurrence. This *can help remove the stigma of being something special,* as that can hinder adoption, especially in an industrial context. Also taking advantage of the regular development of increasing resolution opens up new ways to display stereoscopic and immersive content, leading toward totally new venues that so far shied away from the need for glasses or other head based hardware.

LARGE SCREEN/PROJECTION SYSTEMS

Large screen systems, mostly based on projections (see below for new developments), have become very popular since their inception. They can provide the same 360° fully immersed feeling that HMDs support, but they require much less intrusive and smaller glasses than an HMD; when used with the correct interaction devices, they can be made completely untethered and wireless. Other benefits include the ability for the users to see their own real bodies and the ability for multiple people to see the virtual environment (although in a head-tracked scenario only one of them will get correct stereo images, they can all get an impression of what is displayed). The last aspect is especially interesting in a corporate setting, where many VEs are used for discussions and evaluations. Their disadvantages are the need for large screens to obtain a large field of view, and the need for covering those screens with high resolution pixels. HMDs move with the user's head and can focus their pixel output where he or she is looking. Large screen displays need to have pixels everywhere, in case the user is looking there. This makes the resolution of the display a larger issue than in HMDs.

There has been a lot of effort in trying to achieve higher resolution displays. The two possible approaches are either to use a combination of low resolution projectors or to use higher resolution components in the projector.

The first approach is attractive due to the ability to use low cost COTS components to create very high resolution display systems (Kresse, Reiners, & Knöpfle, 2003). This is alleviated by the challenges to make the multiple components match up correctly, both geometrically and in color and brightness. Automated solutions for these problems have been found, and commercial companies are starting to provide them as turn-key solutions. One interesting approach developed by Jaynes, Seales, Calvert, Fei, and Griffioen (2003) and available from Mersive Technologies (2008) uses the DLP TVs mentioned above as the projection base unit. They are as attractive as they are comparatively cheap. In addition, they feature high (HD) resolution and, due to their construction, require only very limited constructive depth, which addresses one of the major problems of most projection displays. Joining the stereo capabilities of latest-generation TVs with intelligent color and geometric correction in a tiled fashion forms an interesting and affordable option in the projective display space.

An alternative to tiling small units is to design systems with a high intrinsic resolution. This requires a very large engineering effort from the main projector units down to the actual LCD panel or DLP chips. The VE market is not big enough to sustain this kind of development. However, VE systems can benefit from developments in the general projection system market. Movie theatres are making the move to a digital distribution and display scheme, and they need very high resolution displays. As a consequence, such projector manufacturers as Sony and JVC are producing projectors capable of displaying images at resolutions of $4,096 \times 2,160$ (also known as 4K). This is a significant step over older systems, and at normal projection sizes and viewing distances can reach the resolution of the human eye. While not exactly cheap, these systems allow a fairly painless entry into very high resolution, out-of-the-box displays.

In addition to increasing resolutions, cinemas are also moving toward stereoscopic displays. In the near future, combinations of very high resolution and stereoscopic displays can be expected to come out of the commercial cinema domain and be applicable directly in a VE context.

Another interesting new avenue is a combination of the autostereoscopic display idea described above with projective displays. Using a holographic screen and an array of standard projectors it is possible to provide large-screen autostereoscopy, which allows multiple users to have their individual views on the 3-D scene at the same time (Holografika image). These displays provide very interesting opportunities, but at a very high cost, as a large number of pixels need to be calculated and generated.

Large-screen systems continue to be a very active area of development. In the recent past commercial developments have stepped up to reach resolution and stereo capabilities that used to be a specialized topic in VE research and development and bring them into the COTS space, which will help bring costs down and make them more available to a wider variety of application domains.

There are still research avenues, especially in going beyond single projector systems and into autostereoscopic systems, and many other quality aspects, such as brightness, color accuracy, and dynamic range, that are not fully addressed by current systems.

CONCLUSION

Displays have been a defining component of VEs since the very beginning. There has been a little bit of a lull in the field in the past when detailed refinements to existing technology where made, but no real breakthroughs. In the recent past interest has picked up, and new developments have opened new avenues in very high resolution displays, making autostereoscopic displays more available and approachable.

There are many interesting avenues to follow for research in higher resolution and easier use, multiuser systems and to open new avenues in terms of new aspects of display technologies, such as higher dynamic ranges and better color precision, continuing on the quest to create the ultimate display that will allow us to display a virtual environment that is indistinguishable from reality.

REFERENCES

Bolas, M., & McDowell, I. (2006). *The Wide5 HMD.* Retrieved April 26, 2008, from http://www.fakespacelabs.com/Wide5.html

Cruz-Neira, C., Defanti, T. A., Sandin, D. J., Hart, J., & Kenyon, R. (1992). The CAVE audio visual experience automatic virtual environment. *Communications of the ACM 35*(6), 64–72.

Deering, M. (1993). Making virtual reality more real: Experience with the virtual portal. *Proceedings of Graphics Interface '93* (pp. 219–226). New York: ACM.

Fisher, S. S., McGreevy, M., Humphries, J., & Robinett, W. (1986, October). *Virtual environment display system.* Paper presented at the ACM Workshop on Interactive 3D Graphics, Chapel Hill, NC.

iZ3D Monitor. (2008). *Stereoscopic 3D and iZ3D Perception.* Retrieved March 12, 2008, from http://www.iz3d.com/download/iZ3D_Whitepaper.pdf

Jaynes, C., Seales, B., Calvert, K., Fei, Z., & Griffioen, J. (2003, May). *The Metaverse—A collection of inexpensive, self-configuring, immersive environments.* Paper presented at the 7th International Workshop on Immersive Projection Technology/Eurographics Workshop on Virtual Environments, Zurich, Switzerland.

Kresse, W., Reiners, D., & Knöpfle, C. (2003, May). *Color consistency for digital multi-projector stereo display systems: The HEyeWall and the digital cave.* Paper presented at the 7th International Workshop on Immersive Projection Technology/Eurographics Workshop on Virtual Environments, Zurich, Switzerland.

Krueger, W., & Fröehlich, B. (1994). The responsive workbench. *IEEE Computer Graphics and Applications, 3*(3), 12–15.

Mersive Technologies. (2008). *The m-Series displays.* Retrieved on April 15, 2008, from http://www.mersive.com/mSeries_About.html

Sensics. (2008). *The sensics PiSight HMD.* Retrieved February 23, 2008, from http://www.sensics.com/technology

Sutherland, I. E. (1965). The ultimate display. *Proceedings of IFIPS Congress, 2,* 506–508.

Sutherland, I. E. (1968). Head-mounted three-dimensional display. *Proceedings of the Fall Joint Computer Conference, 33,* 757–64.

Ware, C., Arthur, K., & Booth, K. S. (1993). Fish tank virtual reality. *Proceedings of the INTERACT '93 and CHI '93 Conference on Human Factors in Computing Systems* (pp. 37–42). New York: ACM.

Zimmermann, T. G., Lanier, J., Blanchard, C., Bryson, S., & Harvill, Y. (1987). A hand gesture interface device. *Proceedings of the ACM Conference on Human Factors in Computing Systems and Graphics Interface* (pp. 189–192). New York: ACM.

Chapter 41

MINDSCAPE RETUNING AND BRAIN REORGANIZATION WITH HYBRID UNIVERSES: THE FUTURE OF VIRTUAL REHABILITATION

Cali Fidopiastis and Mark Wiederhold

The hallmark of virtual reality (VR) technologies is the capacity to deliver real time simulations of real world contexts that allow for user interaction through multimodal sensory stimulation (Burdea & Coiffet, 2004, p. 3). As a rehabilitation tool, VR augments the therapist's capability to provide such essential therapeutic elements as programmable systematic practice, engaging interaction, immediate feedback, safe nondistracting patient environments, simulations of objects or events difficult to replicate in real life, and environmental manipulations not capable in the real world (Holden, 2005; Riva, 2005; Rose, Brooks, & Rizzo, 2005). More importantly, the ability to create personalized environments that match the disability level of the patient is a major step forward in designing successful rehabilitation protocols (Wilson, Foreman, & Stanton, 1997).

The capability of VR to support a broad range of therapies has driven over 15 years of research, encompassing both cognitive and physical rehabilitation. Researchers have successfully applied advanced VR simulations to improve motor and sensorimotor functions, to assess the extent of disability, to treat anxiety disorders, and to offer clinician-assisted telerehabilitation (Rizzo, Brooks, Sharkey, & Merrick, 2006; Wiederhold, 2006). While effective therapy is the goal of VR based rehabilitation, we predict that the more enduring and empowering legacy of this approach will be to advance our understanding of how the brain functions when persons perform real world tasks (Tarr & Warren, 2002).

While postulating future technologies that can diagnose and deliver exacting therapies is more colorful, the unanswered questions of how the brain performs everyday tasks leave these fanciful ideas as pale as the promise of affordable individual flying transporters. Thus, our more pragmatic reflection on the future of simulation based rehabilitation is that such technologies coupled with portable, noninvasive, unobtrusive neurosensing devices (for example, encephalography, EEG) will provide a unique window into the realistic workings of the behaving brain. In the context of rehabilitation, understanding the retuning and the reorganizing capabilities of the human brain is necessary for determining therapeutic efficacy as well as understanding prognoses and long-term care needs. This aim of this chapter is to highlight the emerging capabilities of virtual worlds and couple them with advances in our understanding of brain function and recovery.

EXTENSIBLE REALITIES AND THE THERAPEUTIC ADVANTAGE

As available technologies for presenting three-dimensional (3-D) graphics continue to improve and expand, virtual reality has come to mean a particular method of simulating virtual worlds. Milgram and Kishino (1994) describe the virtuality continuum as a range with the real environment on one end and purely virtual environments on the other. Mixed reality (MR) falls between the two extremes and merges the physical real world with the virtual world in an interactive setting within the same visual display environment, either with the real environment augmented by virtual objects (augmented reality) or the converse (augmented virtuality). This flexibility allows for the creation of hybrid therapies that employ a mixture of virtual elements and traditional protocols.

For example, exposure therapy where the patient actively confronts the feared object (for example, spiders) or situation (for example, public speaking) is recognized as a necessity for successful phobia treatment; however, there are many situations (for example, car accident) for which direct or graded confrontation is impractical and unacceptable (Wiederhold & Wiederhold, 2005). VR based therapy in these cases fills a necessary void. In fact, comparison studies between traditional and virtual reality exposure therapy show that VR therapy is equally as effective for treating social phobia (Klinger et al., 2005) and fear of flying (Rothbaum et al., 2006). There is also evidence that persons with specific types of phobias (for example, arachnophobia or fear of spiders) are more willing to seek VR based treatment than engaging in traditional exposure therapy (Garcia-Palacios, Botella, Hoffman, & Fabregat, 2007).

While advantageous in phobia therapy, VR is limited by the technological challenges of creating realistic feedback afforded by the physical properties of natural objects (for example, solidness of a cup). MR, however, incorporates the inherent advantages of immediate multisensory feedback from the physical world while retaining the advantages of VR. The real strength in this approach is its ability to re-create the human experience in a manner that potentially facilitates cognitive processing capabilities of the patient. This capability of MR is especially important for physical and cognitive neurorehabilitation applications.

Fidopiastis et al. (2006) demonstrated the importance of MR based therapy for facilitating activities of daily living training. In their single case study, the researchers replicated the kitchen of a patient with severe memory impairment in MR and demonstrated that this contextualized environment could assist in training breakfast-making skills, such as locating breakfast items. An important result of this study was that after training, the participant was able to prepare breakfast in his own kitchen without cuing, which he was unable to do before training. The skills learned in this contextualized (ecologically valid), controlled, and safe environment, therefore, transferred to the patient's own home. Thus, the MR based setting provided the patient with an interactive and flexible environment in which he could "safely explore his functional capabilities" (Fidopiastis et al., 2006, p. 186).

Both therapy examples above illustrate that the use of simulated real world environments is effective as rehabilitation tools for their respective application areas. While anxiety therapies provide a real world corollary for VR based treatment comparison, the heterogeneity of treatment methods for both simulation based and traditional cognitive rehabilitation precludes cross-comparisons (Cicerone et al., 2005; Fidopiastis, 2006). Judging treatment suitability for clinical adoption is an ongoing concern, especially for cognitive rehabilitation (Carney et al., 1999).

Standard therapeutic goals and outcome measures remain clinically undefined for both traditional and virtual cognitive rehabilitation treatments (Weiss, Kedar, & Shahar, 2006; Whyte & Hart, 2003). There is an estimated total lifetime expense of approximately $60 billion for persons with traumatic brain injury (TBI) in the United States (Finkelstein, Corso, Miller, & Associates, 2006). Thus, there is an urgent need to develop the methods and the metrics to soundly assess the impact of rehabilitation therapies and assistive devices today (Whyte, 2006).

Virtual rehabilitation technologies provide a therapeutic advantage in the development of methods and metrics for evaluating the effectiveness of rehabilitation treatments (Rizzo & Kim, 2005). First, virtual systems can be designed as a testbed where multiple technologies (for example, haptic devices and 3-D sound) and therapies (for example, functional training and neurofeedback) can be tested simultaneously or separately (Fidopiastis et al., 2005). Second, psychophysiological measures such as EEG can be integrated into the data output for real time data analysis (Fidopiastis, Hughes, Smith, & Nicholson, 2007). These two features allow for replicable systems that can provide cross-facility testing, quantify the level of therapeutic benefit, as well as characterize brain changes (both temporal and spatial), and lead to modular setups that are more appropriate for private use. More importantly, a single system can provide physical, cognitive, and psychological therapies allowing for programmatic level assessment. These advanced protocols will ultimately provide candid guidance for the development of future clinical technologies, such as brain-driven interfaces for the computer or other remote devices.

MINDSCAPE RETUNING AND BRAIN REORGANIZATION USING VIRTUAL REHABILITATION

Brian plasticity involves the capacity of the nervous system to modify its organization either structurally (for example, changes in neural connections) or functionally (for example, changes in neural patterns) due to experience, maturation, or injury (Ormerod & Galea, 2001). The essence of rehabilitation for both anxiety disorders and recovery from brain injury is to positively affect these neural changes and to subsequently enhance the patient's long-term quality of life (Bremner, Elzinga, Schmahl, & Vermetten, 2008; Kelly, Foxe, & Garavan,

2006). Identifying the types of therapies that most successfully meet these aims is the question of the rehabilitation sciences. An inherent problem in determining efficacy of any rehabilitation approach is the issue of individual differences. We propose that virtual rehabilitation tools coupled with state-of-the-art biosensing technologies are a means to not only characterize individual responses to therapy, but to extend our understanding of how persons physically, cognitively, and socially may perform during real world interactions, such as in vocational settings (Carney et al., 1999).

Neuroimaging studies using such devices as functional magnetic resonance imaging (fMRI) may suggest structural and functional brain differences that underlie the various psychiatric disorders and brain changes due to injury. For example, Rauch and Shin (2002) reviewed the neuroimaging literature across the spectrum of anxiety and stress disorders. The authors contend that while there are brain similarities among persons presenting with these disorders, there are features that are unique to each disorder phenotype.

These results support the idea that persistent cognitive impairments (for example, memory deficits) due to post-traumatic stress disorder (PTSD) may be attributable to maladaptive changes in the hippocampus and associated circuitry (for example, amygdala; Nemeroff et al., 2006). Although hyperarousal of the hippocampal circuitry is also seen in persons presenting with social phobias, these heightened activations are usually present only when the patient is faced with fear-evoking stimuli (Mataix-Cols & Phillips, 2007). Persons diagnosed with either disorder type, regardless of these brain differences, are most likely treated with similar psychotherapy (that is, exposure therapy). Yet, the specifics of the treatment protocol (for example, duration and type of medicines) are unsettled questions. This lack of specificity may lend to the chronicity or persistence of the disorder, especially for patients with PTSD. Issues such as these also affect neurorehabilitation treatment selection.

Neuroimaging technologies are also emerging methods for evaluating neurorehabilitation therapies that may lead to beneficial brain plasticity: neuroplasticity that results in cognitive or motoric improvement (Boyd, Vidoni, & Daly, 2007; Kelly et al., 2006). Thus, the emphasis in neurorehabilitation is to apply treatments that facilitate functional neural reorganization or regeneration (for example, Chen, Abrams, & D'Esposito, 2006). In contrast, modifying fear-related memory structures, what we call "mindscape retuning," is a primary goal of exposure therapies (for example, Foa & Kozack, 1986). The emphasis of these treatments may change as more information about the interconnectivity of brain areas and their contributions to cognitive processing (for example, learning) are elucidated.

Regional cerebral blood flow as measured by positron emission tomography does show attenuation or retuning of the neural network associated with fear responding (hippocampus, amygdala, and medial prefrontal cortex) after traditional psychotherapy (De Raedt, 2006). Similar studies are in progress using fMRI for VR based exposure therapy to better understand its underlying neural mechanisms (Hoffman, Richards, Coda, Richards, & Sharar, 2003). There is also fMRI evidence that VR based physical therapy leads to practice-induced cortical reorganization in the primary sensory motor cortex and subsequent improvement in motor behavior (You et al., 2005). The utility of specifying neural responses that underlie any treatment is the capacity to predict therapeutic outcomes on an individual basis (Paulus, 2008; Chen et al., 2006).

As the current review suggests, informing treatment specificity and outcomes with neuroimaging is a nascent field of research. Evidence based analyses of treatments currently depend upon traditional experimental designs (for example, specified control groups and outcome metrics). Mahncke, Bronstone, and Merzenich (2006) provide an example of an evidence based therapeutic framework, Posit Science's Brain Fitness training program, for remediating plasticity processes with negative consequences (for example, weakened neuromodulatory control and negative learning). Outcomes of such training are correlated with improved neuropsychological measures of memory in healthy older adults (Mahncke, Connor, et al., 2006). Extensive training and practice involving these processes have immense potential for improving higher order cognitive functions, such as memory. More importantly, the treatment offers a means to measure individual gains.

Kelly et al. (2006) and Mahncke, Connor, et al. (2006) both utilized computer based programs to target practice-induced plasticity within cognitive processing networks to achieve positive functional results. Such computer based programs, however, do not offer engagement of senses beyond visual and audio. These types of therapies may not benefit brain areas with more complex functions, such as the prefrontal cortex (PFC).

Chen et al. (2006) suggest that restoring function to the PFC and associated networks should be a priority when considering cognitive rehabilitation treatments for persons with TBI. The PFC is thought to exert executive control over integrated cognitive processes, such as attention and working memory, which are necessary to perform goal-directed tasks. This brain area and related network is particularly susceptible to damage from head injury. Thus, training tasks that target and engage the PFC networks may enhance PFC functioning and the resulting emergent executive processing. Virtual rehabilitation therapy, using mixed reality in particular, enables multimodal stimulation and programmable virtual environments that meet the criteria for stimulating, engaging, and training integrated neural processes underlying higher cognitive functions. The opportunity for MR based therapy is to integrate portable, unobtrusive neurosensing devices that can be comfortably worn in the 3-D virtual rehabilitation environment.

NEW SCOPE AND FOCUS

Under the Virtual Technologies and Environments initiative sponsored by the Office of Naval Research, the Media Convergence Laboratory (MCL) and the Institute for Simulation and Training at the University of Central Florida created the human experience modeler (HEM) testbed to evaluate technologies of virtual, augmented, and mixed reality that may enhance cognitive rehabilitation effectiveness (Fidopiastis et al., 2006). This reconfigurable, portable MR based training solution was extended to provide therapy for persons experiencing PTSD, TBI, and loss of limb. Currently, the MCL team has partnered with The Virtual Reality Medical Center to produce an MR system that assists physical rehabilitation for stroke-disabled patients with upper-extremity hemiplegia.

The field of augmented cognition has developed technologies and protocols that can improve the information processing capabilities of learners within their operational environment. The use of biosensing devices, including functional near-infrared imaging, to determine real time changes in the cognitive state of the learner is one innovation of the field. Another innovation is the coupling of cognitive state measures to adaptive system changes within the simulation based training environment. This training system flexibly modifies the information exchange between the learner and the training material such that the learning state of the operator is optimized. Incorporating these features within the HEM will allow for real time assessment of the patient's behavioral and cognitive changes within a contextualized environment. More importantly, this merger offers a greater potential for determining an effective rehabilitation strategy that not only shows promise in the clinic, but also transfers such successes to the home. The potential of demonstrating improved quality of life and overall functional outcomes for persons with anxiety disorders or impairments due to trauma is a true advance forward for the rehabilitation sciences.

REFERENCES

Bremner, J. D., Elzinga, B., Schmahl, C., & Vermetten, E. (2008). Structural and functional plasticity of the human brain in posttraumatic stress disorder. *Progress in Brain Research, 167,* 171–186.

Boyd, L. A., Vidoni, E. D., & Daly, J. J. (2007). Answering the call: The influence of neuroimaging and electrophysiological evidence on rehabilitation. *Physical Therapy, 87* (6), 684–703.

Burdea, G. C., & Coiffet, P. (2004). *Virtual reality technology* (2nd ed.). New York: John Wiley & Sons.

Carney, N., Chestnut, R. M., Maynard, H., Mann, N. C., Paterson, P., & Helfand, M. (1999). Effect of cognitive rehabilitation on outcomes for persons with traumatic brain injury: A systematic review. *Journal of Head Trauma Rehabilitation, 14,* 277–307.

Chen, A., Abrams, G. M., & D'Esposito, M. (2006). Functional reintegration of prefrontal neural networks for enhancing recovery after brain injury. *Journal of Head Trauma Rehabilitation, 21*(2), 107–118.

Cicerone, K. D., Dahlberg, C., Malec, J. F., Langenbahn, D. M., Felicetti, T., Kneipp, S., et al. (2005). Evidence-based cognitive rehabilitation: Updated review of the literature from 1998 through 2002. *Archives of Physical Medicine and Rehabilitation, 86*(8), 1681–1682.

De Raedt, R. (2006). Does neuroscience hold promise for the further development of behavior therapy? The case of emotional change after exposure in anxiety and depression. *Scandinavian Journal of Psychology, 47*(3), 225–236.

Fidopiastis, C. M. (2006). *User-centered virtual environment assessment and design for cognitive rehabilitation applications.* Ph.D. dissertation, University of Central Florida,

Orlando, FL. Retrieved March 5, 2008, from Dissertations & Theses database. (Publication No. AAT 3233649).

Fidopiastis, C. M., Hughes, C. E., Smith, E. M., & Nicholson, D. M. (2007, September). Assessing virtual rehabilitation design with biophysiological metrics [Electronic version]. *Proceedings of Virtual Rehabilitation 2007*, 89.

Fidopiastis, C. M., Stapleton, C. B., Whiteside, J. D., Hughes, C. E., Fiore, S. M., Martin, G. M., et al. (2005, September). *Human experience modeler: Context-driven cognitive retraining to facilitate transfer of learning.* Paper presented at the 4th International Workshop on Virtual Rehabilitation (IWVR), Catalina Island, CA.

Fidopiastis, C. M., Stapleton, C. B., Whiteside, J. D., Hughes, C. E., Fiore, S. M., Martin, G. M., et al. (2006). Human experience modeler: Context-driven cognitive retraining to facilitate transfer of learning. *CyberPsychology & Behavior, 9*(2), 183–187.

Finkelstein, E. A., Corso, P. S., Miller, T. R., & Associates. (2006). *The incidence and economic burden of injuries in the United States.* New York: Oxford University Press.

Foa, E. B., & Kozak, M. J. (1986). Emotional processing of fear: Exposure to corrective information. *Psychological Bulletin, 99*, 20–35.

Garcia-Palacios, A., Botella, C., Hoffman, H., & Fabregat, S. (2007). Comparing acceptance and refusal rates of virtual reality exposure vs. in vivo exposure by patients with specific phobias. *Cyberpsychology & Behavior, 10*(5), 722–724.

Hoffman, H. G., Richards, T., Coda, B., Richards, A., & Sharar, S. R. (2003). The illusion of presence in immersive virtual reality during an fMRI brain scan. *Cyberpsychology & Behavior, 6*(2), 127–123.

Holden, M. K. (2005). Virtual environments for motor rehabilitation: Review. *CyberPsychology & Behavior, 8*(3), 187–211.

Kelly, C., Foxe, J. J., & Garavan, H. (2006). Patterns of normal human brain plasticity after practice and their implications for neurorehabilitation. *Archives of Physical Medicine and Rehabilitation, 87*(2), S20–S29.

Klinger, E., Bouchard, S., Legeron, P., Roy, S., Lauer, F., Chemin, I., et al. (2005). Virtual reality therapy versus cognitive behavior therapy for social phobia: A preliminary controlled study. *CyberPsychology & Behavior, 8*(1), 76–88.

Mahncke, H. W., Bronstone, A., & Merzenich, M. M. (2006). Brain plasticity and functional losses in the aged: scientific bases for a novel intervention. *Progress in Brain Research, 157*, 81–109.

Mahncke, H. W., Connor, B. B., Appelman, J., Ahsanuddin, O. N., Hardy, J. L, Wood, R., et al. (2006). Memory enhancement in healthy older adults using a brain plasticity-based training program: A randomized, controlled study. *Medical Sciences, 33*, 12523–12528.

Maitix-Cols, D., & Phillips, M. L. (2007). Psychophysiological and functional neuroimaging techniques in the study of anxiety disorders. *Psychiatry, 6*(4), 156–160.

Milgram, P., & Kishino, A. F. (1994). Taxonomy of mixed reality visual displays. *IEICE Transactions on Information and Systems, E77-D(12),* 1321–1329.

Nemeroff, C. B., Bremner, J. D., Foa, E. B., Mayberg, H. S., North, C. S., & Stein, M. B. (2006). Posttraumatic stress disorder: A state-of-the-science review. *Journal of Psychiatric Research, 40*(1), 1–21.

Ormerod, B. K., & Galea, L. (2001). Mechanisms and function of adult neurogenesis. In C. A. Shaw & J. C. McEachern (Eds.), *Toward a theory of neuroplasticity* (pp. 85–100). Lillington, NC: Taylor & Francis.

Paulus, M. P. (2008). The role of neuroimaging for the diagnosis and treatment of anxiety disorders. *Depression and Anxiety, 25,* 348–356.

Rauch, S. L., & Shin, L. M. (2002). Structural and functional imaging of anxiety and stress disorders. In K. L. Davis, D. Charney, J. T. Coyle, & C. Nemeroff (Eds.), *Neuropsychopharmacology: The fifth generation of progress* (pp. 953–966). Philadelphia: Lippincott Williams & Wilkins.

Riva, G. (2005). Virtual reality in psychotherapy: Review. *CyberPsychology & Behavior, 8*(3), 220–230.

Rizzo, A. A., & Kim, G. J. (2005). A SWOT analysis of the field of virtual rehabilitation and therapy. *Presence: Teleoperators and Virtual Environments, 14*(2), 119–146.

Rizzo, A. S., Brooks, T., Sharkey, P. M., & Merrick, J. (2006). Advances in virtual reality therapy and rehabilitation. *International Journal on Disability and Human Development, 5*(3), 203–204.

Rose, F. D., Brooks, B. M., & Rizzo, A. A. (2005). Virtual reality in brain damage rehabilitation: Review. *CyberPsychology & Behavior, 8*(3), 241–262.

Rothbaum, B. O., Anderson, P., Zimand, E., Hodges, L., Lang, D., & Wilson, J. (2006). Virtual reality exposure therapy and standard (in vivo) exposure therapy in the treatment of fear of flying. *Behavior Therapy, 37,* 80–90.

Tarr, M. J., & Warren, W. H. (2002). Virtual reality in behavioral neuroscience and beyond. *Nature, Neuroscience Supplement, 5,* 1089–1092.

Weiss, P. L., Kedar, R., & Shahar, M. (2006). Ties that bind: An introduction to domain mapping as a visualization tool for virtual rehabilitation. *CyberPsychology & Behavior, 9*(2), 114–122.

Whyte, J. (2006). Using treatment theories to refine the designs of brain injury rehabilitation treatment effectiveness studies. *Journal of Head Trauma and Rehabilitation, 21*(2), 99–106.

Whyte, J., & Hart, T. (2003). It's more than a black box, it's a Russian doll: Defining rehabilitation treatments. *American Journal of Physical Medicine and Rehabilitation, 82*(8), 639–652.

Wiederhold, B. (2006). CyberTherapy 2006. *CyberPsychology & Behavior, 9*(6), 651–652.

Wiederhold, B. K., & Wiederhold, M. D. (2005). *Virtual reality therapy for anxiety disorders: Advances in evaluation and treatment.* Washington, DC: American Psychological Association.

Wilson, P. N., Foreman, N., & Stanton, D. (1997). Virtual reality, disability and rehabilitation. *Disability and Rehabilitation, 19*(6), 213–220.

You, S. H., Jang, S. H., Kim, Y. H., Hallett, M., Ahn, S. H., Kwon, Y. H., et al. (2005). Virtual reality-induced cortical reorganization and associated locomotor recovery in chronic stroke: An experimenter-blind randomized study. *Stroke, 36,* 1166–1171.

Chapter 42

PERSONAL LEARNING ASSOCIATES AND THE NEW LEARNING ENVIRONMENT

J. D. Fletcher

Much education, training, problem solving, performance aiding, decision aiding, and the like, may, in the not-distant future, rely on dialogues or conversations with personalized computer based devices, which might be called personal learning associates (PLAs). It further seems likely that these devices will be used as portals into virtual worlds and virtual environments where these dialogues will continue in combination with other experiences, contexts, and conditions.

Functionally, such a PLA-inhabited world might rely on three key components:

1. A global information infrastructure, such as today's World Wide Web, populated by sharable digital objects. These objects could be content for display, such as text, video, virtual "islands," and avatars. They could also be nondisplay materials, such as algorithms, instructional strategies, software tools, and databases.

2. Servers to locate and retrieve these digital objects and assemble them to support interactions with users and learners.

3. Devices that serve as PLAs for users and learners. They could be handhelds and laptops so that they are available on demand, anytime, anywhere. They could also be hosted on platforms ranging from integrated circuits to mainframes. The PLAs could be linked for use by groups of geographically dispersed learners working collaboratively. They will be personal accessories, but they need not be limited to individual uses.

TRENDS

There are historical and technological trends in education, training, and elsewhere that point to the likelihood, if not inevitability, of PLA devices and capabilities. In discussing these trends we need a generic term for education, training, performance aiding, problem solving, decision aiding, and similar capabilities. For convenience they are lumped together here and called "learning."

SOME HISTORICAL TRENDS IN LEARNING

In the primordial beginnings and for perhaps 100,000 years thereafter, learning involved direct, in-person interactions between learners and a sage. Seven thousand or so years ago we learned how to write, which effected a major revolution in learning. People with enough time and resources could study the words of sages without having to rely on face-to-face interaction or the vagaries of human memory. Learning began to move in an on-demand, anytime, anywhere direction.

The next step was the development of books (that is, something beyond mud or stone tablets). As discussed by Kilgour (1998), books were based on papyrus and parchment rolls until about 300 B.C. when the Romans began to sew sheets of parchment together into codices. These were cheaper to produce because they were based on locally available parchment made from animal skin and allowed content to be placed on both sides of the sheets.

Use of paper prepared from linen and cotton in about A.D. 100 (China) and A.D. 1200 (Europe) made books even less expensive. Their lowered costs made them more available to a literate and growing middle-class who, in turn, increased the demand for more cost reductions, more books, and more of the learning they provided. This demand led to the introduction of books printed from moveable type, first in China around A.D. 1000, and later in Europe in the mid-1400s (Kilgour, 1998). Learning then continued to become more widely and inexpensively available on demand, anytime, anywhere.

Next, after about 500 years, comes the computer. With its ability to adapt the sequence and type of operations based on conditions of the moment—or microsecond—computer technology may effect yet another revolution in learning. While preserving the capabilities of writing and books to present learning content on demand, it can also provide guidance and tutorial interactions as needed by individual learners. This combination of learning and individually tailored interactivity is not something books, movies, television, or videotape technologies can do to any appreciable degree. It is a new and significant capability for learning.

In short, the progression of learning across human history appears to be toward increased on-demand, anytime, anywhere access to learning. Aided by computer technology, it seems likely to continue. At least that is the argument presented here.

TECHNOLOGY

Many technologies evolve in directions that no one foresees. We had steam engines before railways, wireless telegraph before radio, microwave transmitters before microwave ovens, the Internet before the Web, and so forth. Still, there may be value in trying to envision where our technologies may be taking us. Knowing in advance where we are going can help us get there—or avoid doing so, should that seem more prudent.

It has been suggested that the future is already here, but unrecognized and unevenly distributed. When it comes to learning and learning environments we might ask what is currently unrecognized and unevenly distributed to see where these environments, and we, may be headed. We might begin by hazarding a list of possibly relevant trends and capabilities that are already at hand. Such a list could include the following:

Moore's Law. In 1965 Gordon Moore, a co-founder of Intel Corporation, noted casually that engineers were doubling the number of electronic devices on chips every year. If we expand Moore's time estimate to 18 months, our expectations fit reality quite closely (Brenner, 1997). This pace of development seems likely to continue. Gorbis and Pescovitz (2006) found that about 70 percent of IEEE (Institute of Electrical and Electronic Engineers) Fellows expect Moore's Law to continue holding for at least 10 more years. About 35 percent of them expect it to continue beyond that, up to 20 years. The major consequence of Moore's Law for PLAs is that the technology needed to support them will become increasingly more compact and affordable.

Computer Communications and Networking. The most dramatic and globally pervasive manifestations of computing in our daily lives seem to be the Internet and the World Wide Web. Web use grew about 266 percent between 2000 and 2007, with more than 1.3 billion learners and users of all sorts worldwide as of December 2007 (Internet World Stats, 2007). The Web and the evolving global information infrastructure have made vast amounts of human information—and misinformation— globally accessible. Tens of thousands of people can participate in massively multiplayer online games, such as *EverQuest, Final Fantasy, RuneScape,* and *World of Warcraft*. Similar multitudes of globally dispersed learners may soon be participating in virtual environments through PLAs.

The Semantic Web. The Semantic Web (Berners-Lee, Hendler, & Lassila, 2001), which is being developed under the auspices of the World Wide Web Consortium, should improve cooperation between computers and human beings by imbuing Web information with meaning and ontological connections. These connections are expected to expose semantic linkages between disparate bodies of knowledge regardless of how different they may appear to be at first (for example, Chandrasekaran, Josephson, & Benjamins, 1999). They will make it possible to develop increasingly powerful, accurate, and comprehensive models of learners for use in tailoring learning environments and their interactions to individual needs and interests (Dodds & Fletcher, 2004). They may add substantially to the adaptability and realism of virtual environments.

Computer Graphics, Video, and Animation. The validity of the multimedia principle, which states that people can absorb more information from words and pictures presented together than from words alone, seems well established by research and ensuing cognitive theory (Fletcher & Tobias, 2005). Enhancements in multimedia capabilities (for example, graphics, video, and animation) now available in virtual environments, and therefore available to PLAs, increase the power, flexibility, and functional range of learning environments and, thanks to the multimedia principle, the retention and transfer of what is learned from them.

Learning Objects. Object-oriented applications are becoming ubiquitous. The development of specifications to make learning objects accessible, interoperable, reusable, and durable is an integral part of this trend. These specifications have been described elsewhere (for example, Fletcher, Tobias, & Wisher, 2007; Wiley, 2000). The objects are packaged in metadata, which describes what is in the package, and are being made available on the global information infrastructure, allowing object-oriented applications, such as we might find in PLAs, to identify, locate, and access them, thereby enhancing the flexibility, responsiveness, and adaptability of leaning environments.

Natural Language Processing. The steadily growing capabilities of computer technology to participate in natural language conversations (for example, Graesser, Gernsbacher, & Goldman, 2003) will significantly enhance the mixed initiative dialogues in which participants, both computer generated and real, participating in learning environments can initiate interactions. One can imagine turning an avatar loose on the global information infrastructure to find advice or to answer a question by locating relevant learning objects and/or engaging humans and other avatars in conversations and returning to report when it judges itself ready. Language barriers should diminish in virtual environments as avatars and human participants become increasingly able to interact using a variety of languages (for example, Chatham, in press). Given the economic windfall promised by reliable natural language understanding by computers, it seems likely that these capabilities will continue to develop.

Individualized, Computer-Assisted Learning. Major improvements over classroom instruction occur when education and training can be presented in tutorial, individualized interactions. The difference can amount to two standard deviations as, for instance, Bloom (1984) found. However, we cannot afford a single human instructor for every learner nor a single advisor for every problem solver. A solution to this problem may be found, as Fletcher (1992) and Corbett (2001) have suggested, by using computers to make affordable the substantial benefits of individualized, tutorial learning suggested by Bloom's research.

Computer technology captured these benefits early on. Since the 1960s they have tailored (a) rate of progress for individual learners, (b) sequences of instructional content and interactions to match each learner's needs, (c) content itself—providing different learners with different content depending on what they have mastered, and (d) difficultly levels to ensure that the tasks for the learner are not so easy as to be boring or so difficult as to seem impossible. These capabilities have been available and used in computer based instruction from its inception (for example, Coulson, 1962; Galanter, 1959; Suppes, Jerman, & Brian, 1968).

By the early 1970s, the effectiveness of using computer technology to individualize learning was generally recognized (for example, Ford, Slough, & Hurlock, 1972; Vinsonhaler & Bass, 1972). Findings from many studies comparing the use of computers in learning to standard classroom practice may be summarized, statistically, by a "rule of thirds." This rule suggests that the learning capabilities we would expect to find on computer based devices, such as PLAs, can reduce the cost of delivering instruction by about one-third and, beyond that, either reduce instructional time to reach instructional goals by about one-third (holding learning constant) or increase the skills and knowledge acquired by about one-third while holding instructional time constant.

As a statistical summary that is silent about cause, the rule of thirds is compatible with Clark's (1983) often-cited point that it is not technology itself, but what we do with it that matters. Still, the demonstrably attainable savings that the rule of thirds reports in time to learn can be expected to be found in the use of PLAs and could reduce the costs of specialized skill training in the Department of Defense by as much as 25 percent (Fletcher, 2006). Similar cost savings are attainable through the use of PLAs as performance aids in equipment maintenance (Fletcher & Johnston, 2002).

Intelligent Tutoring Systems. The key and historical difference between computer-assisted instruction and intelligent tutoring systems is a substantive matter and more than a marketing term. When intelligent tutoring was first introduced into computer-assisted instruction, it concerned quite specific goals that were first targeted in the 1960s (Carbonell, 1970; Fletcher & Rockway, 1986; Sleeman & Brown, 1982).

Two defining capabilities were that intelligent tutoring systems should

- Allow either the system or the learner to ask open-ended questions and initiate a "mixed-initiative" dialogue as needed or desired for learning. Mixed-initiative dialogue requires a language that is shared by both the system and the learner. Natural language has been a frequent and continuing choice for this capability (for example, Brown, Burton, & DeKleer, 1982; Collins, Warnock, & Passfiume, 1974; Graesser, Person, & Magliano, 1995; Graesser, Gernsbacher, & Goldman, 2003), but the language of mathematics, mathematical logic, electronics, and other well-structured communication systems have also been used (Barr, Beard, & Atkinson, 1975; Suppes, 1981; Sleeman & Brown, 1982; Psotka, Massey, & Mutter, 1988).

- Generate learning material and interactions on demand rather than require developers to foresee and prestore all such materials and interactions needed to meet all possible eventualities. This capability involves not just generating problems tailored to each learner's needs, but also providing coaching, hints, critiques of completed solutions, appropriate and effective teaching strategies, and, overall, the interactions and presentations characteristic of individualized, tutorial learning environments. Generative capability remains key to the full range of PLA capabilities envisioned here.

Early applications such as BIP in computer programming (Barr, Beard, & Atkinson, 1975), BUGGY in subtraction (Brown & Burton, 1978), EXCHECK in mathematical logic (Suppes, 1981), SOPHIE in electronic troubleshooting (Brown, Burton, & DeKleer, 1982), and others demonstrated that the necessary capabilities to model subject matter and match it with models of the learner and generate interactions on demand and in real time are within our technical grasp. Development of these capabilities has continued to improve their performance (for example, Luckin, Koedinger, & Greer, 2007; McCalla, Looi, Bredeweg, & Breuker, 2005; Polson & Richardson, 1988; Psotka, Massey, & Mutter, 1988).

PLA OPERATIONS, FUNCTIONALITIES, AND CAPABILITIES

What happens as we begin to combine the above technologies, among others, into learning applications? What might we expect a PLA to be and do?

A PLA might be carried in a pocket or on a belt, worn as a shirt, or even implanted. It will operate wirelessly, accessing the global information infrastructure. It will include all the eagerly sought and widely used functionalities found on today's mobile telephones—e-mail, games, instant messaging, and even voice communication between people. It will use natural language, speech and/or text, to communicate—although other modes, including the language of science, mathematics, and engineering, will be available. It will provide a full range of media for interactions, including graphics, photographics, animation, video, and the like.

An important feature of PLAs will be their ability to allow participation in virtual environments and simulations, which could be used as virtual laboratories, mimicking equipment, situations, markets, and so forth. Virtual laboratories will allow the learner to test different hypotheses concerning the subject matter, try out different problem solving strategies and solutions, participate in collaborative learning and problem solving, and examine the effects and implications of different decisions. Because PLAs will be able to link to other PLAs, they will be able to contact experts and engage with other learners in virtual environments using software tools (for example, Soller & Lesgold, 2003) that identify and assemble potential communities of interest and enhance communication and collaboration within them.

PLAs will become intensely personal accessories. Through explicit and/or implicit means, they will develop, test, and modify models of the learner(s). These models will reflect each learner's knowledge, skills, abilities, interests, values, objectives, and style of encoding information. By using this information to access the global information infrastructure, PLAs will be able to collect and assemble precisely the learning objects that an individual needs to learn, solve a problem, or make a decision. In effect, PLAs may provide a polymath in every pocket, accessing the whole of human knowledge and information, filtering and adapting it for relevance and accuracy, and supplying it, on demand, in a form and level of difficulty that an individual learner is prepared to understand and apply.

By incorporating natural language understanding, PLAs may provide the goal-directed, on-demand, interactive conversations that have long been the goal of automated learning (for example, Uttal, 1962). The foundation of these interactions would be a mixed-initiative conversation between the learner and the PLA to achieve targeted objectives.

PLA INFRASTRUCTURE: PROGRESS

It seems reasonable to anticipate the ready availability of devices that can support PLA functions. Moore's Law, computer communications, wireless infrastructure, and the development of handheld computing should all help ensure this outcome.

The sharable learning objects required by PLAs are achievable, but not so easily assumed. These require effort and agreement among developers more than scientific breakthroughs. The global information infrastructure, currently instantiated as the World Wide Web, is obviously in place. It needs to be complemented by capabilities that automatically and precisely locate digital objects that will operate on most, perhaps all, of the PLA platforms to which they might be delivered.

Objects that meet these criteria have been specified by the Sharable Content Object Reference Model (SCORM) developed by the Advanced Distributed

Learning (ADL) initiative (Dodds & Fletcher, 2004; Fletcher, Tobias, & Wisher, 2007). SCORM ensures that learning objects developed in accord with its specifications allow them to be interoperable across computing platforms of many types, durable across different versions of underlying system support software, and reusable across multiple environments and applications. SCORM has received global acceptance as a specification and is progressing through the steps needed to be certified as an international standard. Whether or not SCORM is the ultimate specification for supporting PLAs remains to be seen, but it is an essential beginning. It has demonstrated the feasibility and acceptability of sharable learning objects.

The issue of access remains. Even if the global information infrastructure is well populated with interoperable, reusable, and durable objects, the problem of finding precisely correct objects to meet PLA user requirements remains. The Content Object Registry/Repository Discovery and Resolution Architecture (CORDRA) and the accompanying ADL Registry infrastructure have made substantial advances toward this goal (Dodds & Fletcher, 2004; Fletcher, Tobias, & Wisher, 2007). CORDRA uses metadata packaging and ontologies to allow substantially more precise location of digital objects than the text crawling techniques of many current search engines. Its precision can be expected to continue improving with the development of the Semantic Web and other emerging capabilities. As with SCORM, the eventual tool used by PLAs may or may not be CORDRA based, but its functionalities are likely to remain quite similar.

SCORM and CORDRA give us the means to populate the global information infrastructure with PLA-usable objects. We have only to create them. That appears to be happening. A survey of learning materials developed for industry and government found that over 4 million SCORM objects had been produced (Rehak, 2006). More have been appearing steadily since that survey was made.

The critical part of PLA functioning, then, remains the capability of servers to assemble learning material on demand, in real time, and in accord with learners' needs. PLAs will implement the generative, dialogue based, information-structured capabilities called for by intelligent tutoring systems. In effect, PLAs must participate in the design and development of the learning environment—virtual and otherwise—in addition to presenting it. They can become more than just delivery systems. This goal has not yet been reached, but it appears achievable. Still, there is much that can be done in the interim.

FINAL WORD

It seems likely that the technological trends and capabilities discussed here and carried forward by the ancient and continuing trend toward on-demand, anytime, anywhere learning will lead to the appearance of something very much like PLA based learning environments. We may reasonably expect substantially increased, globally available learning opportunities through enhanced access to education, training, problem solving, performance aiding, and decision aiding—or learning—made possible by PLAs. We can expect learning to become more responsive and effective through the continuous assessment, learner modeling, and interactions tailored on demand to learner needs that our technologies are making feasible. Finally, we can expect PLAs to vastly enhance access to and use of virtual environments for learning.

REFERENCES

Barr, A., Beard, M., & Atkinson, R. C. (1975). A rationale and description of a CAI program to teach the BASIC programming language. *Instructional Science, 4,* 1–31.

Berners-Lee, T., Hendler, J., & Lassila, O. (2001). The semantic web. *Scientific American, 284,* 34–43.

Bloom, B. S. (1984). The 2 sigma problem: The search for methods of group instruction as effective as one-to-one tutoring. *Educational Researcher, 13,* 4–16.

Brenner, A. E. (1997). Moore's Law. *Science, 275,* 1551.

Brown, J. S., & Burton, R. R. (1978). Diagnostic models for procedural bugs in basic mathematical skills. *Cognitive Science, 2*(2), 155–192.

Brown, J. S., Burton, R. R., & DeKleer, J. (1982). Pedagogical, natural language and knowledge engineering in SOPHIE I, II, and III. In D. Sleeman & J. S. Brown (Eds.), *Intelligent Tutoring Systems* (pp. 227–282). New York: Academic Press.

Carbonell, J. R. (1970). AI in CAI: An artificial intelligence approach to computer-assisted instruction. *IEEE Transactions on Man-Machine Systems, 11,* 190–202.

Chandrasekaran, B., Josephson, J. R., & Benjamins V. R. (1999). Ontologies: What are they? Why do we need them? *IEEE Intelligent Systems and Their Applications, 14,* 20–26.

Chatham, R. E. (in press). Toward a second training revolution: Promise and pitfalls of digital experiential training. In K. A. Ericcson (Ed.), *Development of professional expertise: Toward measurement of expert performance and design of optimal learning environments.* Cambridge, United Kingdom: Cambridge University Press.

Clark, R. E. (1983). Reconsidering research on learning from media. *Review of Educational Research, 53,* 445–459.

Collins, A., Warnock, E. H., & Passfiume, J. J. (1974). *Analysis and synthesis of tutorial dialogues* (BBN Rep. No. 2789). Cambridge, MA: Bolt Beranek and Newman. (ERIC ED 088 512)

Corbett, A. (2001). Cognitive computer tutors: Solving the two-sigma problem. In M. Bauer, P. J. Gmytrasiewicz, & Y. Vassileva (Eds.), *User Modeling* (pp. 137–147). Berlin: Springer-Verlag.

Coulson, J. E. (Ed.). (1962). *Programmed learning and computer-based instruction.* New York: John Wiley and Sons.

Dodds, P. V. W., & Fletcher, J. D. (2004). Opportunities for new "smart" learning environments enabled by next generation web capabilities. *Journal of Education Multimedia and Hypermedia, 13*(4), 391–404.

Fletcher, J. D. (1992). Individualized systems of instruction. In M.C. Alkin (Ed.), *Encyclopedia of educational research* (6th ed., pp. 613–620). New York: Macmillan.

Fletcher, J. D. (2006). A polymath in every pocket. *Educational Technology, 46,* 7–18.

Fletcher, J. D., & Johnston, R. (2002). Effectiveness and cost benefits of computer-based aids for maintenance operations. *Computers in Human Behavior, 18,* 717–728.

Fletcher, J. D., & Rockway, M. R. (1986). Computer-based training in the military. In J. A. Ellis (Ed.), *Military contributions to instructional technology* (pp. 171–222). New York: Praeger Publishers.

Fletcher, J. D., & Tobias, S. (2005). The multimedia principle. In R. E. Mayer (Ed.), *The Cambridge handbook of multimedia learning* (pp. 117–133). New York: Cambridge University Press.

Fletcher, J. D., Tobias, S., & Wisher, R. L. (2007). Learning anytime, anywhere: Advanced distributed learning and the changing face of education. *Educational Researcher, 36*(2), 96–102.

Ford, J. D., Slough, D. A., & Hurlock, R. E. (1972). *Computer assisted instruction in Navy technical training using a small dedicated computer system: Final report* (Research Rep. No. SRR 73-13). San Diego, CA: Navy Personnel Research and Development Center.

Galanter, E. (Ed.). (1959). *Automatic teaching: The state of the art.* New York: John Wiley & Sons.

Gorbis, M., & Pescovitz, D. (2006). IEEE Fellows survey: Bursting tech bubbles before they balloon. *IEEE Spectrum, 43*(9), 50–55.

Graesser, A. C., Gernsbacher, M. A., & Goldman, S. (Eds.). (2003). *Handbook of discourse processes.* Mahwah, NJ: Lawrence Erlbaum.

Graesser, A. C., Person, N. K., & Magliano, J. P. (1995). Collaborative dialogue patterns in naturalistic one-on-one tutoring. *Applied Cognitive Psychology, 9,* 495–522.

Internet World Stats. (2007). *Usage and population statistics.* Retrieved January 30, 2008, from http://www.internetworldstats.com

Kilgour, F. G. (1998). *The evolution of the book.* New York: Oxford University Press.

Luckin, R., Koedinger, K. R., & Greer, J. (Eds.). (2007). *Artificial intelligence in education.* Amsterdam: IOS Press.

McCalla, G., Looi, C. K., Bredeweg, B., & Breuker, J. (Eds.). (2005). *Artificial intelligence in education.* Amsterdam: IOS Press.

Polson, M. C., & Richardson, J. J. (Eds.). (1988). *Intelligent tutoring systems.* Mahwah, NJ: Lawrence Erlbaum.

Psotka, J., Massey, L. D., & Mutter, S. A. (Eds.). (1988). *Intelligent tutoring systems: Lessons learned.* Hillsdale, NJ: Lawrence Erlbaum.

Rehak, D. R. (2006). Challenges for ubiquitous learning and learning technology. *Educational Technology, 46,* 43–49.

Sleeman, D., & Brown, J. S. (Eds.). (1982). *Intelligent tutoring systems.* New York: Academic Press.

Soller, A., & Lesgold, A. (2003). A computational approach to analyzing online knowledge sharing interaction. In U. Hoppe, F. Verdejo, & J. Kay (Eds.), *Proceedings of Artificial Intelligence in Education 2003* (pp. 253–260). Amsterdam: IOS Press.

Suppes, P. (Ed.). (1981). *University-level computer assisted instruction at Stanford: 1968–1980.* Stanford, CA: Institute for Mathematical Studies in the Social Sciences.

Suppes, P., Jerman, M., & Brian, D. (1968). *Computer-assisted instruction: The 1965–66 Stanford arithmetic program.* New York: Academic Press.

Uttal, W. R. (1962). On conversational interaction. In J. E. Coulson (Ed.), *Programmed learning and computer-based instruction* (pp. 171–190). New York: John Wiley and Sons.

Vinsonhaler, J. F., & Bass, R. K. (1972). A summary of ten major studies on CAI drill and practice. *Educational Technology, 12,* 29–32.

Wiley, D. (2000). *The instructional use of learning objects.* Retrieved January 30, 2007, from http://www.reusability.org/read

Chapter 43

THE FUTURE OF MUSEUM EXPERIENCES

Lori Walters, Eileen Smith, and Charles Hughes

THE UBIQUITOUS MUSEUM

The evolution of the museum will be tightly woven with the desires of what noted educator Marc Prensky defines as the "digital native" generation (Prensky, 2001). It is a generation that has no direct link to the vacuum tube, the rotary phone, or a world before satellites. To these individuals born after 1990, the personal computer is as much a natural component of life as a television for those born after 1960. Museums of the future have the incredible opportunity to merge the interactivity and visualization that the digital generation enjoys, while still maintaining one of the museum's paramount missions—the intergenerational transfer of cultural memory. Traditionally we have provided this intergenerational transfer as if free-choice learning centers were isolated points of interconnectivity. The links to exploration and learning should not weaken simply because an individual has left the brick and mortar confines of the museum. Future museums will have available to them a web of connectivity that provides for a seamless transfer of knowledge between visitor, facility, and beyond.

Current technology can permit learning data and experiences to travel with individual learners through time and space. Their museum experiences can be captured onto a personal learning "journal" that follows them after they depart the facility. Radio frequency identification (RFID) technology, personal digital assistants (PDAs) and cell phones, allow customizable exploration of the museum's learning experiences. As visitors make choices during their exploration on priority community issues at a science center, or examine artifacts and artwork of interest at a historical or art museum, their decisions (implicit and explicit) are logged into their journal. Data gathered at the museum on field trips can follow students back into the classroom, giving educators a tool that can link the discovery learning at the museum with curricular learning at their students' particular grade level. Families can access data gathered at the museum once they return home, allowing deeper and longer exploration than is possible at a museum exhibit.

Museums are just beginning to explore the potential of the digital revolution. While most maintain a Web presence, these Web sites offer little more than reformatted versions of their marketing information that a potential visitor could acquire at any tourist bureau kiosk. Museums that have ventured beyond the brochure phase and have developed supplemental materials for their larger exhibits have, to a large extent, yet to tap into the Internet's full interactive potential. Most sites offer generic experiences where everyone begins at the same point and few drill down through layers to locate information of particular interest to those individuals. Any interactivity comes not from how the visitor interacted with exhibits at the museum; it is derived from interaction with the Web site only—reinforcing the separation of the museum and its Web site.

The next generation of the museum Web site can provide interactive adventures that are customized to the interests of individual visitors. Museum experiences are a combination of personal, social, and physical contexts for learning as noted by Falk and Dierking (2000). This multifaceted context allows for rich exploration and connection to everyday life. While technology continues to evolve and will ultimately provide new solutions, the Media Convergence Laboratory at the University of Central Florida is examining this concept with current RFID technology at three partner museums. At each facility a network of RFID transceivers will be positioned throughout a specific exhibit hall. Visitors receive a pre-encoded disposable wristband, and as they walk through the exhibit the wristband sends data to a transceiver recording the length of time that a visitor interacts with a particular artifact/exhibit. Each exhibit is identified with a multitude of potential interests. The acquired data are used to create personalized online experiences that can be accessed by logging on to the facility's Web site and entering the number printed on each visitor's wristband. Another primary use of RFID is to customize any particular visitor's learning while in the museum facility itself. Exhibits can allow visitors to answer polls, prioritize issues, explore "what if?" scenarios, and keep the decisions made and the simulations experienced in their digital "files" throughout their visits. The RFID experience is completely anonymous as the numbered wristbands have no connection to individual names—thus there are no privacy issues to concern visitors. The personalized interactive Web site will allow further, deeper exploration of a topic of specific interest, stemming from the in-facility experiences that the individual or group had, and enhance the overall individual learning experience.

A significant secondary use for the captured data is providing a new tool to museum exhibit and education professionals by providing an exhibit layout and usage analysis for them. The captured RFID data will be available for aggregation, analysis, and display for the museum's development use. Animations of visitor paths can demonstrate how individuals travel through the on-site exhibit hall. Dwell time on specific exhibits can be revealed. The data can uncover general patron patterns that can be used to improve the current gallery in regard to flow and assist in the design of future learning experiences.

Today, museums are often resistant to displaying large portions of their exhibits online due to fears of the Web accessibility reducing the number of on-site visitors. This parochial thought does a disservice to two of the basic tenets of a museum—dissemination of learning experiences and stimulating curiosity to explore further. It also fails to address the desires of individuals who simply are unable to physically travel to that museum. The treasures entrusted to any historical or art museum, for example, are outside the physical reach of a large percentage of the world's population. This is not an issue of not wishing to visit—it is a true issue of inability. Museums of the future should embrace the connectivity of the Internet—not only for sharing their exhibits and collections with everyone, but for their own economic survival. The recent introduction of the 16 billion pixel digital image of Leonardo da Vinci's "The Last Supper" available on the Internet at www.haltadefinizione.com has the potential to significantly alter the relationship between the museum and the Internet.

Every 15 minutes, groups of 25 visitors are permitted to view the "The Last Supper"—which provides an annual visitation of 320,000 per year. With its placement on the Internet, anyone has the ability to view the masterpiece to a degree beyond that of even an in-person experience. Granted it would be cost and time prohibitive to digitize all historical artifacts and works of art to such a great degree of detail as the "The Last Supper," but there are multiple possibilities for potential sponsorship and other funding opportunities. As connectivity expands, museums will want to evolve as how they serve their communities, and rich discussion will occur on how they might serve the global community. We can easily imagine the installation of a series of high definition cameras within museums where virtual visitors could peruse a museum at any time of the day from any location. There are many possible operational issues to address, such as having live cameras in exhibit halls and what level of viewing would be possible for exhibits and floor programs, but those issues are beginning to be discussed by multiple industries as connectivity continues to expand.

The argument that offering Web based experiences reduces attendance figures can be continually debated. One could argue that an inviting Web portal would encourage many individuals to visit the facility in person. Others could claim that many will refrain from an in-person visit when an online experience could be accomplished from the confines of their lounge chair. The fact is that the museum experience and the online experience, if they are well designed, could not be more different. The museum experience is inherently social, where groups of people visit together, and their discussion, along with interaction with other museum visitors, makes a unique imprint when they leave. Online experiences are inherently personal and customized and serve to allow deeper exploration to a level not possible on a museum exhibit floor. The two experiences can indeed seed each other for ongoing learning and excitement.

In addition to connectivity for learners between learning environments, the future will be one where increased networking between museums might be seen. Imagine if many of your museum visits from this moment on could be interconnected. Each exhibit that you find interesting is noted, where museum exhibits are networked and your PDA or cell phone alerts you to exhibits within a self-determined distance based on your interests. As you pass through an unfamiliar town you are notified that a small museum contains an artifact or artwork of interest about which you may never have known. Networking is critical to museums in the future—in particular, to smaller museums where collections are often eclectic. Travelers may pick up a brochure that provides a broad overview of a facility and have no idea that within its walls is an artifact that is of particular interest to them. If smaller museums created a national artifact database, your PDA or phone could cross-check the database with your location and alert you to the artifact and guide you directly to the front door of the facility. This could be achieved with automatic alerts, even making you aware of objects of interest as you are driving down the highway or by explicit request with the choice being up to the user

THE EVOLVING MUSEUM

Museums have always been places with unique personalities. The uniqueness generally comes from a theme and a centerpiece that expresses that theme. Having a centerpiece can, however, be a mixed blessing. In general, the venue that has the greatest initial impact comes with a high price tag. The cost of the exhibit and the organization's identification with its implicit message make it difficult to replace the exhibit, even years after its novelty and the public's perception of its relevance have dissipated. The consequence is low attendance and minimal dwell time in a venue that typically consumes a large portion of the museum's overall real estate.

It is in this context of a highly valued but underused and largely static exhibit that the technology of mixed reality can and will make a positive difference. By mixed reality (MR), we are referring to simulation based experiences where the user is placed in an immersive setting that is either real with virtual asset augmentation or virtual with real world augmentation (Milgram & Kishino, 1994). Additionally, in the model proposed in Stapleton and Hughes (2003), the underlying story must draw on the user's imagination. This latter requirement is needed if the experience is to leave a lasting impression, a clear objective of museums (Stapleton & Hughes, 2006).

It is important to differentiate mixed from virtual reality. In a purely virtual experience, a user's visual system is dominated. This removes the experience from the current physical context. That is, clearly, undesirable when we want to maintain the venue's existing context. It is also counter to one of a museum's most important attributes, that of encouraging discussion and social interactions among visitors, especially among family members. In contrast, an MR experience does not exclude the current context; rather, it enhances that context. Visual contact with other humans still exists, and dialog is encouraged as the visitors are collectively surrounded by virtual and physical objects that seem to interact with each other.

As the above is vague, an example is in order. In its simplest form, mixed reality is a visual overlay, where virtual objects and virtual signage are placed in front of real objects. The mixed real/virtual information is seen either by wearing a head-mounted display (HMD) or viewing the scene through the mediation of a monitor. In the case of an HMD, there are two choices—video see-through and optical see-through (Rolland & Fuchs, 2000, Uchiyama, Takemoto, Satoh, Yamamoto, & Tamura, 2002). In the former, the real world is captured by cameras mounted on the outside of the HMD and the mixed reality is delivered on liquid crystal displays mounted on the inside and aligned with the user's eye. In the latter, the real world is seen through transparent lenses with the virtual content projected into the user's field of view. Using a monitor is a less expensive approach and usually involves mounting a camera on the back (nondisplay) side of the monitor. The camera captures reality, and that is merged with virtual content; the merged scene is then rendered onto the display. This paradigm can be extended to use large screens, such a dome screens, so the experience feels immersive. Audio, with real world sounds being heard naturally and synthetic ones being introduced through carefully placed speakers or via headsets, can augment the experience. Speakers are more hygienic and currently more capable of delivering three-dimensional precisely placed sounds (Hughes, 2005).

We noted that the easy case is when virtual content overlays physical objects, but MR can also deliver experiences where the real and virtual are intertwined with real objects partially occluding virtual ones and vice versa. This is the kind of MR that we believe can be used to reinvigorate wonderful, but stale content pieces in museums. The goal of the MR community is to achieve the holodeck from *Star Trek* fame. Such a future is quite probable (see Bimber, 2006, for work already in use at museums), given advances in the field of optics. However, its availability in the museums not directly associated with mixed reality researchers and at a price affordable to smaller- and medium-sized venues is still in the future, and the problem of unchanging exhibits is already being faced by museums that must compete in today's media centric world.

In 2004, to test out the concept of MR bringing an exhibit back to life, our lab developed the Sea Creatures experience at the Orlando Science Center's Dino-Digs exhibition hall (Hughes, Stapleton, Hughes, & Smith, 2005). This venue contains fossils of marine reptiles and fish in a clean, inviting environment. In the midst of this, we brought in a large dome screen, outfitted with a camera on the nonviewing side, which faced the main exhibits. Speakers were added above and around the viewing area. Prior to this, we had digitally scanned the area, acquiring a three-dimensional model that was perfectly registered with (overlaid on) the room's fixed objects (artifacts, exhibit cases, fossils, and support columns). See Figures 43.1(a) and (b)

When visitors walked around to the display side of the dome screen, they were greeted by a virtual guide who informed them of the kinds of sea life that they would have encountered in the Cretaceous Period. This virtual guide appears to be standing on the museum floor in front of all the real activities currently taking place in the hall. Typical young visitors look around to the other side seeing real people who also show up on the screen, but no sign of our helpful guide. Surprising to their family and/or friends, the curious youth is now part of the mixed

Figure 43.1. (a) shows the DinoDigs venue with the dome screen nonviewing side that contains a mounted camera to capture the real world.

reality. As the virtual guide leaves, the visitors hear rushing water and the entire venue appears to be experiencing a flood and then the Cretaceous Period reptiles appear. They swim in front of and behind the columns and display cases in the museum, and yet a peek around the side of the dome screen reveals nothing but normal activity. The visitors are encouraged to explore this enhanced world and, through story, seduced to return to the physical space and look more closely once the experience ends.

MR has the ability to enhance and attract; its inclusion adds new information and activities to the exhibit. The attraction is that each subsequent visit to the venue has the potential to expose new knowledge. Thus, a museum's long-term

Figure 43.1.   (b) shows the viewing side of the dome screen with mixed real and virtual content.

centerpiece experience can evolve and remain relevant, encouraging repeat visits and memberships that are so central to a healthy museum. Sea Creatures is just one example of how MR can reinvigorate museum experiences, allowing them to evolve with the new scientific knowledge and new interaction paradigms. Other examples include Geller (2006); Liu, Fernando, Cheok, Wijesena, and Tan (2007); and the extensive work done by the Augmented Reality Group at Bauhaus-University Weimar (Bimber & Raskar, 2005).

REFERENCES

Bimber, O. (2006). Augmenting holograms. *IEEE Computer Graphics and Applications, 26*(5), 12–17.

Bimber, O., & Raskar, R. (2005). *Spatial augmented reality: Merging real and virtual worlds.* Wellesley, MA: A K Peters.

Falk, J., & Dierking, L. (2000). *Learning from museums: Visitor experiences and the making of meaning.* Lanham, MD: Altamira Press.

Geller, T. (2006). Interactive tabletop exhibits in museums and galleries. *IEEE Computer Graphics and Applications, 26*(5), 6–11.

Hughes, C. E., Stapleton, C. B., Hughes, D. E., & Smith, E. (2005). Mixed reality in education, entertainment and training: An interdisciplinary approach. *IEEE Computer Graphics and Applications, 26*(6), 24–30.

Hughes, D. E. (2005, July). Defining an audio pipeline for mixed reality. In *Proceedings of Human Computer Interfaces International 2005-HCII'05* [CDROM]. Mahwah, NJ: Lawrence Erlbaum.

Kondo, T., Inaba, R., Arita-Kikutani, H., Shibasaki, J., Mizuki, A., & Minabi, M. (2007). Mixed reality technology at a natural history museum. In J. Trant & D. Bearman (Eds.), *Museums and the Web 2007*. Toronto, Canada: Archives & Museum Informatics. Available from http://www.archimuse.com/mw2007/papers/kondo/kondo.html

Liu, W., Fernando, O., Cheok, A., Wijesena, J., & Tan, R. (2007). Science museum mixed reality digital media exhibitions for children. In *2nd Workshop in Digital Media and Its Application in Museum & Heritage* (pp. 389–394.) Washington, DC: IEEE Computer Society.

Milgram, P., & Kishino, A. F. (1994). Taxonomy of mixed reality visual displays. *IEICE Transactions on Information and Systems, E77-D (12),* 1321–1329.

Prensky, M. (2001). Digital natives, digital immigrants. *On the Horizon, 9*(5), 1–2.

Rolland, J. P., & Fuchs, H. (2000). Optical versus video see-through head-mounted displays in medical visualization. *Presence: Teleoperators and Virtual Environments, 9* (3), 287–309.

Stapleton, C. B., & Hughes, C. E. (2003). Interactive imagination: Tapping the emotions through interactive story for compelling simulations. *IEEE Computer Graphics and Applications, 24*(5), 11–15.

Stapleton, C. B., & Hughes, C. E. (2006). Believing is seeing. *IEEE Computer Graphics and Applications, 27*(1), 80–85.

Uchiyama, S., Takemoto, K., Satoh, K., Yamamoto, H., & Tamura, H. (2002). MR platform: A basic body on which mixed reality applications are built. In *International Symposium on Mixed and Augmented Reality 2002-ISMAR2002* (pp. 246–256). Washington, DC: IEEE Computer Society.

ACRONYMS

AA	*America's Army*
AAR	after action review
AAV	amphibious assault vehicle
ACT-R	adaptive control of thought-Rational
ADL	Advanced Distributed Learning
ADW	air defense warfare
AFIST	Abrams full crew interactive simulator trainer
A4I	advanced active analysis adjunct for interactive multisensor analysis training
AI	artificial intelligence
AMIRE	authoring mixed reality
AO	area of operation
APE	auxiliary physics engine
API	application programming interface
AQT	advanced qualification training
AR	augmented reality
ARCI	acoustic rapid commercial off-the-shelf insertion
ARI	Army Research Institute
ARNG	Army National Guard
ASL	advanced scripting language
ASRA	Advanced Systems Research Aircraft
ASTD	American Society for Training and Development
ASW	antisubmarine warfare
ATELIER	architecture and technologies for inspirational learning environments
ATM	automated teller machine
AVs	autonomous vehicles
AW	aviation warfare
AWAVS	aviation wide-angle visual system
BARS	Battlefield Augmented Reality System
BCG	Brogden-Cronbach-Gleser
BiLAT	*Bilateral Negotiation*
BIP	basic instructional program
BL	blended learning

CAD	computer-aided design
CAN	Combined Arms Network of DVTE
CAPT	Combined Arms Planning Tool
CARP	Computerized Airborne Research Platform
CAS	close air support
CATS	cognitive avionics tool set
CAVE	cave automatic virtual environment
CCTT	close combat tactical trainer
CD	compact disc
CDMTS	common distributed mission training system
CFF	call for fire
CFn	composable FORCEnet
C4I	command, control, communications, computers, and intelligence
C4ISR	command, control, communications, computers, intelligence, surveillance, and reconnaissance
CGF	computer-generated forces
CGI	computer-generated imagery
ChrAVE	Chromakey Augmented Virtual Environment
CI	confidence interval
CLS	combat lifesaver
CO	commanding officer
COC	combat operations center
COD	Cognitive Delfin
COMPTUEXs	Composite Training Unit Exercises
CORDRA	Content Object Registry/Repository Discovery and Resolution Architecture
COTS	commercial off-the-shelf
COVE	Conning Officer Virtual Environment
CRM	customer relationship management
CRS4	Center for Advanced Studies, Research, and Development in Sardinia
CRT	cathode ray tube
CTA	cognitive task analysis
CTER	cumulative transfer effectiveness ratio
CTPS	combat trauma patient simulation
C2	command and control
C2V	command and control vehicle
CVE	collaborative virtual environment
DAGGERS	distributed advanced graphics generator and embedded rehearsal system
DARPA	Defense Advanced Research Projects Agency
DDKE	data-driven knowledge engineering
Desdemona	desoriëntatie demonstrator amst
DESRON	destroyer squadron
DI	dismounted infantrymen
DIS	distributed interactive simulation
DI-SAF	Dismounted Infantry Semi-Automated Force
DLP	digital light processing (trademark owned by Texas Instruments)

DMX	digital multiplex
DoD	Department of Defense
DoDD	Department of Defense directive
DoDI	Department of Defense instruction
DOF	degrees of freedom
DMO	distributed mission operations
DMT	distributed mission training
DSP	digital signal processing
DVTE	Deployable Virtual Training Environment
DVTE-CAN	Deployable Virtual Training Environment–Combined Arms Network
EAX	environmental audio extensions
ECATT/MR	embedded combined arms team training and mission rehearsal
ECG	electrocardiogram
ECS	emergency care simulator
EEG	electroencephalogram
EFL	English as a foreign language
ESN	European Simulation Network
ET	embedded training
ETDS	embedded training for dismounted soldiers
EU	European Union
EWS	Expeditionary Warfare School (USMC)
FAC	forward air controller
FATS	firearms training system
FBW	fly-by-wire
FCS	future combat system
FFW	future force warrior
FiST	Fire Support Team
FMB	full mission bridge
fMRI	functional magnetic resonance imaging
FMT	full mission trainer
FO	forward observer
FOM	federation object model
FOV	field of view
FPS	first-person shooter
FRS	Fleet Replacement Squadron
FRSS	forward resuscitative surgical system
FY	fiscal year
GAO	General Accounting Office
GFI	goodness-of-fit index
GIST	Gwangju Institute of Science and Technology
GM	General Motors Corporation
GOTS	government off-the-shelf
GPS	global positioning system
GUI	graphical user interface
HARDMAN	hardware versus manpower
HCI	human-computer interface
HD	high definition

HDTV	high definition television
HEM	human experience modeler
HFM	Human Factors and Medicine
HLA	high level architecture
HMD	head/helmet-mounted display
HMX-1	Marine Helicopter Squadron One, the Presidential Helicopter Squadron
HPSM	human performance systems model
HSI	human-systems integration/interaction
HVI	high value individual
IAAPA	International Association of Amusement Parks and Attractions
ICALT	International Conference on Advanced Learning Technologies
ICS	intercockpit communications system
ICV	infantry carrier vehicle
IED	improvised explosive device
IEEE	Institute of Electrical and Electronics Engineers
IG	image generator
I/ITSEC	Interservice/Industry Training, Simulation, and Education Conference
ILUMA	illumination under realistic weather conditions
IMAT	Interactive Multisensor Analysis Training
INVEST	intervehicle embedded simulation and training
IOC	Infantry Officer's Course (USMC)
IOS	instructor operator station
IPTs	integrated product teams
IR	infrared
IRS	internal referencing strategy
ISMAR	International Symposium on Mixed and Augmented Reality
ISMT	Indoor Simulated Marksmanship Trainer
ISNS	integrated shipboard network systems
IT	information technology
ITER	incremental transfer effectiveness ratio
ITK	Infantry Tool Kit
ITS	intelligent tutoring systems
J&J	Johnson & Johnson
J&JPRD	Johnson & Johnson Pharmaceutical Research & Development
JFCOM	Joint Forces Command
JO	junior officer
JSAF	Joint Semi-Automated Forces
JSF	joint strike fighter
JTAC	joint terminal air controller
JTEN	Joint Training and Experimentation Network
KR	knowledge of results
KSAs	knowledge, skills, and abilities/attitudes
LAN	local area network
LCAC	landing craft, air cushion
LCD	liquid crystal display
LCS	littoral combat ship

LED	light emitting diode
LRU	line replaceable unit
LSP	learning support package
LTC	lieutenant colonel
LVC	live, virtual, and constructive
LW	land warrior
MAGTF	Marine Air-Ground Task Force
MANPRINT	manpower and personnel integration
MCL	Media Convergence Laboratory
MGS	mobile gun system
MIDI	musical-instrument digital interface
MILES	multiple integrated laser engagement system
MITAC	map interpretation and terrain association course
MITAVES	map interpretation and terrain association virtual environment system
MMOG	massively multiplayer online game
MMORPGs	massively multiplayer online role-playing games
MMP	massively multiplayer
ModSAF	modular semi-automated forces
MOS	military occupational specialty
MOT^{2}IVE	Multi-Platform Operational Team Training Immersive Virtual Environment
MOUT	military operations on urban terrain
MPT	manpower, personnel, and training
MR	mission rehearsal
MR	mixed reality
MRC	Marmara Research Center
MRMC	Medical Research and Materiel Command
MSTC	medical simulation training centers
MTC	mission training center
MX	mixed reality toolkit
NASA	National Aeronautics and Space Administration
NATO	North Atlantic Treaty Organization
NAVAIR	Naval Air Systems Command
NCTE	Navy Continuous Training Environment
NFL	National Football League
NMCI	Navy Marine Corps Internet
NOPF	Naval Ocean Processing Facility
NPS	Naval Postgraduate School
NRC-FRL	National Research Council Flight Research Laboratory
NVG	night vision goggle
OCONUS	outside contiguous United States
OFET	Objective Force Embedded Training
OIF	Operation Iraqi Freedom
ONEnet	OCONUS Navy Enterprise Network
OneSAF	One Semi-Automated Forces
ONR	Office of Naval Research
ONS	operational needs statement

OOD	officers of the deck
OPL	Operator Performance Laboratory
ParaSim	parachute simulator
PC	personal computer
PCC	pre-command course
PDA	personal digital assistant
PFC	prefrontal cortex
PLA	personal learning associate
PTSD	post-traumatic stress disorder
QTEA	quality of training effectiveness assessment
radar	radio detection and ranging
RAM	random access memory
RAP	Ready Aircrew Program
R&D	research & development
RDECOM	Research, Development, and Engineering Command
RDECOM STTC	Research, Development and Engineering Command, Simulation and Training Technology Center
RFID	radio frequency identification
RGB	red, green, and blue
ROM	rough order of magnitude
RTI	run-time infrastructures
SAF	semi-automated forces
SAGAT	situation awareness global assessment technique
S&T	science and technology
SAPS	stand-alone patient simulator
SBIR	Small Business Innovative Research
SCORM	Sharable Content Object Reference Model
SCP	School for Command Preparation
SE	standard error
SE	systems engineering
SET	sonar employment trainer
SET-MR	scalable ET and mission rehearsal
SGI	Silicon Graphics, Inc.
SIGGRAPH	Special Interest Group on Graphics and Intermixed Techniques
SIMILAR	state, investigate, model, integrate, launch, assess, and reevaluate
SIMNET	simulation network
6 DOF	6 degrees of freedom
SLEP	service life extension program
SLOC	source lines of code
SMART	simulated mission and rehearsal training
SME	subject matter expert
SMMTT	submarine multimission team trainer
SORTS	status of resources and training system
SO3	study of organizational opinion
SPAWAR	space and naval warfare
SSP	sound speed profile
STA	sensory task analysis

STDA	sonar tactical decision aid
STE	synthetic task environment
STI	Systems Technology, Inc.
STL	stereolithography
STO	science and technology objective
STOW	synthetic theater of war
STTC	Simulation and Training Technology Center
SVS	soldier visualization station
SWAT	special weapons and tactics
SWOS	Surface Warfare Officers School
TacOpsMC	*Tactical Operations Marine Corps*
TBI	traumatic brain injury
TC3	tactical combat casualty care
TD	training device
TDS	tactical decision simulation
TECOM	Training and Education Command (USMC)
TEE	training effectiveness evaluation
TER	training effectiveness ratio
3-D	three-dimensional
3G	third generation
TM&SMP	Training Modeling and Simulation Master Plan
TNO	Netherlands Organisation for Applied Scientific Research
TNA	training needs analysis
TOPS	tera operations per second
ToT	transfer of training
TRADOC	Training and Doctrine Command
TRPPM	training planning process methodology
TTE	tactical training equipment
UA	utility analysis
UAS	unmanned aerial system
UAV	unmanned aerial vehicle
UCD	user-centered training system design
UCF	University of Central Florida
UCF-IST	University of Central Florida Institute for Simulation and Training
UNREP	underway replenishment
USACOM	U.S. Atlantic Command (now JFCOM)
USAF	United States Air Force
USE	user scrutiny event
USMC	U.S. Marine Corps
UV	ultraviolet
UV	unmanned vehicle
V&V	verification and validation
VBS-1	*Virtual Battlefield System 1*
VCSA	Vice Chief of Staff of the Army
VE	virtual environment
VEAAAV	Virtual Environment Advanced Amphibious Assault Vehicle
VEHELO	Virtual Environment Helicopter

VEL	Virtual Environment Laboratory
VELCAC	Virtual Environment Landing Craft, Air Cushion
VESUB	Virtual Environment Submarine
VIRTE	Virtual Technologies and Environments
VLNET	virtual life network
VMS	voyage management system
VMT	virtual maneuver trainers
VR	virtual reality
VRLab	Virtual Reality Lab
VRVis	Virtual Reality and Visualization
XML	Extensible Markup Language

INDEX

ABOUT THE
EDITORS AND CONTRIBUTORS

THE EDITORS

JOSEPH COHN, Ph.D., is a Lieutenant Commander in the U.S. Navy, a full member of the Human Factors and Ergonomics Society, the American Psychological Association, and the Aerospace Medical Association. Selected as the Potomac Institute for Policy Studies' 2006 Lewis and Clark Fellow, Cohn has more than 60 publications in scientific journals, edited books, and conference proceedings and has given numerous invited lectures and presentations.

DENISE NICHOLSON, Ph.D., is Director of Applied Cognition and Training in the Immersive Virtual Environments Laboratory at the University of Central Florida's Institute for Simulation and Training. She holds joint appointments in UCF's Modeling and Simulation Graduate Program, Industrial Engineering and Management Department, and the College of Optics and Photonics. In recognition of her contributions to the field of Virtual Environments, Nicholson received the Innovation Award in Science and Technology from the Naval Air Warfare Center and has served as an appointed member of the international NATO Panel on "Advances of Virtual Environments for Human Systems Interaction." She joined UCF in 2005, with more than 18 years of government experience ranging from bench level research at the Air Force Research Lab to leadership as Deputy Director for Science and Technology at NAVAIR Training Systems Division.

DYLAN SCHMORROW, Ph.D., is an international leader in advancing virtual environment science and technology for training and education applications. He has received both the Human Factors and Ergonomics Society Leland S. Kollmorgen Spirit of Innovation Award for his contributions to the field of Augmented Cognition, and the Society of United States Naval Flight Surgeons Sonny Carter Memorial Award in recognition of his career improving the health, safety, and welfare of military operational forces. Schmorrow is a Commander in the U.S. Navy and has served at the Office of the Secretary of Defense, the Office of Naval Research, the Defense Advanced Research Projects Agency, the Naval Research Laboratory, the Naval Air Systems Command, and the Naval

Postgraduate School. He is the only naval officer to have received the Navy's Top Scientist and Engineers Award.

THE CONTRIBUTORS

ALI AHMAD is a Lead Researcher at Design Interactive, Inc. He holds a Ph.D. in Industrial Engineering from the University of Central Florida. His research interests include multimodal interaction design, audio interfaces, and advanced application of statistical techniques. Ali is a Certified Simulation Analyst and a Six Sigma Black Belt.

G. VINCENT AMICO, Ph.D., is one of the pioneers of simulation—with over 50 years of involvement in the industry. He is one of the principal agents behind the growth of the simulation industry, both in Central Florida and nationwide. He began his simulation career in 1948 as a project engineer in the flight trainers branch of the Special Devices Center, a facility now known as NAVAIR Orlando. During this time, he made significant contributions to simulation science. He was one of the first to use commercial digital computers for simulation, and in 1966, he chaired the first I/ITSEC Conference, the now well-established annual simulation, training, and education meeting. By the time he retired in 1981, he had held both the Director of Engineering and the Direct of Research positions within NAVAIR Orlando. Amico has been the recipient of many professional honors, including the I/ITSEC Lifetime Achievement Award, the Society for Computer Simulation Presidential Award, and an honorary Ph.D. in Modeling and Simulation from the University of Central Florida. The NCS created "The Vince Amico Scholarship" for deserving high school seniors interested in pursuing study in simulation, and in 2001, in recognition of his unselfish commitment to simulation technology and training, Orlando mayor Glenda Hood designated December 12, 2001, as "Vince Amico Day."

DEE ANDREWS, Ph.D., is a Senior Scientist with the Human Effectiveness Directorate of the Air Force Research Laboratory. His Ph.D. is in Instructional Systems from Florida State University. His research interests include distributed simulation training, aircrew training, and cyberoperations training.

RICHARD ARNOLD is President of Human Performance Architects, a human factors consulting firm based in Orlando, Florida, specializing in military personnel, training, and safety research. Prior to establishing the company, he served in the U.S. Navy as a designated Aerospace Experimental Psychologist.

ED BACHELDER received his Ph.D. from the Massachusetts Institute of Technology subsequent to flying the SH-60B as a Naval Aviator. His areas of research at Systems Technology, Inc., include (1) augmented reality, (2) optimized control guidance for helicopter autorotation, (3) system identification, and (4) 3-D helicopter cueing for precision hover and nap-of-earth flight.

JOHN BARNETT is a Research Psychologist with the U.S. Army, whose research interests include human-automation interaction, aviation, training, and human performance in extreme environments. He holds a Ph.D. in Applied Experimental and Human Factors Psychology from the University of Central Florida and is a former U.S. Air Force officer.

KATHLEEN BARTLETT, Technical Writer, earned her MA in English at the University of Central Florida (UCF). After teaching for Orange County Public Schools, Bartlett held several instructional and administrative positions at UCF. In addition to writing for UCF's Institute for Simulation and Training, she teaches at Florida Institute of Technology.

WILLIAM BECKER, Ph.D., is research faculty in the MOVES Institute at the Naval Postgraduate School. His specialty is the development of hardware and software to support advanced training for military personnel. He is currently working with the Marine Corps.

HERBERT BELL, Ph.D., is Technical Advisor for the Warfighter Readiness Division, Human Effectiveness Directorate of the Air Force Research Laboratory. His research interests include distributed simulation, training effectiveness, and research methodology. He received a Ph.D. in experimental psychology from Vanderbilt University.

NOAH BRICKMAN graduated from University of California at Santa Cruz (UCSC) in 1995 with a bachelor's degree in Computer Science. He has 12 years of experience writing virtual reality, aerospace simulation, AI, and gaming software. He is currently working for Systems Technology, Inc., and pursuing a computer science master's degree at UCSC.

C. SHAWN BURKE is a Research Scientist at the Institute for Simulation and Training, University of Central Florida. She is currently investigating team adaptability, multicultural team performance, multiteam systems, and leadership, measurement, and training of such teams. Dr. Burke received her doctorate in Industrial/Organizational Psychology from George Mason University in 2000.

MEREDITH BELL CARROLL is a Senior Research Associate at Design Interactive, Inc. She is currently a Doctoral Candidate in Human Factors and Applied Experimental Psychology at the University of Central Florida. Her research interests include human/team performance and training in complex systems with focuses on performance measurement and virtual training technology.

ROBERTO CHAMPNEY is a Senior Research Associate at Design Interactive, Inc. He is currently a Doctoral Candidate in Industrial Engineering at the University of Central Florida. His research interests include the design, development, and evaluation of human-interactive systems and emotions in user experience.

NICOLE COEYMAN is a Science and Technology Manager for the Asymmetric Warfare–Virtual Training Technologies (AW-VTT) program at the U.S. Army RDECOM STTC. She received a B.S. degree in Computer Engineering from the University of Central Florida (UCF) and is currently pursuing a master's degree in Industrial Engineering at UCF.

JOSEPH COHN received his Ph.D. in Neuroscience from Brandeis University's Ashton Graybiel Spatial Orientation Laboratory and continued his postdoctoral studies with Dr. J. A. Scott Kelso. His research interests focus on maintaining human performance/human effectiveness in real world environments by optimizing the symbiosis of humans and machines.

CAROLINA CRUZ-NEIRA is the Executive Director and Chief Scientist of the Louisiana Immersive Technologies Enterprise (LITE). Her interests are in interdisciplinary applications for immersive Virtual Environments in combination with supercomputing and high speed networking. She was the technical developer of the original CAVE and is the holder of the IEEE VGTC Virtual Reality Technical Achievement Award 2007.

RUDOLPH DARKEN is Professor of Computer Science at the Naval Postgraduate School. He is also the Director of Research for the Center for Homeland Defense and Security and is the former Director of the MOVES Institute for modeling and simulation.

DAVID DORSEY is employed by the National Security Agency. Dr. Dorsey holds a Ph.D. in Industrial-Organizational Psychology and a graduate minor in Computer Science from the University of South Florida. His professional interests include performance measurement, testing and assessment, training and training technologies, and computational modeling.

JAMES DUNNE, CDR, M.D., is chief of trauma and surgical critical care and surgical director of intensive care at the National Naval Medical Center in Bethesda, Maryland. He has completed two postdoctoral research fellowships focused on blood transfusion and its effect on morbidity and mortality in trauma.

CALI FIDOPIASTIS, Ph.D., is the Associate Director for Applied Cognition in the ACTIVE Lab at the Institute for Simulation and Training at the University of Central Florida. Cali studies human brain plasticity in naturalistic environments employing biosensing devices, such as fNIR and EEG, along with biomathematical modeling techniques.

NEAL FINKELSTEIN, Ph.D., is a graduate of Florida Atlantic University with a degree in Electrical Engineering. He also holds a Doctorate in Industrial Engineering from the University of Central Florida. At the STTC (Simulation and Training Technology Center) in Orlando, Florida, he serves as the technical advisor on technical, programmatic, and organizational issues that cut across the organization.

J. D. FLETCHER is a member of the senior research staff at the Institute for Defense Analyses, where he specializes in personnel and human performance issues. His research interests include design and evaluation of education and training using technology, cost-effectiveness analysis, and the development of human performance and expertise.

GEORGE GALANIS is head of the Training and Preparedness group with the Defence Science and Technology Organisation in Australia. His research includes investigating the effectiveness of simulation for individual and collective training, and learning at the organizational level. He holds a Ph.D. in Engineering and Human Factors from the RMIT (Royal Melbourne Institute of Technology).

PAT GARRITY is a principal investigator at U.S. Army Research, Development and Engineering Command (RDECOM), Simulation and Training Technology Center (STTC). He currently works in Dismounted Simulation conducting R&D in the area of dismounted soldier training and simulation where he was the Army's Science and Technology Objective Manager for the Embedded Training for Dismounted Soldiers program.

ROBERT GEHORSAM is President of Forterra Systems, a provider of enterprise virtual world solutions. He has more than 25 years of management experience in the online games and digital media world and has held senior positions at Sony Online Entertainment, Viacom, Scholastic, and Prodigy Services Company.

KEVIN GEISS, Ph.D., is a principal scientist in the Human Effectiveness Directorate of the Air Force Research Laboratory. Dr. Geiss was the Associate Chief Scientist for the Directorate from 2003 to 2006. Dr. Geiss holds a Ph.D. in Zoology and an M.S. in chemistry, both from Miami University, Ohio.

STEPHEN GOLDBERG, Ph.D., is the Chief of the Orlando Research Unit of the U.S. Army Research Institute. He received a doctorate in Cognitive Psychology from the State University of New York at Buffalo. He supervises a research program focused on feedback processes and training in virtual simulations and games.

BRIAN GOLDIEZ holds a Ph.D. in Modeling & Simulation and an M.S. in Computer Engineering from the University of Central Florida (UCF). He is Deputy Director at UCF's Institute for Simulation and Training and has a joint appointment in Industrial Engineering and Management Systems. He has 30 years of experience in M&S.

STUART GRANT is a Defence Scientist at Defence Research and Development Canada, where he leads the Learning and Training Group. His research addresses interfaces to virtual environments for dismounted combatants, simulation for direct fire training, and distributed simulation. He received a Ph.D. in Cognitive Psychology from the University of Toronto.

GARY GREEN was a Principal Investigator for IST's Embedded Simulation Technology Lab that is conducting research in support of the U.S. Army Research, Development and Engineering Command (RDECOM) Simulation and Training Technology Center (STTC). He has an M.S. in Operations Research from the U.S. Naval Postgraduate School.

MATT GUIBERT has focused on human/machine interfacing problems at STI. Mr. Guibert was responsible for the creation of the system architecture and software development of STI's Driver Assessment and Training System (DATS), which serves as a training and feedback module for STI's commercially available driving simulator STISIM Drive.

ALFRED HARMS JR., Vice Admiral, U.S. Navy (Ret), is currently serving on the staff of the University of Central Florida (UCF). Admiral Harms joined UCF following his final active-duty assignment where he served as Commander, Naval Education and Training Command where he was responsible for accession, professional and warfare training for all naval personnel.

JOHN HART is the Chief of Creative Learning Technologies Division and the Learning with Adaptive Simulation and Training Army Technology Objective Manager at U.S. Army RDECOM-STTC. He manages research and development of immersive technologies to create effective learning environments for military training that include the use of game based technologies.

CARL HOBSON is the founder and President of Oasis Advanced Engineering Inc. and has been involved in the simulation industry since 1985. Oasis has led the U.S. Army's Embedded Training R&D for ground combat systems for the past decade. Prior to forming Oasis, Mr. Hobson managed the General Dynamics Land Systems engineering labs for eight years.

ADAM HOOVER, Ph.D., earned his B.S. (1992) and M.S. (1993) in Computer Engineering and his Ph.D. (1996) in Computer Science and Engineering from the University of South Florida. Adam is currently an Associate Professor in Electrical and Computer Engineering at Clemson University. His research focuses on tracking, embedded systems, and physiological monitoring.

CHARLES HUGHES is Professor and Associate Director of the School of Electrical Engineering and Computer Science at the University of Central Florida. He is also Director of the Media Convergence Laboratory at the Institute for

Simulation and Training. His research interests are in mixed reality and interactive computer graphics.

DARIN HUGHES is a research faculty member at the Media Convergence Laboratory within the Institute for Simulation and Training, University of Central Florida. His research interests include sound in simulation, auditory perception, and audio engines.

CHRISTIAN JEROME is a research psychologist at the U.S. Army Research Institute. He received his Ph.D. in applied experimental psychology from the University of Central Florida in 2006. He has performed research with attention, situation awareness, decision making, presence, human performance cognitive modeling, driver distraction, training using virtual/augmented reality, and game based simulation.

DAVID JONES is a Senior Research Associate at Design Interactive, Inc. He received his M.S. degree in Industrial Engineering from the University of Central Florida, where he focused on multimodal design science. At Design Interactive, he has performed training evaluations on a number of government funded training systems.

PHILLIP JONES has over 22 years of professional experience as an Army combat-arms officer. He has extensive army and joint operational and training experience. More recently, Mr. Jones has led a series of studies on training effectiveness in army and joint units.

ROBERT C. KENNEDY is a doctoral candidate in I/O Psychology at the University of Central Florida. His research experience includes NASA Space Adaptation, NSF Cognitive Effects of Stress, and DoD Sensory/Perceptual Performance, training transfer, and criterion development and measurement.

ROBERT S. KENNEDY, Ph.D., has been a Human Factors Psychologist for over 48 years and has conducted projects with numerous agencies, including DoD, NASA, NSF, DOT, and NIH, on training and adaptation, human performance, and motion/VE sickness. He is also an Adjunct Professor at the University of Central Florida.

BRUCE KNERR, Ph.D., is a team leader at the U.S. Army Research Institute, where he conducts research on the use of virtual simulations for soldier training. He received a B.S. in Psychology from The Pennsylvania State University and M.S. and Ph.D. degrees in Engineering Psychology from the University of Maryland.

STEPHANIE LACKEY, Ph.D., is the Deputy Director of the Concept Development and Integration Laboratory at the Naval Air Warfare Center Training

Systems Division. Stephanie earned a B.S. in Mathematics from Methodist University and M.S. and Ph.D. degrees from the Industrial Engineering and Management Systems Department, University of Central Florida.

FOTIS LIAROKAPIS, Ph.D., holds a Ph.D. in Computer Engineering (University of Sussex), an MSc in Computer Graphics and Virtual Environments (University of Hull), and a BEng in Computer Systems Engineering (University of Sussex). He is employed by Coventry University as a Senior Lecturer and University of Sussex as a Visiting Lecturer.

RODNEY LONG is the program lead for research using Massively Multiplayer Online Games (MMOG) at the SFC Paul Ray Smith Simulation and Training Technology Center. Mr. Long has a breadth of simulation and training experience that spans more than 20 years in the Department of Defense.

TODD MACUDA, Ph.D., is currently Vice President of Business Development and Operations at Gladstone Aerospace Consulting. Dr. Macuda is a graduate of the University of Western Ontario and holds an undergraduate and master's degrees in Psychology and a Ph.D. in Neuroscience. He is an adjunct professor of several universities and a qualified instructor of human factors and related aerospace medicine courses.

HENRY MARSHALL's 25-plus years with the government have been spent primarily in leading-edge simulation technologies, such as embedded training technology development, semi-automated forces (SAFs), and software acquisition. He received a BSE in Electrical Engineering and an M.S. in Systems Simulation from the University of Central Florida.

THOMAS MASTAGLIO, Ph.D., has a career that includes 22 years of service as a U.S. Army Officer, a Senior Engineer, and a Program Manager in industry, as an independent consultant, a university faculty membership, and a business owner. His educational, research, and industry technical background includes computer and cognitive science, simulation design and development, training development and technology, and program development.

MICHELLE MAYO is currently a Science and Technology Manager for the Tactical Digital Holograms project for RDECOM-STTC. Previously, the manager of the Combat Trauma Patient Simulation Program, Ms. Mayo has over eight years of experience in military simulation and training programs. She has a B.S. degree in Computer Engineering.

CLAUDIA MCDONALD, Ph.D., leads the Center for Virtual Medical Education at Texas A&M University–Corpus Christi, specializing in research and development of sophisticated learning platforms utilizing virtual-world technologies. McDonald originated Pulse!! The Virtual Clinical Learning Lab, which has received more than $12 million in federal funding since March 2005.

JAMES MCDONOUGH, MAJ, is an artillery officer and a graduate of the Naval Postgraduate School (NPS) in Modeling, Virtual Environments, and Simulations. Upon graduation from NPS he served as the Modeling and Simulation Officer for Training and Education Command Technology Division at Quantico, Virginia. He is presently assigned to 3d Battalion, 12th Marine Regiment, 3d Marine Division, Okinawa, Japan.

GERALD MERSTEN, Ph.D., is the director of Technology Division of Marine Corps Training and Education Command at Quantico, Virginia. He has extensive executive experience in the aerospace industry.

LAURA MILHAM received her doctorate from the Applied Experimental and Human Factors Psychology program at the University of Central Florida. At Design Interactive, she is the Training Systems Director and Principal Investigator of numerous projects in support of the development and assessment of the effectiveness of training systems and training management systems.

JEFFREY MOSS is a retired U.S. Army warrant officer and UH-60 Black Hawk instructor pilot with nearly 2,000 hours of flight time as an instructor. After completing his army career, he began work in the civilian sector participating in joint, PC based simulation exercises and leading PC based simulation systems integration.

PETE MULLER is President of Potomac Training Corporation and the Systems Engineer for ONR's Human Performance, Training & Education (HPT&E) Thrust area. He served the same role for ONR's Virtual Environments and Technologies (VIRTE). He has worked in the aerospace industry in systems engineering and program management for both large and small companies.

ERIC MUTH, Ph.D., earned a B.A. from Hartwick College in 1991. He earned his M.S. and Ph.D. degrees in Psychology from The Pennsylvania State University in 1993 and 1997, respectively. Eric is currently a Professor of Psychology at Clemson University, where his work focuses on performance in high stress/workload environments.

LONG NGUYEN is an Electronics Engineer at the Naval Air Warfare Center Training Systems Division (NAWCTSD). He manages NAWCTSD's Applied Modeling and Simulation Branch. He holds an M.S. in Electrical Engineering from the University of Central Florida and is pursuing a Ph.D. in Industrial Engineering from the same.

DENISE NICHOLSON, Ph.D., is the Director of the Applied Cognition and Training in Immersive Virtual Environments Laboratory at the University of Central Florida's Institute for Simulation and Training (IST). Her additional UCF appointments include the Modeling and Simulation Graduate Program, the

Department of Industrial Engineering and Management Systems, and the College of Optics and Photonics/CREOL.

JACK NORFLEET heads the Medical Simulation Technologies group at the Army's RDECOM-STTC. He has 24 years of experience in military simulation with experience in medical simulations, live training, and in instrumentation systems. He has a BSEE from UCF and an MBA from Webster University. He has also trained as an EMT.

JOHN OWEN serves as Head of the Weapon Systems HSI Branch at the Naval Air Warfare Center Training Systems Division. He holds a B.S. in Electrical Engineering. Mr. Owen previously served as the Training Program Manager for LPD 17 and is the former manger for SEAPRINT.

DANIEL PATTON is the Deputy Director for Surface and Expeditionary Warfare Projects at the Naval Air Warfare Center, Training Systems Division (NAWC-TSD) in Orlando, Florida. He is a retired Naval Officer and holds an M.A. in Instructional Systems Technology.

M. BETH PETTITT is the Division Chief for Soldier Simulation Environments (SSE), Simulation and Training Technology Center, RDECOM. Prior to this position, she was instrumental in establishing STRICOM's CTPS and Advanced Trauma Patient Simulation (ATPS) DTO programs. Ms. Pettitt has over 19 years of experience in military modeling and simulation.

JAMES PHARMER, Ph.D., works as a Senior Research Psychologist at the Naval Air Warfare Center Training Systems Division (NAWCTSD) applying human systems integration (HSI) principles into navy system acquisition programs. He holds a Ph.D. in Applied Experimental Human Factors Psychology and an M.S. in Engineering Psychology.

WILLIAM PIKE is a Science and Technology Manager at U.S. Army RDECOM-STTC, where he leads research and development efforts in medical simulations to train combat medics. He has also led research efforts on PC game based simulations to determine their most effective use to support military training.

DIRK REINERS is an Assistant Professor at the University of Louisiana, Lafayette. His interests are in interactive 3-D graphics for complex scenes and software systems for real time rendering applications, as well as computer game design and development. He is the project lead for the OpenSG Open Source scenegraph project (http://opensg.vrsource.org/trac).

KATRINA RICCI, Ph.D., is a Senior Research Psychologist with the Naval Air Warfare Center Training Systems Division in Orlando, Florida. Dr. Ricci earned an M.S. in Industrial Organizational Psychology and a Ph.D. in Human Factors

Psychology from the University of Central Florida. Dr. Ricci has over 20 years of experience in training, human performance, and HSI.

DAVID ROLSTON is Chairman and CEO of Forterra Systems and has over 35 years of experience in the high technology industry spanning a broad spectrum of industries, applications, and technologies, including extensive involvement in simulation and training, graphics applications, imagery, gaming, artificial intelligence, entertainment, and early versions of the Internet.

STEVEN RUSSELL is a Research Scientist at Personnel Decisions Research Institutes in Arlington, Virginia. He holds a Ph.D. in Industrial-Organizational Psychology from Bowling Green State University. His professional interests include the design and evaluation of training programs, criterion measurement, and test development and validation, including item response theory (IRT) techniques.

RICHARD SCHAFFER is a Principal Investigator at Lockheed Martin's Advanced Simulation Center in Burlington, Massachusetts. He has been a simulation technology researcher since joining the DARPA SIMNET team in 1985 and has served as PI or Lead Simulation Integrator for numerous defense modeling and simulation programs. Richard is a Lockheed Martin Fellow.

DYLAN SCHMORROW, Ph.D., is an international leader in advancing virtual environment science and technology for training and education applications. Dr. Schmorrow is a Commander in the U.S. Navy and has served at the Office of the Secretary of Defense, the Office of Naval Research, the Defense Advanced Research Projects Agency, and the Naval Research Laboratory.

TOM SCHNELL is an Associate Professor in Industrial Engineering at the University of Iowa. He is the Director of the Operator Performance Laboratory (OPL). Tom has degrees in electrical (BSEE) and in industrial engineering (MS, Ph.D.). He is a commercial pilot, flight instructor, and helicopter and glider pilot with jet-type ratings.

LEE SCIARINI is a doctoral candidate the University of Central Florida. His research interests include training system development and effectiveness, human performance, human systems integration, team performance, unmanned systems, neuronergonomics, augmented cognition, and how all of these areas can be leveraged to enhance future systems.

RANDALL SHUMAKER, Ph.D., is the Director of the Institute for Simulation and Training (IST). Previous assignments include Superintendent for Information Technology at the U.S. Naval Research Laboratory and Director of the Navy Center for Applied Research in Artificial Intelligence. His research interests include artificial intelligence, biomorphic computing, and human-agent collaboration.

ALEXANDER SINGER has been a Motion Picture Director for 40 years, directing over 280 TV shows in all forms and genres, five feature films, and a short film for DARPA. Three projects with the NRC led to an award as Lifetime National Associate of the National Academies (of Science).

JUDITH SINGER has published two novels, written a Columbia Pictures feature screenplay and various daytime and prime-time TV screenplays, has contracted treatments for TV and feature films, and shared conceptualizing with her husband's science-driven explorations, including this project. For a decade she has been a professional Film Script Supervisor.

EILEEN SMITH is the Associate Director of the Media Convergence Laboratory and an Instructor in the Digital Media Department at the University of Central Florida. Her research interests are using mixed reality to enhance learning experiences.

ROGER SMITH, Ph.D., is the Chief Technology Officer for U.S. Army Simulation, Training and Instrumentation. He holds degrees in computer science (Ph.D.), statistics (M.S.), mathematics (B.S.), and management (MBA and M.S.).

ROBERT SOTTILARE is the Deputy Director for the U.S. Army Research, Development and Engineering Command's Simulation and Training Technology Center in Orlando, Florida. He has an M.S. in Simulation and is currently a Ph.D. candidate in the Modeling & Simulation program at the University of Central Florida.

KAY STANNEY is President of Design Interactive, Inc. She received her Ph.D. in Industrial Engineering from Purdue University, after which time she spent 15 years as a professor at the University of Central Florida. She has over 15 years of experience in the design, development, and evaluation of human-interactive systems.

ROY STRIPLING, Ph.D., is the program manager for the Office of Naval Research Human Performance, Training, and Education thrust area. He previously served as the head of the Warfighter Human-Systems Integration Laboratory at the Naval Research Laboratory. Dr. Stripling received his Ph.D. in neuroscience from the University of Illinois.

JOSEPH SULLIVAN, CDR, is a Permanent Military Professor in the Computer Science Department at the Naval Postgraduate School (NPS). Prior to assignment to NPS, CDR Sullivan completed numerous operational tours as helicopter pilot.

FRED SWITZER III, Ph.D., earned a B.A. from University of Texas at Austin in 1975. He earned his M.S. and Ph.D. degrees in Industrial/Organizational Psychology from Lamar University in 1982 and University of Illinois at

Urbana-Champaign in 1988, respectively. Fred is currently a Professor of Psychology at Clemson University.

Dr. **JACK THORPE** is a consultant involved in the definition and planning of advanced technology development projects. His expertise is in Distributed Simulation, and he was the Program Manager at the Defense Advanced Research Projects Agency that created SIMNET, micro-travel, video arcade trainers, the electronic sand table, and seamless simulation.

JUAN VAQUERIZO has been an industry leader in VST for the past 25 years. He founded two high technology visual simulation companies: Soft Reality and Advanced Simulation Research. He has been at the forefront of the design, development, and delivery of hundreds of training systems including the U.S. Army's RDECOM DAGGERS.

DENNIS VINCENZI, Ph.D., is a Research Psychologist for the NAWCTSD in Orlando, Florida. He earned his Ph.D. in Human Factors Psychology from the University of Central Florida in 1998. Dr. Vincenzi is currently the lead researcher on an ONR-funded research project involving Next Generation Helmet-Mounted Display Systems.

DANIEL WALKER, COL, is Chief, Warfighter Readiness Research Division, Air Force Research Laboratory. He holds a B.S. from the USAF Academy, four master's degrees, and is pursuing a Ph.D. He is a Master Navigator with experience in the B-1 and the B-52, including commanding the B-1 Division of the USAF Weapons School.

WILLIAM WALKER is a Visual Systems Engineer with HMD, collimated, domes, and large cylindrical display system experience. He has a BSEE from Auburn University and is a Navy Surface Warfare Officer. Prior work includes C-130H, F-14A & D, and E-2C aviation visual systems. He is the visual engineer on the COVE and VESUB projects.

LORI WALTERS, Ph.D., is joint faculty with the Institute of Simulation and Training and Department of History at the University of Central Florida. Her research interests are the use of virtual reality to enhance the story of history and technology in the museum.

TIMOTHY WANSBURY is a Technology Transition Officer at U.S. Army RDECOM-STTC, where he leads efforts in transitioning tools, technologies, and prototypes developed through a variety of research and development projects. He has led research efforts focused on developing a better understanding of how to design, develop, and use PC game based simulations to support military training.

SANDRA WETZEL-SMITH is a Senior Research Psychologist at the Space and Naval Warfare Systems Center in San Diego, California. She also serves as Director, Tactical Systems, at the Naval Mine and ASW Command, San Diego. Her most recent awards include the SSC-SDLauritzen-Bennett award for Scientific Excellence and the Federal Computer Week Federal 100.

MICHAEL WHITE is a Certified Modeling and Simulation Professional (CMSP) with Alion Science and Technology and has over 10 years of experience in Modeling and Simulation. Mr. White holds a B.S. in Professional Aeronautics and an MBA/A from Embry-Riddle Aeronautical University and is pursuing a Ph.D. at Old Dominion University.

SUSAN WHITE is an independent consultant in the Washington, D.C., area. Dr. White holds a Ph.D. in Industrial-Organizational Psychology from the University of Maryland. Her professional interests include performance appraisal, training, services marketing and management, and climate and culture.

MARK WIEDERHOLD Ph.D., is President and Director of Virtual Reality Medical Center in San Diego. Dr. Wiederhold and his team have been treating patients with VR therapy for the past 12 years. He has served on several advisory, editorial, and technical boards, and he has more than 150 scientific publications.

WALLACE WULFECK is a Senior Research Psychologist at the Space and Naval Warfare Systems Center where he serves as Co-Principal Investigator and Project Scientist on the Interactive Multisensor Analysis Training (IMAT) project. He previously directed the Instructional Simulations Division and Training Research Computing Facility and served at the Office of Naval Research.

WILLIAM YATES, LT COL, is an artillery officer and a graduate of the Naval Postgraduate School in Modeling, Virtual Environments, and Simulations. He was the director of the Battle Simulation Center at MAGTF Training Command, Twentynine Palms prior to being assigned as the M&S officer for the Program Manager for Training systems.